# POPULATION DYNAMICS

# POPULATION DYNAMICS

The 20th Symposium of
The British Ecological Society
London 1978

EDITED BY

R. M. ANDERSON
Imperial College, London

B. D. TURNER
King's College, London

AND

L. R. TAYLOR
Rothamsted
Experimental Station

BLACKWELL SCIENTIFIC PUBLICATIONS
OXFORD LONDON EDINBURGH
MELBOURNE
1979

Editorial offices:
Osney Mead, Oxford, OX2 0EL
8 John Street, London, WC1N 2ES
9 Forrest Road, Edinburgh, EH1 2QH
214 Berkeley Street, Carlton
Victoria 3053, Australia



First published 1979

Printed and bound
in Great Britain by
William Clowes (Beccles) Limited
Beccles and London

DISTRIBUTORS

USA and Canada
Halsted Press,
a Division of
John Wiley & Sons Inc.
New York

Australia
Blackwell Scientific Book
Distributors
214 Berkeley Street, Carlton
Victoria 3053

British Library
Cataloguing in Publication Data

Population dynamics.—(British
Ecological Society. Symposia; 20th).
1. Population biology—Congresses
I. Anderson, R M II. Turner, B D
III. Taylor, Lionel Roy IV. Series
574.5'24 QH352

ISBN 0-632-00184-4

# CONTENTS

# EDITORS' PREFACE

Over the last two decades the study of the population dynamics of animals and plants has assumed a new vitality.

Our aim as organizers of this, the twentieth symposium of the British Ecological Society, was to attempt to provide an up-to-date review of past work in this rapidly expanding field and also to capture the flavour of recent developments. We believed the time was right for such an attempt because of increasing interest in the role played by dynamics in providing a common conceptual framework for the interpretation of ecological problems and also in their application. The need for a sound theoretical base for the development of an applied dynamics is everywhere evident in problems as varied as the harvesting of animals and plants, the epidemiology of pests and infectious diseases, the use of predators, parasitoids and parasites as biological control agents and, conversely, in the problems arising from man's increasing influence on his own, as well as on other animal and plant, community structures.

No single volume can provide a comprehensive treatment of the numerous areas of ecology that fall beneath the umbrella of population dynamics, but the papers presented here range in approach from theoretical to field study and, we hope, provide an indication of a developing trend in population biology. That is, the increasing emphasis on distillation to find common ecological and evolutionary threads running through the observed population biologies of animal and plant species. Ecologists of a new generation are searching for basic similarities, rather than differences, and trying to establish ecological principles to bring order into a mass of data built up by past generations.

In this recent development, theory has played an increasing role. Fortunately, however, and perhaps in contrast to fields such as population genetics, its development has not overshadowed the fundamental importance of experiment and field study. A trend common to these contributions is, therefore, the integration of theory, experiment and field observation.

This symposium volume is organized as follows. The first two papers set the stage for a recurrent theme, namely the effect of heterogeneity on the behaviour of individuals that results in the spatial relationships which are such potent determinants of population dynamics. The introductory paper (Taylor & Taylor) stresses the relevance of spatial inhomogeneity in the vital role of individual movement which, in response to population density, generates overall persistence in populations whose separate patches are ephemeral. The topic of heterogeneity is later expanded by Connell, Hassell and Anderson in the respective contexts of spatial and temporal instability of complex communities, non-random search of parasitoids and predators and the distribution of parasites within populations of hosts.

The second paper (Harper and Bell) provides a striking example of convergence in the way botanists and zoologists now view the population biology of organisms. Harper and Bell demonstrate how the dynamics of plant growth depend upon the behavior of modular units and so determine the resulting spatial architecture of plant structure and form; this is another aspect of behavioural motion but at a different time-scale from that commonly observed in animals. Single species' studies are still the theme of the next three papers; Berry bridges the gap between dynamics and genetics by illustrating the way in which population dynamics are always confounded by population genetics; Law focuses on the evolution of plant life history strategies, while Dixon indicates the significance of climatic factors on the fluid dynamics of aphid populations.

In the next three papers competition provides the point of focus, discussing its role in moulding plant communities, determining bird faunas on islands, and controlling complex community structure within tropical rain forests and coral reefs (Law, Diamond, Connell). By their studies on optimal foraging Krebs & Cowie, like Taylor & Taylor, focus on that important but somewhat neglected area of population biology, the effect of integrating the behaviour of individuals into the overall dynamics of populations.

The interaction between organisms in different trophic levels forms the theme of the next five papers. Arthropod herbivores and plants provide the topic of the papers by Whittaker and Lawton & McNeill. These contributions clearly indicate that plant structure, both in its architectural or morphological sense and in its chemistry, is of utmost significance in determining arthropod dynamics. The subject of predator–prey, host–parasitoid and host–pathogen associations has developed rapidly and become increasingly sophisticated in recent years and, as shown by the papers of Hassell and Anderson, theory is based on a firm empirical framework. These papers also emphasize the stabilizing impact of spatial heterogeneity, or within-patch distribution, on the overall population dynamics of two-species interactions between patches. Beddington continues this theme with a review of recent developments in the dynamics of harvesting animals.

The last section focuses on multispecies interactions with contributions on fish and insect systems. It discusses the concepts of resilience and the existence of multiple stable states (Peterman *et al.*), marine systems where spatial and temporal heterogeneity are of major significance (Steele), microarthropods in decomposer-communities (Usher *et al.*), whilst the last paper (May) provides an overview of past and current developments in the theoretical richness of dynamical behaviour latent in multiple stable states and the relationship with trophic structure, and in the effects of spatial patchiness.

March 1979

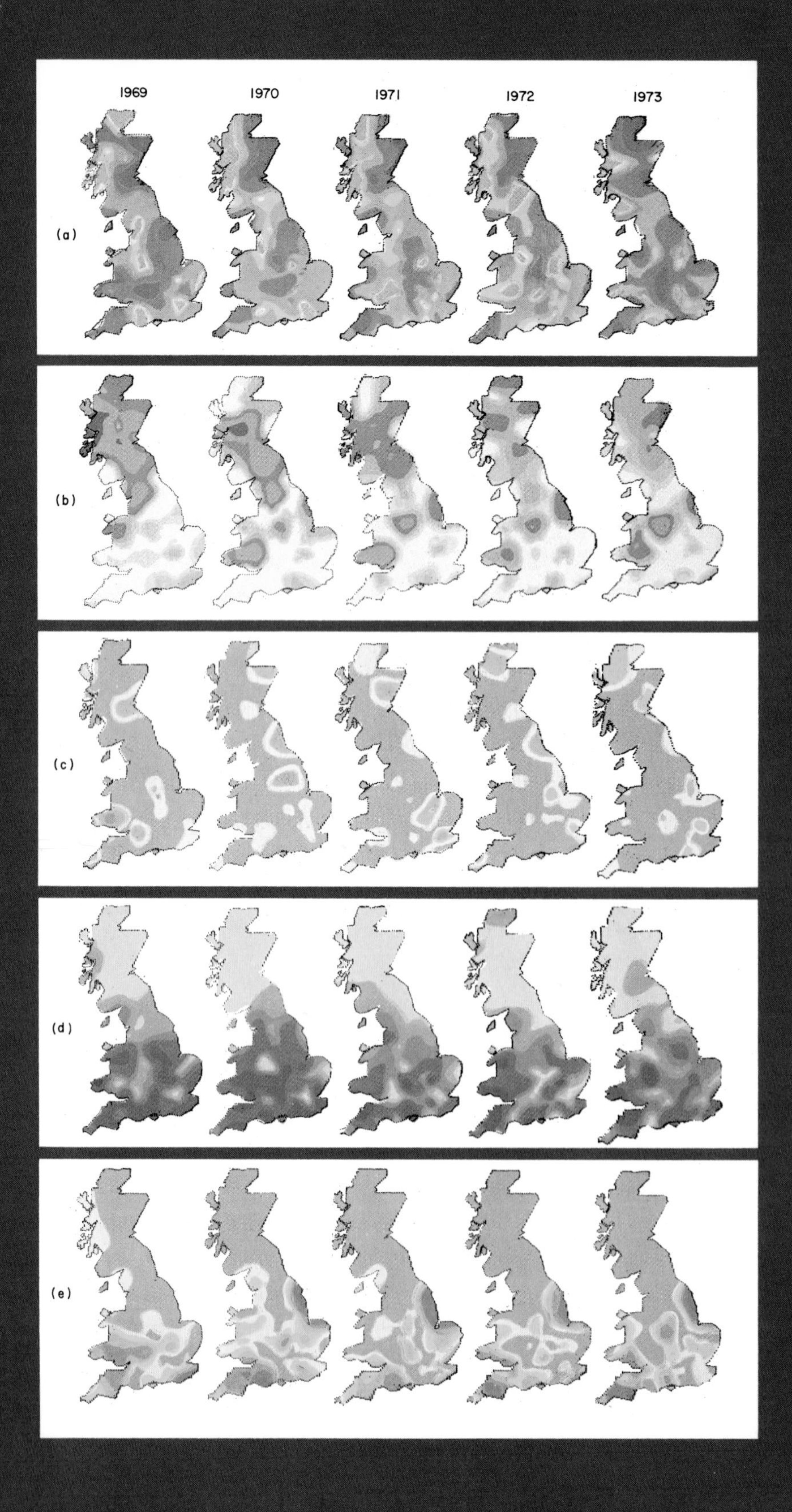
1969
1970
1971
1972
1973
(a)
(b)
(c)
(d)
(e)

# 1. A BEHAVIOURAL MODEL FOR THE EVOLUTION OF SPATIAL DYNAMICS

R. A. J. TAYLOR[1] AND L. R. TAYLOR[2]

[1] *Department of Zoology and Applied Entomology, Imperial College of Science and Technology, Silwood Park, Ascot, Berkshire*

[2] *Department of Entomology, Rothamsted Experimental Station, Harpenden, Hertfordshire*

*The Ecological Theatre and the Evolutionary Play*, Evelyn Hutchinson's (1965) illuminating analogy, likens ecology to a theatre where a play is endlessly performed by players whose behaviour evolves as the parts develop. The script adapts continuously to an ever-changing environmental set. Our concern at this meeting is less with the script, which is largely the province of population genetics, than with the stage directions, which are provided by population dynamics. Yet much population dynamics as well as genetics underestimates adaptive behaviour, and as a result a paradox has become apparent as the subjects have grown.

It has come to be realized over the last few decades that the negative feedback systems for population regulation required by theory, and seemingly found in laboratory experiments, depended on resource limitations for which there is scant field evidence. Furthermore, evidence has accumulated for epideictic (Wynne-Edwards 1962) and social (Wilson 1975) behaviour at all levels of organization that is obviously an integral part of population dynamics in the field, but is difficult to incorporate into the theory. This difficulty exists because behaviour involves space and the original theory was in one dimension

---

PLATE 1. Maps showing characteristic features of the changes in geographical distribution of moth populations between 1969 and 1973.

a *Xanthorhoë fluctuata* (Garden Carpet moth) is a widespread species, numerically fairly stable but with mobile 'holes' in its non-random spatial distribution which has no fixed population centre.

b *Cerapteryx graminis* (Antler moth) has a distribution limited by its plant hosts, which are fine upland grasses but, within these limits, the distribution of density is in constant motion.

c *Euxoa nigricans* (Garden Dart moth) has a spotted distribution that is spatially and temporally chaotic (see Fig. 1.2a).

d *Spilosoma lutea* (Buff Ermine moth) has a numerically very stable population (see Fig. 1.2a) with an unstable northern limit.

e *Callimorpha jacobaeae* (Cinnabar moth) is highly concentrated in suitable areas but, even so, the distribution is spatially mobile.

only, time. Hence the theory made implicit assumptions of linearity in behavioural responses, and randomness in both motion and distribution which were not valid (R.A.J. Taylor 1978; Taylor, Woiwod & Perry 1978). The behavioural sciences have made these assumptions increasingly untenable and this is elegantly demonstrated in the step-by-step increase in sophistication of the predator–prey models (e.g. Nicholson & Bailey 1935; Holling 1959; Hassell & Varley 1969; Rogers 1972; May 1978; see also papers by Anderson, Lawton & McNeill and Hassell this volume) as they have been made successively more realistic, incorporating behavioural responses (see Whittaker, Beddington and Peterman, Clark & Holling this volume) often verifiable by experimentation (e.g. Cowie & Krebs this volume). The earlier solution of a collective categorization of all interactive behaviour under the umbrella of 'competition' (with its subjective overtones) is fading (Grime, Connell and Diamond this volume), while a new approach to dynamics is developing in which spatial relationships are being examined for potential influences on the interactions between individuals. Such relationships, which were formerly treated in separate disciplines, mainly by statisticians and students of migration, are now being integrated into the fabric of population dynamics. As can be seen from papers in this volume, spatial relationships at all levels of organization, from cellular to multispecies populations, are being explored to account for the paradox which, when expressed in its simplest form, amounts to the common experience that while limiting resources are the key to classical theory, resources are hardly ever more than very locally limiting in nature.

The paradox arises from the evolutionary precept of selection for reproductive superiority. In fact, the first premise for natural selection is not maximized fecundity as classical theory assumes, but the survival of the individual, and this is determined by its ability to choose correctly those actions that will maximize its chances of staying alive until it can reproduce. Without this ability its fecundity is of little consequence. Many of its ancestors may have had low fecundity, but none made an error of choice crucial to its own or its offspring's survival. To resolve the paradox and make population dynamics something more than a continuing elaboration of the Verhulst (1838) equation requires this behaviour to be incorporated into the dynamic system. The difficulty lies in defining behaviour numerically.

Categories such as competitive, agonistic and amicable may measure a behavioural pressure building up in a population but they do not provide the movement necessary to find a safer place to live. It is the avoidance and escape mechanisms of negative competition—cowardice, cunning and migration—that produce much of the essential motion leading to the 'niche shift' of Diamond in this volume. Migration has often been suggested as a dynamic property but, without some counter-balancing congregatory mechanism, the movement it creates is randomizing and does not lead to the highly specific dynamic systems that occur. Nevertheless this characteristic, of movement for survival, seems to provide the definition for behaviour most useful in dynamics. In Kennedy's (1969) words, behaviour is 'the integrated function of the whole animal in its environment with special reference to its movements'. These movements are the product of behaviour initiated either by 'closed' or 'open' memory (Mayr 1976); that is, genetically programmed memory which is usually identified with instinctive, compulsive behaviour that anticipates its dynamic function, or short-term memory that leads to behaviour modifiable in response to current environmental stimuli and is commonly called facultative. Even facultative responses may not be immediate but may be delayed for a generation or more (Lees 1975) because survival is determined as much by where an organism is born as by what it is. The parents and grandparents who select the place of birth partly pre-determine an individual's behaviour, and hence its prospects, by their own success in life reflected, say, in size (Krebs & Boonstra 1979), and also by putting the offspring's unique set of responses in a unique place (Łomnicki 1978). In other words, it is not the genotype that is selected but the phenotype which correctly anticipates changes in its environment (Waddington 1975; Berry this volume). Ecologically, the genetic sources of behaviour are relevant only in their effect on its mode of operation, for all behaviour has been 'learned' by the species at some stage in its evolution. Also, as Harper and Bell's growth 'trees' demonstrate so elegantly in this volume, there is a unique historical element in each life history. This uniqueness, reflected in each species' population and in evolution itself, requires a real space-time continuum in which to operate.

In his Presidential Address to this Society, Southwood (1977) further developed his concept of the habitat as a template for ecological strategies. He stressed a recent trend in single-species dynamics, away from classical monolithic populations shaped by hostile external forces in abstract space, towards a more dynamic anatomy of populations of real individuals in a state of flux in real space as well as in real time (Taylor & Taylor 1977). This spatial flexibility adapts populations to their ever-changing environment and so provides both dynamic stability and resistance to evolutionary change. It originates in the behaviour of individuals, selfishly selecting from options offered by an environment that is, in the main, indifferent to their response; behaviour that results in the survival of the best gamesman in the art of being in the right place at the right time.

## OBJECTIVES

In this paper we try to bridge the gap between the naturalists and behaviourists who see what real individuals do, the experimentalists who collect evidence from real populations and deduce what actually must have happened, and the theoreticians whose models suggest how things might happen in an ideal world. To seek a common quantitative component in population structure, we examine the geographical behaviour of populations of a range of species at different spatial and temporal scales. We find that populations are not spatially stable. The instantaneous spatial distributions are not random and the movements of individuals are not random either, nor accountable to instantaneous transport coefficients. To put behaviour into a real multidimensional frame necessitates classifying it on its dynamic outcome rather than on motivational criteria. In doing this we do not intend to suggest that the dynamic outcome of every act is directly evident in its proximate cause. This misunderstanding has been a source of controversy in the study of insects, where much migration is compulsive and repeated attempts have been made to find behavioural criteria for it (Johnson 1969). The sexual, social and feeding behaviours of the more highly organized animals have spatial outcomes in addition to their more obvious functions. For example, pheromones may be regarded as purely sexual physiological mechanisms, but they incidentally stimulate congregation (Usher, Davis, Harris & Longstaff this volume) from very low densities by inducing movements over comparatively great distances. It is impossible now to determine why this behaviour arose, but it greatly increases the ambit of awareness of the individual and affects local population density. So also do other sexual aggregations, territorial displays, common nesting sites and collective feeding, herding or schooling for protection from predators, aerial quartering for food finding; the examples are endless because all behaviour can be expressed as movement affecting redistribution. If, as a result, offspring are born in a different place from their parents, then dynamic migration has successfully taken place whatever other kinds of behaviour have been involved (see Figs 1 and 2, Taylor & Taylor 1978).

We then seek a common evolutionary origin for this spatial behaviour in the environmental pressures on an individual when life first emerged. We suppose that the pressures remain the same, although the individual's responses are now more complex. Finally, we speculate on the functional form of the spatial response of individuals to these pressures and, using a general simulation model on a real space-time frame, show that populations growing under such environmental pressures and responding by movement have characteristics which match those found in real populations. Any particular individual may not move far, but, if it survives, it is that selfish motion which kept it alive and, when integrated throughout the population, keeps the population in motion. Consequently, all species have an element of spatial flexibility. We therefore turn now to the evidence for the stability of populations in space.

## EVIDENCE FROM REAL POPULATIONS

There has been a tendency in the past to characterize species as migratory or non-migratory, and 'patches' as if they were spatially permanent. This is a question of scale, but it is our conviction that a truly spatially stable, autochthonous, population could not survive.

### *Global scale redistribution of populations*

Analyses, notably by G.R. Coope and his colleagues, of the fossiliferous sediments from Quarternary deposits of lake muds, silts and peats throughout Great Britain show a constant turn-over of insect species during the last 100 000 years. Forty per cent of the beetle species in some British assemblages are different now from 11 000 years ago (Coope 1977a), and at previous times the rate of turn-over has been much more rapid despite the fact that the Coleoptera have probably been evolutionarily stable for more than a million years, at least in northern temperate and polar regions (Matthews 1976). Diagnostic characters, such as the sculpturing of the elytra, are well preserved and of about 2 000 species examined all but one or two show no evidence of morphological or behavioural evolution; the same groups of species inhabited the same appropriate habitats then as now (Coope & Angus 1975) although not in the same places.

About 44–45 000 years ago the beetle fauna of Great Britain was dominated by present-day arctic species. About 95% of the species present 43 000 years ago now occur further south than England. Between 40 000 and 42 000 years ago a new fauna invaded and later retreated eastwards beyond Europe into Asia, some to the very far east. Their place was taken for the next 15 000 years by beetles that have now emigrated mainly to the Siberian arctic. About 14 500 years ago a different arctic assemblage invaded Britain but by 12–13 000 years ago the species occurring at Windermere in the English Lakeland were typical of present-day populations. Around 12 000 years ago northern European species were in possession once more. By ten thousand years ago present-day Northern and Eastern Siberian and Arctic American species were in residence and since that time the present-day fauna has become re-established.

This constant large-scale mobility must be an integral part of any dynamic system. To take two named examples, *Boreaphilus henningianus* Sahlb. (Staphylinidae) visited Great Britain about 10 000, 14 000 and 47 000 years ago whilst *Morychus aeneus* F. (Byrrihidae) was here about 43 000, 60 000 and 120 000 years ago and is here again now. 'The fossil record provides abundant evidence that the geographical ranges of even the most sedentary species have changed on an enormous scale . . .' (Coope 1977b). In this the Coleoptera are not unusual, except that they comprise the biggest single order of organisms with over 360 000 known species, and leave behind an excellent historical record over hundreds of thousands of generations. The populations of some of these species have repeatedly migrated thousands of kilometres over enormously varied terrain. The likelihood that the same controlling system of key factors could be maintained for so long and over such a large and varied area is small.

### *Autochthony and the instantaneous state*

*Aphodius holdereri* Reitter (Coleoptera: Scarabaeidae) is a distinctive dung beetle currently endemic and apparently autochthonous only in the high Tibetan plateau between 3 000 and 5 000 metres (Coope 1973). However, during the mid-Weichselian period (26–42 000 years ago) this was the commonest large *Aphodius* species, widespread throughout Midland England (Coope 1975).

Such beetles are not long-distance migrants. Many of them, especially the predatory ground beetles, are flightless. The process of population shift is a gradual, accumulative effect of generations of individual migrants searching over small areas for places to live. So the whole population creeps in an exploratory, amoeboid, manner over the terrain (see Plate 1, facing page 1). Neither is the population a single monolithic block. At any instant in time, isolated segments of the population are present simultaneously at widely separated sites (Coope 1975). Only in space and time do they form a continuum that we have elsewhere likened to a reticulate fern stele (Taylor & Taylor 1977).

### *Redistribution at other scales of space and time*

At the other extreme of the scales of space and time, Huffaker's (1958) experiments provide clear evidence of continuous movement within small population elements, with and without predation. Experimental arenas for small mammals also demonstrate movement and the zoo syndrome that results when it is denied (Lidicker 1975). However, since the organisms in these arenas cannot select their terrain, nor migrate without restraint, the difference from natural populations is difficult to ascertain. Even when an experiment can be conducted in a large enclosure (e.g. Gibb 1977), the survival or extinction of such a local patch is not evidence for the stability of the population as a whole.

It is seemingly almost impossible to measure these two properties, of changing spatial distribution and unrestrained migration, simultaneously for more than one species at a time. There are excellent single-species studies now available like those on the red grouse (Watson 1977) and the great tit (Krebs & Perrins 1978); too many to list but not sufficient to compare parameters and so avoid drawing conclusions from what may turn out to be special cases, for they inevitably tend to be 'non-migratory' species. Taking the two problems separately enables us to consider them comparatively between species.

### *Medium scale density distribution patterns*

In 1959 one of us proposed a project to investigate 'simultaneous changes in time and space' for a range of species from 'total migrants' to 'non-migrants' throughout the island of Great Britain (L.R. Taylor 1974), and there are increasing data from 1960 onwards. The system gives a stratified random sample of relative density for 585 species of moths (Taylor, French & Woiwod 1978) and absolute density for 278 species of aphids (L.R. Taylor 1977a) which show changes in distribution between successive population cycles from 1968 to 1978. We confine our interests here to relative densities. The maps in Plate 1 and Fig. 1.1 give a visual impression of the changing geographical patterns of the log density of real populations in their chosen terrain, typical of the hundreds examined, and show six relevant characteristics.

Firstly, *the geographical distribution invariably changes as the population level changes, there being no permanent population centre.* This apparently simple conclusion does not follow from temporal models in which population change is attributed primarily to natality and mortality.

Secondly, there are outer geographical limits to potential distribution. Within these limits, however, *in no instance is the whole permissible area occupied at any one time by any species.* This also is not a necessary expectation from most temporal models.

Thirdly, *occasional rogue individuals appear far from their natal areas.* These are from the long tail of the density/distance function (see later).

Fourthly, when the population returns to its original level after a fluctuation, its geographical pattern has changed. *Each generation's distribution is spatially unique.* This is the population element of Coope's movements.

Fifthly, some species have less surface modulation than others. In other words, *the population of a species has a characteristic spatial density surface which is not a simple property of the resources.*

Finally, *there is a specific rate of change of the characteristic surface from one generation to the next* (see Fig. 19, L.R. Taylor 1974).

These maps do not provide direct evidence of motion, but with Huffaker's evidence of fine scale individual movement and Coope's evidence of global scale population movement, the probability that these changing patterns involve movement as well as mortality becomes exceedingly high.

We therefore conclude that characteristic individual spatial redistribution and characteristic specific population distribution occur at different organizational levels and it is likely that the one leads to the other.

## THE ROUGHNESS CHARACTERISTIC OF POPULATION DENSITY

The maps themselves are the Harvard Computer Graphics SYMAP programme's interpretation of a log density surface based on a set of stratified random samples in which the two main measurable properties are mean level of the population and surface roughness, except that the sampling arena is unusually large. To interpret this roughness, frequency distributions like the

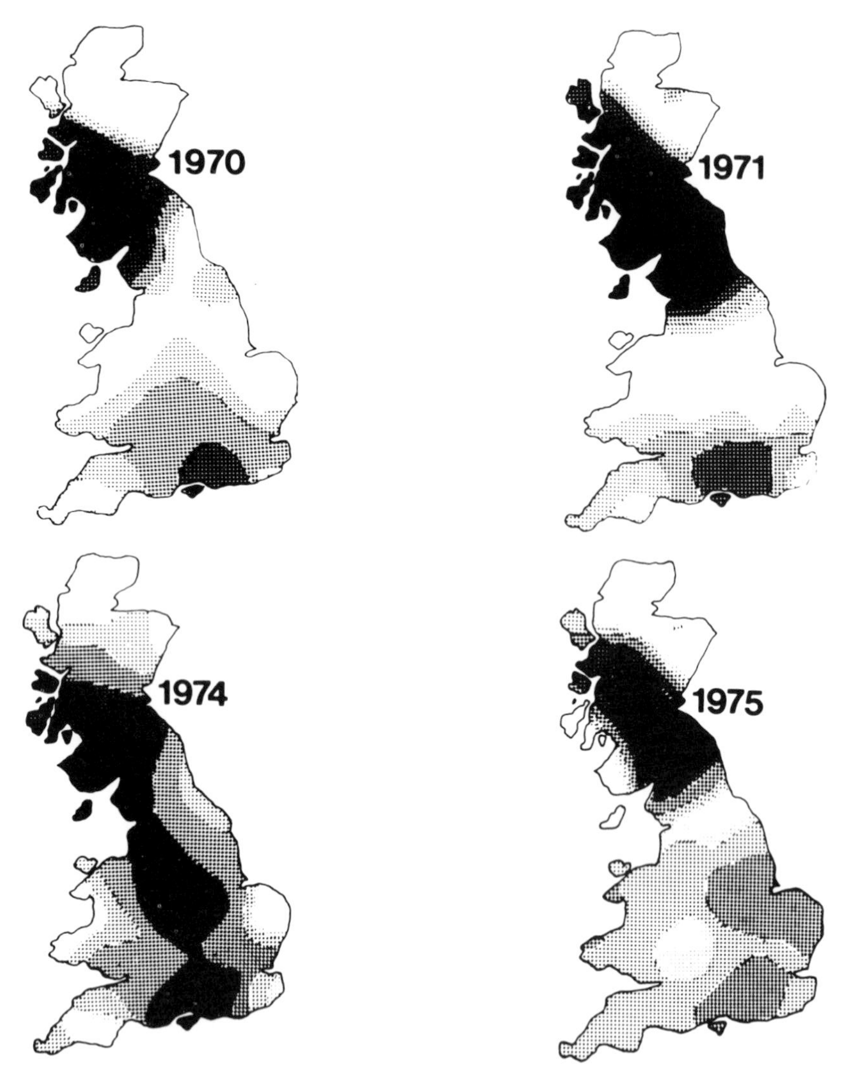


FIG. 1.1 (above and on facing page). Maps showing the changes in geographical distribution of *Aphis sambuci* (Elder aphid) between 1970 and 1977. Elder aphid has winter hosts (*Sambucus*) and summer hosts (*Rumex*) in every part of Great Britain, but its own distribution is totally different each year.

negative binomial have usually been fitted to sample counts. However, there is little ecological evidence to recommend this usage (Taylor, Woiwod & Perry 1979), although considerable mathematical effort has been expended on its justification, and we prefer a simpler, more general method. The modulation, or roughness, of any density surface may be considered as its rate of change

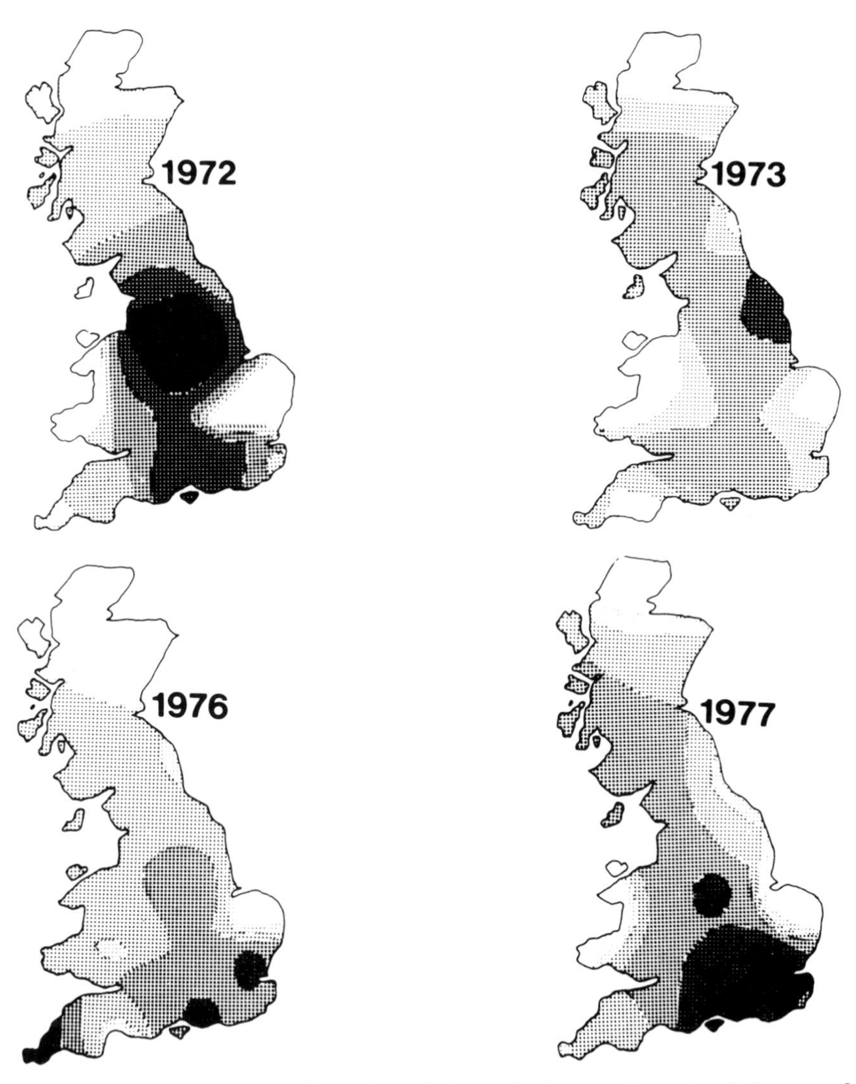


with respect to place at a given time. It is equivalent to the rate of change of density with respect to time at a given place and may be measured in the same way, by mean and variance. For density surfaces, L.R. Taylor (1961) proposed that this variance is the product of mutual attraction and repulsion between individuals and predicted a power model, spatial variance is proportional to a fractional power of mean density ($s^2 = am^b$), to describe it.

In a recent survey of spatial distribution, Taylor, Woiwod & Perry (1978) examined 156 sets of data comprising over 200 000 sample units in 3 840 samples from field populations of 102 species of plants and animals using the generalized linear model programme GLIM (Baker & Nelder 1978). It was found that Iwao's (1977) regression model, which uses Bartlett's (1936) expression ($s^2 = am + bm^2$) and its special case the negative binomial with a common $k$ ($s^2 = m + bm^2$), has the serious structural defect that negative variance could sometimes be expected, and is therefore suspect as an ecological, as distinct from statistical, model. L.R. Taylor's (1961) power model fitted 147 sets of data best, including the random and regular sets, and accounted for most variance.

The data covered a wide range of species, higher taxa, spatial scales and sampling methods. In Figure 1.2a we show variance–mean plots for data of a ciliate protozoon, a tick, an aphid, two moths, a sea urchin, a fish, two birds, modern man and an orchid. The ciliate is epizooic, the tick is an ectoparasite, the tit a predator, the moths phytophagous and man omnivorous. The great tit is territorial, the aphid is migratory in this set of data. In the collared dove and man the population is rising continuously, whilst in the garden dart moth and the great tit the population cycles continuously. In data for the coloured fox from the classical Moravian Mission (later Hudsons Bay Company) fur records, the population cycled about twenty-three times in the ninety-one years during which the samples were collected (see Fig. 6, Taylor, Woiwod & Perry 1978). The buff ermine moth has a numerically stable population, whilst for colonially sedentary bean aphids on individual plant stems (Fig. 6, Taylor, Woiwod & Perry 1978) mean density ranges over five orders of magnitude and variances over eight. The ciliate sampling area is the surface of flat-worms whilst that for man is the United States of America. The moths and the aphid are from the survey of Great Britain that produces the maps in Figure 1.1. Some of the samples are complete census counts, and all kinds of sampling methods are included. Finally, some samples were collected simultaneously at different places whilst, in others, each sample covers the same area but at a different time.

We conclude that, apart from very occasional special cases, such as the wheat-blossom midge (L.R. Taylor 1971), all taxa and nearly all species conform to the model. This generalizes the evidence given in the maps (Fig. 1.1), and we therefore have good reason to expect that, when sequential maps become available for other organisms, such as small mammals, for example, geographical distribution will be found to change in the same way, on a time scale appropriate to their life history.

## COMBINED RESULTS OF MAPS AND ROUGHNESS CHARACTERISTICS

In these log $\times$ log regressions the coefficients $b$, which have a mean and S.D. of $1{\cdot}45 \pm 0{\cdot}39$, are a measure of the linearity of variance with respect to the mean, i.e. they are roughness coefficients. Only when $b = 1$ is the variance, or when

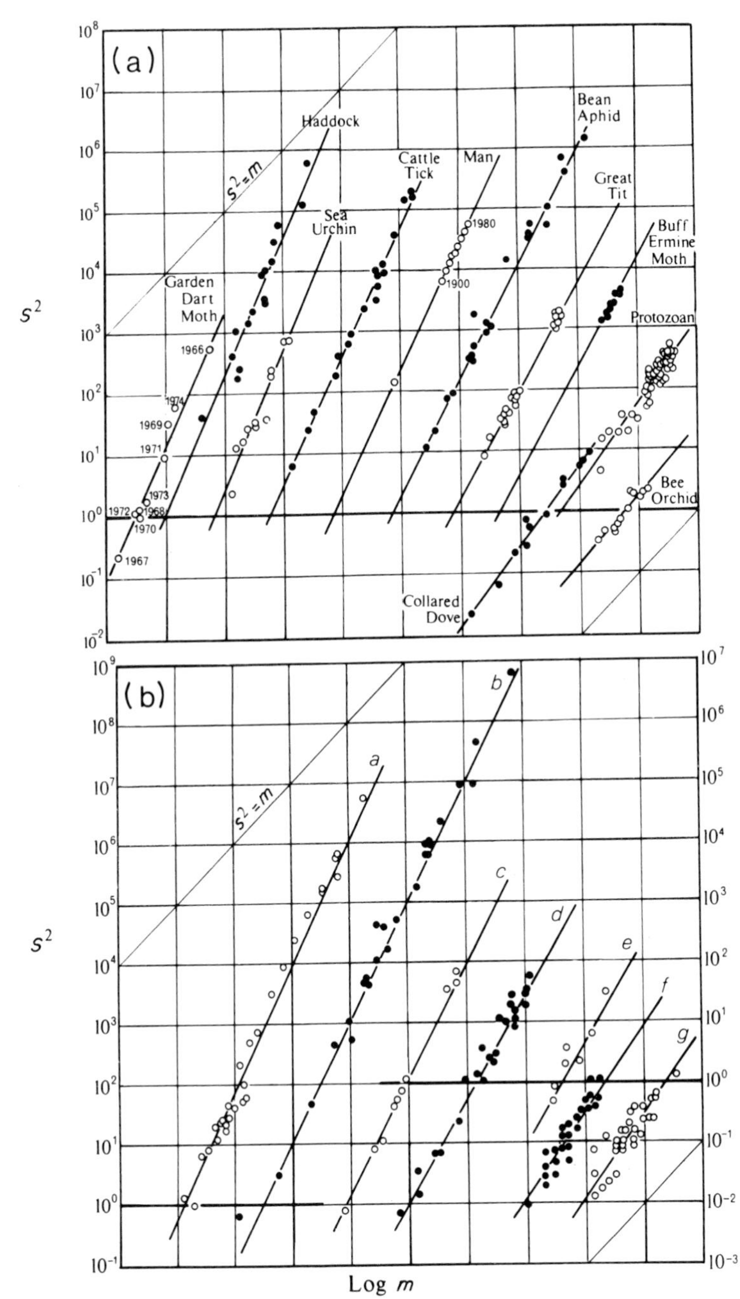


FIG. 1.2. (a) Variance of spatial samples is proportional to a fractional power of mean population density for all species and at all scales, including samples that are mapped in Fig. 1.1 (from Taylor, Woiwod & Perry 1978).
(b) The Δ-function generates spatial distributions with the same variance-mean functional form as natural populations (for parameters see Taylor & Taylor 1978).

$b = 2$ the standard deviation, linearly related to the mean, hence the relationship is non-linear for most species, $b$ usually being significantly different from both one and two and sometimes $> 2$. Data collected by the Rothamsted Insect Survey (L.R. Taylor 1974) of 225 species of moths and 75 species of aphids confirm this. From the inspection of maps (point four, p. 7) we have already established that each instantaneous geographical pattern is unique, even at the same mean density. In sharp contrast, the same spatial variance or roughness coefficient tends to be restored each time the same mean density returns (L.R. Taylor 1977b).

Extrinsic biotic factors affecting birth and death cannot achieve this dual control. It requires a feedback mechanism in which $k$-values (Varley & Gradwell 1960) are the same over all parts of the geographical range and are synchronized to restore the appropriate variance by counter-balancing reductions in density at one place with a corresponding specific non-linear increase in others. Nor can resource limitation do so because it operates as an upper limit only. We conclude that this intrinsic property of the population must be behavioural and the negative feedback mechanism must therefore involve species specific motion.

Any population model must therefore yield this general spatial condition for all organisms, $s^2 = am^b$, as a result of intrinsic motion or migration.

## THE SPECIFIC BEHAVIOUR OF MIGRATION

'Random migration' was a concept applied early to ecology from random physical processes, by Pearson & Blakeman (1906), and later adopted by Fisher who predicted that the spread of genes in space would be found to obey the gas laws (Fisher 1922), and by Wright (Dobzhansky & Wright 1943). Wright's model is a solution of the one-dimensional Fokker–Planck equation and assumes that particles, in this case flies, move a fixed distance in one of two opposite direction along the $X$-axis with a probability of $p\ (= 1 - q)$, during equal time intervals. The probability of the fly being at a distance $x$, from its starting point after time $t$, is given by a form of the normal probability function,

$$\phi(x, t) = \frac{1}{\sqrt{(4\pi Dt)}} \exp\left[\frac{1}{4Dt}(x - 2Ct)^2\right] \tag{1.1}$$

in which there are constants for drift ($C$) and characteristic diffusion rate ($D$). The mean, $\mu = 2Ct$, and variance, $\sigma^2 = 2Dt$, are linearly dependent on time. If there is no directional bias, $p = q = 0{\cdot}5$ and $C =$ zero and, if all observations are made simultaneously at time $t$, then distribution (1.1) can be replaced by the regression equation

$$N = \exp(a + bX^2) \tag{1.2}$$

where $N$ is the number moving along the $X$ axis and $a$ and $b$ are constants, the 'half normal' equation. In two dimensions, particles diffuse fixed distances, $d$, in all directions from a point source in a plane, $XY$, over fixed periods. They have a probability density function in $x$, $y$ and $t$ which gives the distribution of absolute distance $r$ from the origin as

$$\phi(r, t) = \frac{2r}{d^2 t} \exp\left[\frac{-r^2}{d^2 t}\right] \tag{1.3}$$

i.e. similar to eqn (1.1) in which $\sigma^2 = 2Dt \equiv \frac{1}{2}d^2 t$, with $d^2$ the mean square displacement per unit time.

Wright's model (1.1) was tested in the classical one-dimensional 'cross experiment' (Dobzhansky & Wright 1943) on genetically-marked *Drosophila pseudoobscura.* Although the results seemed acceptable at the time, leptokurtosis was detected in the distribution and the estimates for $D$ were not repeatable (Dobzhansky & Powell 1974). However, using the derived two-dimensional model (1.3) and assuming an exponential growth rate in Ulbrich's (1930) muskrat data, Skellam (1951) reached the qualified conclusion that the randomness assumption of the model 'is at least approximately true'.

Skellam's paper is often cited to justify the assumption of random diffusion in ecological models ignoring both Skellam's qualification and the fact that, even for *Drosophila pseudoobscura*, the log of density regressed on the square root of distance gives a better description than the half-normal (eqn (1.2)) (Wallace 1966). For most species these models are a gross over-simplification and the randomness assumption is not valid. Of the many sets of data for various organisms (Wolfenbarger 1946, 1959) few are adequate for testing to discriminate random from non-random processes. Of eight sets found to be suitable, only *Meligethes aeneus* conformed to the random model, the logarithm × square root model fitted three sets best, and the Wadley & Wolfenbarger's (1944) semi-log models and the reciprocal model of Paris (1965) failed to give any best fits. The five models found to fit at least one data set are all special cases of a more general model,

$$N = \exp(a + bX^c) \qquad \text{(R.A.J. Taylor 1978)} \tag{1.4}$$

which provides a better description of all the data used (Fig. 1.3a), generalizes the 'half-normal' model (eqn (1.2)) and is related to the generalized gamma distribution (R.A.J. Taylor 1979).

In Fig. 1.3b the cumulative frequency distribution of the *Drosophila* data is plotted against distance and the 'half-normal' (eqn (1.2)), which is the top half of the familiar normal distribution's sigmoid curve of accumulated frequency, clearly does not fit whilst eqn (1.4) does. If the axes are transposed, we obtain the distance moved away from the centre as a function of the total population sampled (Fig. 1.3c).

The behaviour involved in movement is sophisticated and this is especially

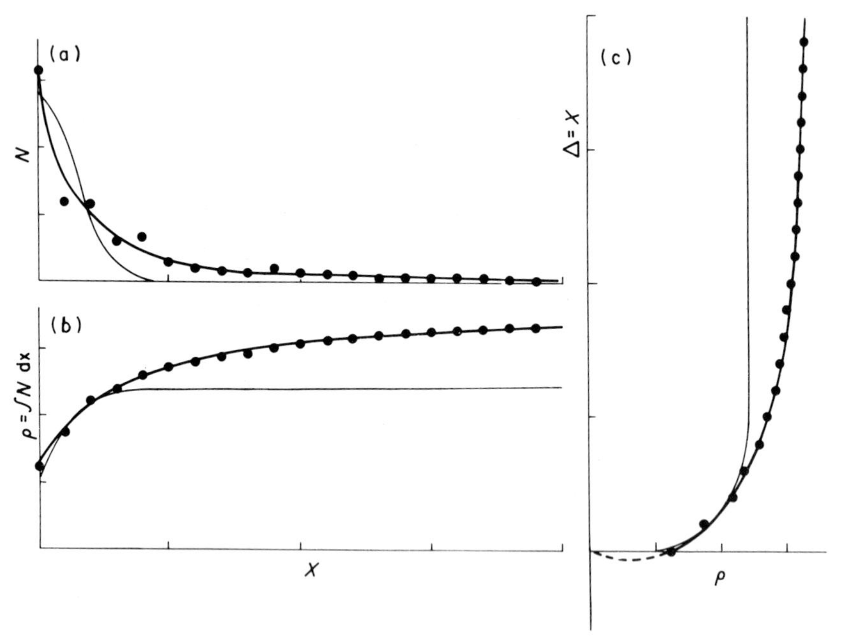


FIG. 1.3. (a) Density × distance data on *Drosophila*, from Dobzhansky & Wright (1943), does not fit the random diffusion model (——) but does fit eqn (1.4) (——).
(b) Cumulative frequency from 1.3a.
(c) Transposed axes give the Δ-function form (eqn (1.6)), distance moved as a function of total population.

evident when different ages and sexes behave differently. Despite the increased flexibility and complexity provided by the additional parameter in eqn (1.4), it often remains inadequate and a still more complex expression is needed. The blackbird data of Greenwood & Harvey (1976) and the great tit data from Wytham Wood (Greenwood, Harvey & Perrins 1979), for example, require the greater flexibility afforded by yet another parameter,

$$N = AX^d \exp(bX^c) \qquad \text{(R.A.J. Taylor 1979)} \qquad (1.5)$$

This specific density × distance behaviour function, or one like it, must also be an outcome of any population model.

## THE ORIGINS OF INTRINSIC MOTION

The dynamic model we consider is based on an evolutionary proposition about the origins of behaviour. For diagrammatic purposes only, we suppose, initially, that the earliest proto-organisms were sub-spherical and grew in a homogeneous environment until they reached a critical size, perhaps when surface-area/volume ratio reached a threshold differential between internal and external environments and division became nutritionally advantageous. At the moment of division, each new proto-cell restored its area/volume ratio but now had an inclusion, the sister cell, in its otherwise homogeneous environment. Competition would not be advantageous, so the environmental *status quo*, and with it the nutritional advantage, would be restored to those proto-cells that were accidentally (randomly) separated. There is therefore an environmental advantage in separation and an evolutionary pressure to acquire intrinsic motion independently of competition. This may be achieved by the minimum migration of one daughter cell (50% in a population) or by both (100%) (for a formal treatment see Hamilton & May 1977). We now have the three basic kinds of behaviour: (1) growth, (2) division and (3) re-distribution (Fig. 1.4a).

In a heterogeneous environment most cells are produced in the most favourable places so that, on average, any continued random motion would result in proto-cells losing environmental status, thus creating an environmental advantage in controlling motion. There is also an evolutionary pressure not to move too far, to remain congregated or, if separated, to restore congregation (Fig. 1.4b). We suppose that these two fundamental environmental pressures—to move but not to move too far—lead to an optimal spacing behaviour that would be selected for during evolution and would become the prime arbiters of environmental status determining reproductive advantage and, if fixed in inherited (closed) memory, would become compulsive or pre-saturation migration, sometimes directional of the kind dealt with by Williams (1930) in insects.

In the most realistic case, where the environment is both heterogeneous and changing (Fig. 1.4c), the environmental pressures would be always for the organism to evolve an optimum distance for movement by a combination of compulsive (closed) and facultative (open) behaviour and so to match the changing environment whilst avoiding too much competition. There are specialist alternatives, such as diapause, aestivation and dormancy, to await the return of the favourable environment or learning the space–time vector and migrating directionally to anticipate it (Fig. 1.4c). With the advent of speciation, each species would experience a different environment [Southwood's (1977) template] and its behaviour would then resolve the problem differently by acquiring a characteristic congregation–migration balance and a characteristic density–distance function to operate it. The net result would be a characteristic spatial distribution with a characteristic rate of temporal and spatial change, space being a frame for evolution, not a resource. The evolutionary prize would go to the organism that 'learned' best how to modify its behaviour so as to be always in the optimum place, by manipulating the 'open' and 'closed' components of behaviour appropriately for its own environment.

There is no reason to suppose that the three kinds of behaviour, (1) growth,

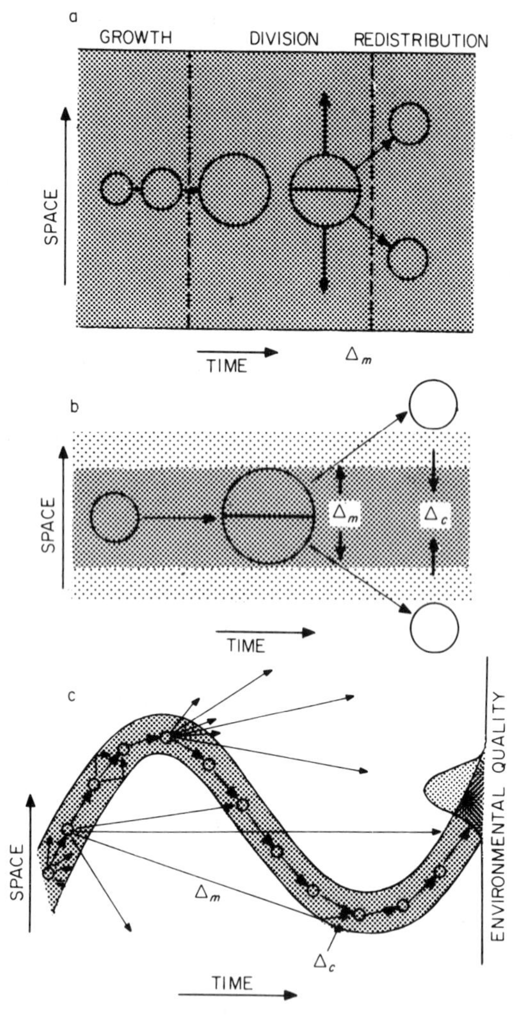


FIG. 1.4. (a) Diagrammatic representation of the three kinds of behaviour—growth, division and redistribution—of a proto-cell in a homogeneous environment shows the environmental pressure to restore the *status quo* by migration.
(b) In a heterogeneous environment there is a counter-balancing pressure not to migrate too far (i.e. into a poorer quality environment) or to restore the *status quo* after migration by congregation. The Δ-function balances these pressures.
(c) In a more realistic environment that is both heterogeneous and mobile there are pressures to anticipate regular environmental cycles by compulsive motion but also to retain flexibility by facultative motion. Each species optimizes its spatial strategy differently, dependent on its own environmental template.

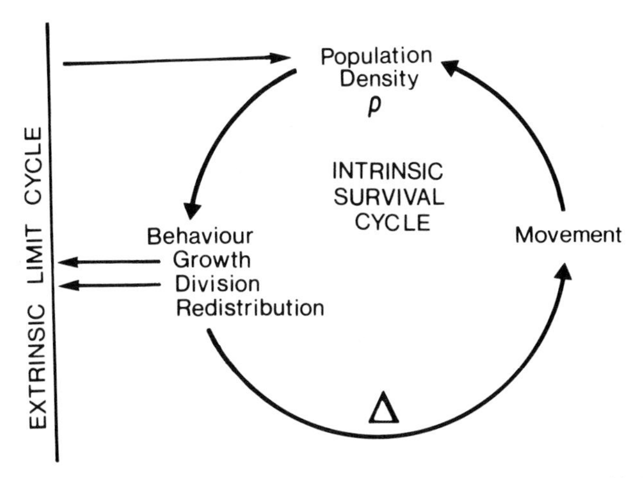


FIG. 1.5. Control is only possible by negative feedback in the three kinds of behaviour. Growth and division can be extrinsically controlled only if non-altruistic, but redistribution can be both intrinsic and selfish.

(2) division and (3) reproduction, are less appropriate now than they were in the beginning, or that the environmental pressures have changed. Population control, or regulation, must be based on negative feedback within this system. Extrinsic control can operate on either growth or division but intrinsic control can, so far as we can see, occur only in behaviour of the third kind, spatial behaviour (Fig. 1.5).

## MODEL REQUIREMENTS

We now attempt to list the requirements for a spatially dynamic model.

The model must define behaviour mathematically, operating in a real three- (or four-) dimensional space–time continuum (see Harper & Bell this volume); it must match the direct evidence from the real world for the decline in density with distance from a source of dispersal, the statistical relationship between spatial (or volumetric) variance and mean density, and the geographical pattern of populations. A model is incomplete if it fails to generate spatially fluid populations with these properties of distribution and abundance.

The model must not limit fitness; if it leads to reduced reproductive potential, this must be compensated by increased survival of offspring; it should explain such phenomena as the zoo syndrome and should be capable of interpreting social and sexual behaviour such as herding and lecking as well as colonial and territorial behaviour; it should apply in predictable and unpredictable, and in spatially and temporally heterogeneous, environments, and for apparently static as well as obviously migratory species. To match observations of natural populations, extinction should be rare and, to ensure evolutionary generality, it should not depend upon sex or environmental response, including predation and parasitism or competition (Lewontin 1970), although it should be capable of generating them. Lastly, and possibly most important, the model must be applicable to all species and to all stages of evolution; it should highlight the conditions existing at the first appearance of life.

We now make the following proposition; that the spacing of individuals is the indirect result of facultative or compulsive behaviour, evolved partly in response to other individuals, which may be treated as the net difference between mutual repulsion and attraction, both proportional to a power function of population density.

## THE MODEL

The model has been expressed as the difference of two power laws (Taylor & Taylor 1977) and therefore has the required property of a negative feedback loop,

$$\Delta = \Delta_m - \Delta_c = G\rho^p - H\rho^q \qquad (1.6)$$

where $\Delta_m$ and $\Delta_c$ are measures of repulsion ($\equiv$ migration) and attraction ($\equiv$ congregation) respectively. In its strictest form the $\Delta$-response relates the behavioural pressure exerted by one individual on another to the distance separating them, but it is easier to measure distance in relation to density, which is a property of average separation, although not a simple one.

In his brilliant paper on the selfishness of spacing behaviour Hamilton (1970) defined a six-sided 'domain of danger' which relates separation to density. If we represent the domain as a circle of radius $R$ instead of a polygon, then $a = \pi R^2$, where $a$ is the area of each domain, and the point density $\rho = 1/\pi R^2$. Thus, we can express density in terms of domain radius with minimum value equal to the radius of the animal, and maximum limited only by its ability to detect other individuals, its ambit of awareness, a property similar to the area of discovery of Nicholson & Bailey's (1935) model, and eqn (1.6) becomes $\Delta = gR^{-2p} - hR^{-2q}$. If both axes are in the same distance units, movements in response to separation and the return to optimum separation ($R_0 = (g/h)^{1/2(p-q)}$ can be plotted (Fig. 1.6).

Depending on the starting values, limit cycle oscillations in density can occur but are easily disrupted by birth and death, and recovery to $R_0$ after disturbance is usually quite rapid. When $p < q$, the feedback loop becomes positive, causing contraction of the population to accelerate as the separation decreases, an evolutionary improbability that leads inevitably to extinction.

Interpretation of $p$ and $q$ is best made in terms of 'life-style'. *Drosophila*

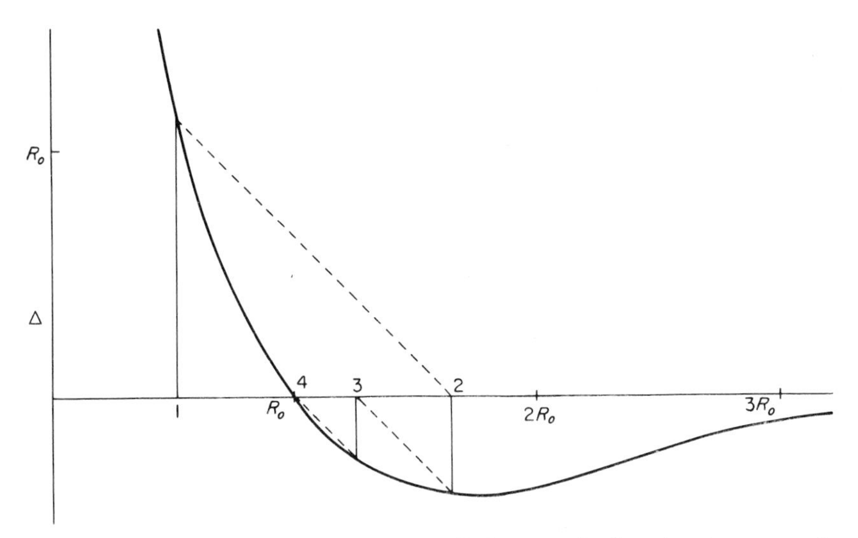


FIG. 1.6. When the Δ-function is plotted as displacement (ordinate) to the same scale as distance of separation (abscissa), $R = (\pi\rho)^{-1/2}$, spacing returns to an optimum $R_0$ (4). Movements in response to excessive natality (decrease in $R$) (1) overshoot $R_0$ to (2), which is equivalent to losses caused by mortality. Movements subsequently cause the separation to hunt (3) back to $R_0$ (4).

has small values for both $p$ and $q$, indicating only a small degree of density dependence and relatively small departure from random diffusion. Blackbirds are territorial and the Δ-function shows very weak attraction at low density, and then, just when density exceeds $\rho_0$ (the density corresponding to $R_0$), a very strong and sudden repulsion causing migration over large distances (Taylor & Taylor 1977). The small increase in density before the sudden change in response can be interpreted as tolerance of a small compression of territories. The small negative component to the Δ-function suggests that blackbirds are prepared to defend territories larger than those corresponding to $R_0$ and this inhibits invasion by immigrants (see the Beau Geste syndrome p. 23).

## SIMULATION EXPERIMENTS

An experimental arena of 15 600 environmental unit areas was first made uniformly hostile by allocating to each area a negative environmental score ($m = -1$) that killed any arriving individual. Areas were then made progressively more favourable by adding one to the environmental score of randomly placed patches of 9 or 25 contiguous units. One increment made the unit neutral ($m = 0$), allowing survival but not reproduction. Successive chance deposition of overlapping patches gave benign areas with $m = +1, 2, 3$, etc.

until the whole arena had an approximately Poisson distribution of non-negative scores with $\mu \simeq 2$. Each simulation run was assigned a basic reproductive rate, $r$, with the net reproductive rate equal to the product of $r$ and the environmental score, $m$, giving a range from $-r$ (death), and 0 (survival) to $r$, $2r$, $3r$, etc. In some experiments the arena was constant, in others it changed between generations, but in all cases the environment was inert; it did not respond in any way to the organisms' presence so there was no extrinsic density-dependent mortality.

Founder individuals, usually 16, were released randomly, except on hostile squares. Their intrinsic reproductive rate, $r$ per unit $m$ (maximum $m = 6$), was usually between 2 and 5, and effective reproductive rate, $mr$, had a Poisson distribution with means usually between 4 and 10, maximum 30. To test extreme pressures on the system, a few runs with $r = 15$ giving a maximum $mr$ of about 90 were performed, resulting in an increase in the number and violence of population explosions.

The arena was divided into 156 population elements of $10 \times 10$ environmental units, and individuals in each element moved towards or away from the 'gravitational' centre of their population element. The distance moved was in response to the average density of the element rather than their own point density, to give a random element of variance in the response. In asexual reproduction experiments, the individuals reproduced according to the unit they reached and there was no random mortality. To simulate sexual reproduction, hermaphrodites were permitted to reproduce only if they arrived in a population element containing another individual. After reproduction these parents died. The individuals that migrated out of the arena were presumed lost, i.e. the arena was an island.

About 150 'experiments' were done, with a total of over 5 000 generations.

### *Results from simulation*

With a range of different starting conditions, each resulting population growth curve was characterized by an increase in numbers to an equilibrium (Fig. 1.7a) from which the numbers fluctuated, more in the sexual than in the asexual populations. With high reproductive rates, populations had the appearance of the outbreaks called by Li & Yorke (1975) 'chaos', and simulations then sometimes took several generations to return to near the endemic level (Fig. 1.7b).

FIG. 1.7. (a) Simulated Δ-model population totals grow and stabilize in a typical logistic form, but the limit is intrinsic (behavioural) not extrinsic (resources).
(b) Some parameter values lead to numerically chaotic total populations with return times which may be prolonged (arrows mark relevant maps in Fig. 1.7c).
(c) When mapped, the geographical distributions range from spotted chaotic (cf. *Euxoa nigricans* Fig. 1.2)
(d) to stable with mobile holes (cf. *Xanthorhoë fluctuata* Fig. 1.2).

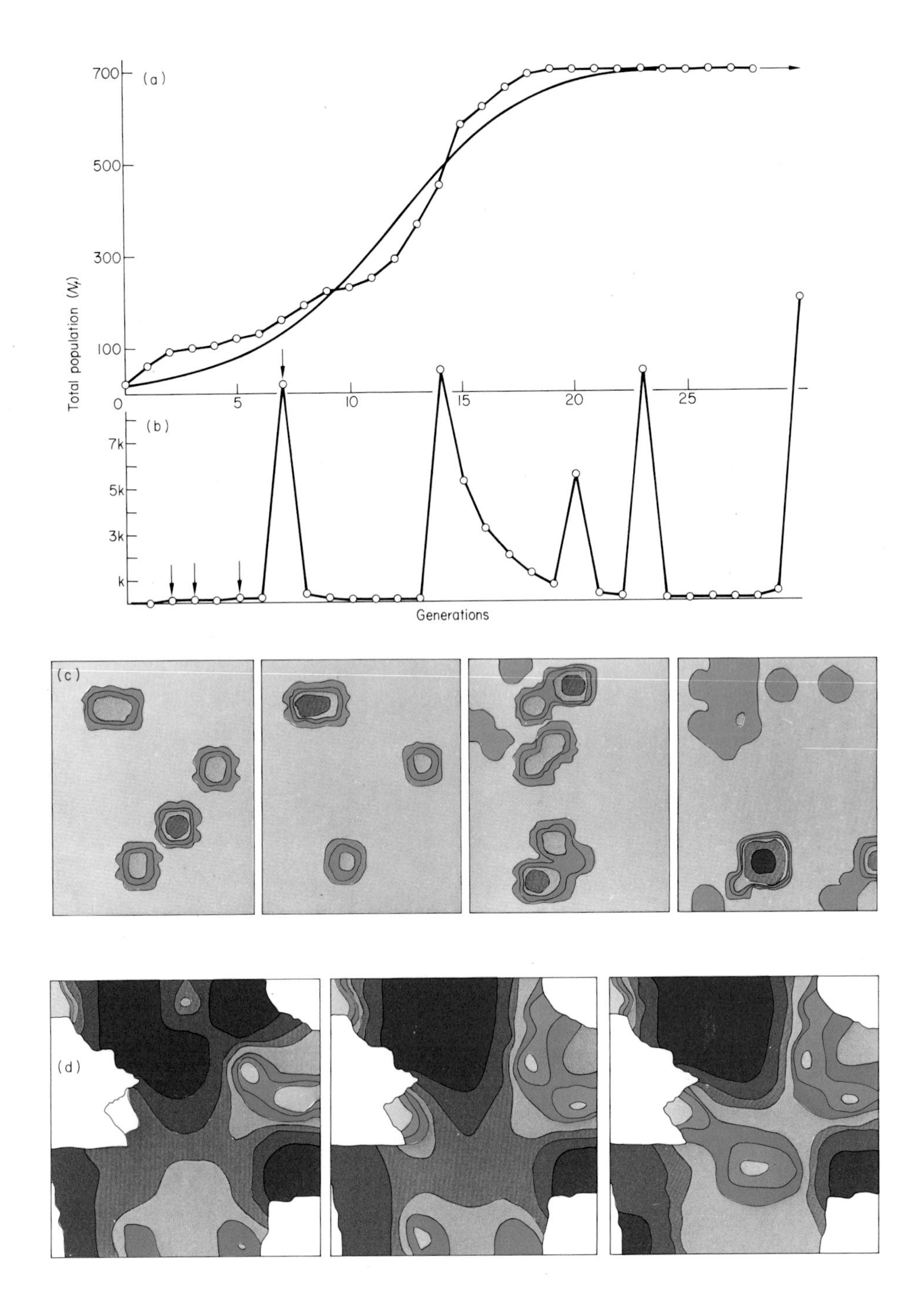

700
500
300
100
0
(a)
Total population ($N_t$)
5
10
15
20
25
(b)
7k
5k
3k
k
Generations
(c)
(d)

This flexibility is a feature of natural populations which is lost in the 'characteristic return time' of deterministic models (May *et al.* 1974). Populations are continuously mobile, in varying degrees (Fig. 1.7c & d) and outbreaks occur unpredictably, rarely at the same place twice, because the net reproductive rate is determined by both the animal and its environment, and migrants find the very benign regions fortuitously. The descendants of chaotic outbreaks often find themselves in less favourable habitats than their parents. In the simulation, evolution would tend to favour those that stayed behind because the habitat does not react and so remains undamaged by the high density it has been forced to support. In reality, such damage and the attraction of predators and parasites would reinforce the pressure to emigrate, or die.

Where the habitat remained fixed, the population stabilized somewhere near to $\rho_0$. In no case was every habitable unit occupied and only rarely were there more animals than units. The outcome was very sensitive to the starting unit of founder females, a very benign unit causing rapid population increase, with stability sometimes being achieved at higher total population number ($N_t$) and more elements being occupied. However, this is variable because the offspring rely for their total fecundity on the neighbouring elements which may be more or less hospitable than their mother's.

Several properties can now be predicted beforehand. The population will never remain spatially static and is unlikely to over-populate its habitat, in accordance with common observation. Also, extinction of species, as opposed to population elements, is comparatively rare but can be forced by making the benign areas too far apart for the animals to find them easily. Variance analysis of the simulation always produces the power function required to match natural populations, with exponents, $b$, ranging from about 1·0 to 3·5 (Fig. 1.2b, lines *a–g*). The relationship between density and distance generated by the $\Delta$-function (eqn (1.6)) conforms to the 3- or 4-parameter regression equations 1.4 and 1.5 which fit all species so far examined (see Fig. 4, Taylor & Taylor 1977). Changing the environment every generation can permit extinction by reducing fecundity and inhibiting movement simultaneously. In other words, a species can survive in a rapidly changing environment only by investing in a large number of highly mobile offspring, again as expected.

Comparison of binary and hermaphrodite simulations shows that the long-term results are surprisingly similar, indicating that reproductive strategy is less important than the degree of spatial cohesion defined by the parameters of the model. However, the rate of spread of a population element is to some extent governed by the reproductive strategy, so that sex appears as an adaptation to manipulate the environment. We have already commented that sexual behaviour may be associated with the process of congregation; we can see now that there is a potential positive feedback in sexual differentiation which may lead to extreme sexual dimorphism.

The model works equally well whether the density-dependent behaviour is immediate in response to the individual's congeners or is delayed in response to the parents'.

## DENSITY-DEPENDENT BEHAVIOUR

The Δ-model proposes density dependence of behaviour, measured as movement, but the simulation covers only immediate and delayed responses because of the limits on computer storage. In nature, anticipated or pre-saturation emigration is the more familiar and the evolutionary logic (p. 15) leads one to expect systematic adversity, such as seasonal or tidal cycles, to create such a response in the closed memory (genetic) system. The open system is more appropriate to unpredictable eventualities but even here the relationship is rarely simple or direct. Increased neural sophistication should make response to the environment more subtle and generate migration at an earlier stage in population growth, before density changes are necessarily obvious. Both systems may operate in the same organism. A non-migrant bean aphid, for example, can be made to produce all its remarkable range of behavioural and morphological emigrants by experimental stimulation after almost 1 000 generations of clonal non-migrants (Taylor & Taylor 1978); but a migrant cannot be stimulated to produce another migrant.

In pea aphids, the appearance of a predator can initiate density-dependent migration (Roitberg, Myers & Frazer 1979) and direct contact between any two individuals can stimulate both of them to produce emigrant offspring in many aphid species (Dixon 1977). Alternatively, the response may be focused in a single individual. In female water voles, it is the cowardly loser in an encounter that emigrates (Leuze 1976), but in social animals the activating cues may be from a single central, dominant individual and it is usually, but not always, the satellite individual who emigrates (Bekoff 1977). In the same way congregatory behaviour can sometimes appear to be largely concentrated on the sexual partner or on a dominant individual of the same sex. Agonistic and amicable behaviour may form a continuum depending on status, so determining who emigrates. Although in all these instances the immediate cue for the individual to move may be concentrated in an isolated event, the number of cues received collectively by all individuals is density-dependent. Closed and open behaviour may be confounded and we cannot know, *a priori*, whether density dependence is delayed, direct or anticipatory or a combination of these. Much migration leads to mortality, especially in juveniles, and any stabilization that results may then be allocated mistakenly to the immediate cause of death. Migration is not in itself a dangerous occupation but the movement may take the individual to a region of higher risk (e.g. predators) so that the outcome of migration may be increased local mortality. Also, interaction between variates confounds analysis and the problem is exacerbated in natural populations by the difficulties of measuring both density and movement simultaneously and by the large element of extraneous variance. Finally, and perhaps most important, because emigration reduces the density that creates it, the more effective the mechanism, the less easily can it be detected.

From the Δ-logic, immigration must also be density-dependent and, in contrast to emigration, this is more easily recognized because it relates largely to current density. In Webber's (1979) model for the Wytham Wood great tit population, for example, immigration is not only density-dependent but, as expected from the Δ-model, it also accounts for a larger proportion of the return time ($T_r$) than does reproduction. Even immigration may interact with density, however, if the Beau Geste effect (Krebs 1977) causes cunning residents to misrepresent their own density and thereby effectively discourage immigrants.

## CONCLUSIONS

The idea of feedback generated by the difference of two power functions is a familiar one in physics, for example the Lennard–Jones potential (Flowers & Mendoza 1970), but we must emphasize that we do not attribute the behaviour we describe to physical properties. Indeed, the essence of our argument is that too much attention has been paid in the past in population dynamics to purely physical processes of accretion and loss, and too little to the prime product of biological evolution, the ability of living organisms to behave as if they are alive and can repeatedly break the Newtonian laws of motion by choosing where and when to start and stop moving. If, as Mayr (1976) suggests, the unique characteristics of open populations are still inadequately interpreted by geneticists in spite of their persistence and mathematical brilliance, this may also be because abstract population genetics theory extrapolates too rapidly beyond the limits of the observed behaviour of real individuals.

In population dynamics a large body of patiently acquired field data and experimental evidence at both individual and population levels (see, for example, Dixon this volume) still lacks the techniques necessary to analyse it and a general theory to unite it. Although we may choose to ignore the problem for convenience in experiments and in small samples, these data are always non-linear when extrapolated over the many orders of magnitude required for prediction, either in practical applications (Steele this volume) or in theoretical modelling (May this volume).

Only very recently has the concept of generalized linear models (Nelder & Wedderburn 1972) made possible the development of programmes such as GLIM (Baker & Nelder 1978) capable of stabilizing variance effectively in such open-ended ecological data. Until these are in general use, in population ecology as well as in population genetics, there is a danger that we may continue to misinterpret moderately good data by finding what we expect to find, due to the inadequate fitting of progressively more sophisticated versions of the same basic Malthusian model based on the concept of resource limitation. In that event, like population genetics, population dynamics could become an appendage to abstract mathematics (Berry 1977). We have attempted to show that resources are never limiting, because they are never entirely depleted. The limiting factor is the ability of the individual to find them—its behaviour-controlled motion—so that premature mortality is always, in an evolutionary sense, attributable to the incompetence of the individual.

Dynamic stability and evolutionary change are the only alternative solutions to the conflicting requirements of survival in a changing world (Law this volume), and, as Thoday (1975), Thorpe (1978) and others have remarked, there is more to either than emerges from the human preoccupation with births and deaths. The principles of maximized reproduction and resource limitation are commonly accepted because, like Lack (1966) and Williams (1971), we can see no logical way for reproductive restraint to be inherited, and yet there must be some property that can perpetuate a species for hundreds of thousands of generations without change in, for example, the sculpture on the wing-case of a beetle (see p. 5). The likelihood that this continuity could be maintained against all the vicissitudes of change in the physical environment, by a beneficient but also constantly changing alliance of parasites, hyperparasites, predators, competitors and resource suppliers from outside the population, seems as Panglossian (Haldane in Lack 1966) as altruistic restraint from within, and an evolutionary risk that is unlikely to have survived.

## ACKNOWLEDGMENTS

We are grateful for this opportunity to thank the following for critical encouragement during the development of these ideas: V.C. Wynne-Edwards, V.B. Wigglesworth, M.J. Way, G.C. Varley, J.M. Thoday, T.R.E. Southwood, G. Murdie, S. McNeill, A.D. Lees, J.S. Kennedy, M.P. Hassell and W.D. Hamilton. The model was developed by one of us (R.A.J.T.) whilst on an A.R.C. grant and using computing facilities provided by the Ford Foundation.

Reference to other papers in this volume is meant to suggest only that we see connecting links with them, not necessarily that the other authors approve of our views, and we appreciate having had the opportunity to make these comments.

## REFERENCES

**Baker R.J. & Nelder J.A. (1978)** *The GLIM system manual, release 3.* Numerical Algorithms Group, Oxford.

**Bartlett M.S. (1936)** Some notes on insecticide tests in the laboratory and in the field. *Supplement to the Journal of the Royal Statistical Society*, **3**, 185–194.

**Bekoff M. (1977)** Mammalian dispersal and the ontogeny of individual behavioural phenotypes. *The American Naturalist*, **111**, 715–732.

**Berry R.J. (1977)** Evolution by natural selection—or not? *Biologist*, **24**, 236–238.

**Coope G.R. (1973)** Tibetan species of dung beetle from late Pleistocene deposits in England. *Nature, London*, **245**, 335–336.

**Coope G.R. (1975)** Mid-Weichselian climatic changes in western Europe, re-interpreted from coleopteran assemblages. *Quaternary Studies* (Ed. by R.P. Suggate & M.M. Cresswell) pp. 101–108. Royal Society of New Zealand, Wellington.

**Coope G.R. (1977a)** Quaternary Coleoptera as aids in the interpretation of environmental history. *British Quaternary Studies* (Ed. by F.W. Shotton) pp. 55–68. Clarendon Press, Oxford.

**Coope G.R. (1977b)** Fossil coleopteran assemblages as sensitive indicators of climatic changes during the Devensian (last) cold stage. *Philosophical Transactions of the Royal Society of London, B*, **280**, 313–340.

**Coope G.R. & Angus R.B. (1975)** An ecological study of a temperate interlude in the middle of the last glaciation based on fossil Coleoptera from Isleworth, Middlesex. *Journal of Animal Ecology*, **44**, 365–389.

**Dixon A.F.G. (1977)** Aphid ecology: life cycles, polymorphism and population regulation. *Annual Review of Ecology and Systematics*, **8**, 329–353.

**Dobzhansky Th. & Powell J.R. (1974)** Rates of dispersal of *Drosophila pseudoobscura* and its relatives. *Proceedings of the Royal Society of London, B*, **187**, 281–298.

**Dobzhansky Th. & Wright S. (1943)** Genetics of natural populations X. Dispersion rates in *Drosophila pseudoobscura*. *Genetics*, **28**, 304–340.

**Fisher R.A. (1922)** On the dominance ratio. *Proceedings of the Royal Society of Edinburgh*, **42**, 321–341.

**Flowers B.H. & Mendoza E. (1970)** *Properties of Matter*. John Wiley, London.

**Gibb J.A. (1977)** Factors affecting population density in the wild rabbit *Oryctolagus cuniculus* (L.), and their relevance to small mammals. *Evolutionary Ecology* (Ed. by B. Stonehouse & C. Perrins) pp. 33–46. Macmillan, London.

**Greenwood P.J. & Harvey P.H. (1976)** The adaptive significance of variation in breeding area fidelity of the blackbird (*Turdus merula* L.). *Journal of Animal Ecology*, **45**, 887–898.

**Greenwood P.J., Harvey P.H. & Perrins C.M. (1979)** The role of dispersal in the great tit (*Parus major*): the causes, consequences and heritability of natal dispersal. *Journal of Animal Ecology*, **48**, 123–142.

**Hamilton W.D. (1970)** Geometry for the selfish herd. *Journal of Theoretical Biology*, **31**, 295–311.

**Hamilton W.D. & May R.M. (1977)** Dispersal in stable habitats. *Nature, London*, **269**, 578–581.

**Hassell M.P. & Varley G.C. (1969)** New inductive population model for insect parasites and its bearing on biological control. *Nature, London*, **223**, 1133–1137.

**Holling C.S. (1959)** Some characteristics of simple types of predation and parasitism. *Canadian Entomologist*, **91**, 385–398.

**Huffaker C.B. (1958)** Experimental studies on predation: dispersion factors and predator–prey oscillations. *Hilgardia*, **27**, 343–383.

**Hutchinson G.E. (1965)** *The Ecological Theatre and the Evolutionary Play*. Yale University Press, New Haven.

**Iwao S. (1977)** The $m^* - m$ statistics as a comprehensive method for analyzing spatial patterns of biological populations and its application to sampling problems. *Japanese International Biological Programme Synthesis*, **17**, 21–46.

**Johnson C.G. (1969)** *Migration and Dispersal of Insects by Flight*. Methuen, London.

**Kennedy J.S. (1969)** *The Relevance of Animal Behaviour*. Imperial College of Science and Technology, London.

**Krebs C.J. & Boonstra R. (1978)** Demography of the spring decline in populations of the vole *Microtus townsendii*. *Journal of Animal Ecology*, **47**, 1007–1015.

**Krebs J.R. (1977)** The significance of song repertoires: the Beau Geste hypothesis. *Animal Behaviour*, **25**, 475–478.

**Krebs J. R. & Perrins C. (1978)** Behaviour and population regulation in the great tit (*Parus major*). *Population Control by Social Behaviour* (Ed. by F.J. Ebling & D.M. Stoddart) pp. 23–47. Institute of Biology, London.

**Lack D. (1966)** *Population Studies of Birds*. Clarendon Press, Oxford.

**Lees A.D. (1975)** Aphid polymorphism and 'Darwin's Demon'. *Proceedings C of the Royal Entomological Society of London*, **39**, 59–64.

**Leuze C.C.K. (1976)** Social behaviour and dispersion in the water vole *Arvicola terrestris*, Lacepede. *Unpublished Ph.D. thesis, University of Aberdeen.*

**Lewontin R.C. (1970)** The units of selection. *Annual Review of Ecology and Systematics*, **1**, 1–18.

**Li T.-Y. & Yorke J.A. (1975)** Period three implies chaos. *American Mathematical Monthly*, **82**, 985–992.

**Lidicker W.Z. Jr (1975)** The role of dispersal in the demography of small mammals. *Small Mammals: Their Productivity and Population Dynamics* (Ed. by F.B. Golley, K. Petrusewicz & L. Ryszkowski) pp. 103–128. University Press, Cambridge.

**Łomnicki A. (1978)** Individual differences between animals and the natural regulation of their numbers. *Journal of Animal Ecology*, **47**, 461–475.

**Matthews J.V. (1976)** Evolution of the sub-genus *Cyphelophorus* (Genus *Helophorus*, Hydrophilidae, Coleoptera): description of two new fossil species and discussion of *Helophorus tuberculatus* Gyll. *Canadian Journal of Zoology*, **54**, 652–673.

**May R.M. (1978)** Host-parasitoid systems in patchy environments: a phenomenological model. *Journal of Animal Ecology*, **47**, 833–844.

**May R.M., Conway G.R., Hassell M.P. & Southwood T.R.E. (1974)** Time delays, density-dependence and single species oscillations. *Journal of Animal Ecology*, **43**, 747–770.

**Mayr E. (1976)** *Evolution and the Diversity of Life*. Belknap Press, Cambridge, Massachusetts.

**Nelder J.A. & Wedderburn R.W.M. (1972)** Generalised linear models. *Journal of the Royal Statistical Society, A*, **135**, 370–384.

**Nicholson A.J. & Bailey V.A. (1935)** The balance of animal populations. Part I. *Proceedings of the Zoological Society of London*, **105**, 551–598.

**Paris O.H. (1965)** The vagility of $P^{32}$-labelled Isopods in grassland. *Ecology*, **46**, 635–648.

**Pearson K. & Blakeman J. (1906)** Mathematical contributions to the theory of evolution. XV. A mathematical theory of random migration. *Drapers Company Research Memoirs Biomathematical Series*, **3**, 1–54.

**Rogers D.J. (1972)** Random search and insect population models. *Journal of Animal Ecology*, **41**, 369–383.

**Roitberg B.D., Myers J.H. & Frazer B.D. (1979)** The influence of predators on the movement of apterous pea aphids between plants. *Journal of Animal Ecology*, **48**, 111–122.

**Skellam J.G. (1951)** Random dispersal in theoretical populations. *Biometrika*, **38**, 196–218.

**Southwood T.R.E. (1977)** Habitat, the templet for ecological strategies? *Journal of Animal Ecology*, **46**, 337–365.

**Taylor L.R. (1961)** Aggregation, variance and the mean. *Nature, London*, **189**, 732–735.

**Taylor L.R. (1971)** Aggregation as a species characteristic. *Statistical Ecology*, **1** (Ed. by G.P. Patil, E.C. Pielou & W.E. Waters) pp. 357–377. Pennsylvania State University Press, University Park.

**Taylor L.R. (1974)** Monitoring change in the distribution and abundance of insects. *Report of the Rothamsted Experimental Station for 1973, Part 2*, 202–239.

**Taylor L.R. (1977a)** Aphid forecasting and the Rothamsted Insect Survey. *Journal of the Royal Agricultural Society of England*, **138**, 75–97.

**Taylor L.R. (1977b)** Migration and the spatial dynamics of an aphid. *Journal of Animal Ecology*, **46**, 411–423.

**Taylor L.R., French R.A. & Woiwod I.P. (1978)** The Rothamsted Insect Survey and the urbanization of land in Great Britain. *Perspectives in Urban Entomology* (Ed. by G.W. Frankie & C.S. Koehler) pp. 31–65. Academic Press, New York (in press).

**Taylor L.R. & Taylor R.A.J. (1977)** Aggregation, migration and population mechanics. *Nature, London*, **265**, 415–421.

**Taylor L.R. & Taylor R.A.J. (1978)** The dynamics of spatial behaviour. *Population Control by Social Behaviour* (Ed. by F.J. Ebling & D.M. Stoddart) pp. 181–212. Institute of Biology, London.

**Taylor L.R., Woiwod I.P. & Perry J.N. (1978)** The density-dependence of spatial behaviour and the rarity of randomness. *Journal of Animal Ecology*, **47**, 383–406.

**Taylor L.R., Woiwod I.P. & Perry J.N. (1979)** The negative binomial as a dynamic ecological model for aggregation, and the density-dependence of *k*. *Journal of Animal Ecology*, **48**, 289–304.

**Taylor R.A.J. (1978)** The relationship between density and distance of dispersing insects. *Ecological Entomology*, **3**, 63–70.

**Taylor R.A.J. (1979)** Simulation studies and analysis of migration dynamics. *Unpublished Ph.D. thesis, University of London.*

**Thoday J.M. (1975)** Non-Darwinian 'evolution' and biological progress. *Nature, London*, **255**, 675–677.

**Thorpe W.H. (1978)** *Purpose in a World of Chance.* Oxford University Press, Oxford.

**Ulbrich J. (1930)** *Die Bisamratte.* Heinrich: Dresden.

**Varley G.C. & Gradwell G.R. (1960)** Key factors in population studies. *Journal of Animal Ecology*, **29**, 399–401.

**Verhulst P.-F. (1838)** Notice sur la loi que la population suit dans son accroisement. *Correspondance Mathematique et Physique*, **10**, 113–121.

**Waddington C.H. (1975)** *The Evolution of an Evolutionist.* Edinburgh University Press, Edinburgh.

**Wadley F.M. & Wolfenbarger D.O. (1944)** Regression of insect density on distance from centre of dispersion as shown by a study of the smaller European elm bark beetle. *Journal of Agricultural Research*, **69**, 299–308.

**Wallace B. (1966)** On the dispersal of *Drosophila. American Naturalist*, **100**, 551–563.

**Watson A. (1977)** Population limitation and the adaptive value of territorial behaviour in Scottish red grouse, *Lagopus l. scoticus. Evolutionary Ecology* (Ed. by B. Stonehouse & C. Perrins) pp. 19–26. Macmillan, London.

**Webber M.I. (1979)** A population model for the great tit *Parus major* L. in Wytham Wood. *Journal of Animal Ecology*, **48** (in press).

**Williams C.B. (1930)** *The Migration of Butterflies.* Oliver & Boyd, Edinburgh.

**Williams G.C. (Ed.) (1971)** *Group Selection.* Aldine-Atherton, Chicago.

**Wilson E.O. (1975)** *Sociobiology: The New Synthesis.* Belknap Press, Cambridge, Massachusetts.

**Wolfenbarger D.O. (1946)** Dispersion of small organisms, distance dispersion rates of bacteria, spores, seeds, pollen and insects; incidence rates of diseases and injuries. *American Midland Naturalist*, **35**, 1–152.

**Wolfenbarger D.O. (1959)** Dispersion of small organisms. Incidence of viruses and pollen; dispersion of fungus spores and insects. *Lloydia*, **22**, 1–106.

**Wynne-Edwards V.C. (1962)** *Animal Dispersion in Relation to Social Behaviour.* Oliver & Boyd, Edinburgh.

# 2. THE POPULATION DYNAMICS OF GROWTH FORM IN ORGANISMS WITH MODULAR CONSTRUCTION

JOHN L. HARPER AND A. D. BELL

*School of Plant Biology, University College of North Wales, Bangor, Wales*

Almost all the studies that make up the literature of population dynamics have been concerned with a rather specialized type of organism. Voles, hares, *Drosophila*, *Tribolium* and man exemplify life histories in which the zygote develops to a unitary organism, the genotype of which specifies a unitary morphology and a life cycle that proceeds remorselessly from juvenility through a reproductive phase to senility and death. In most plants, however, and a great many animals the genotype is expressed in a quite different way. The zygote develops to a modular organism in which a basic structural unit (or several forms of a unit) is iterated.

In the animal kingdom iterative modular form is characteristic of hydroids, bryozoans, colonial ascidians as well as many protozoa. The iterated modules may remain joined to give a characteristic morphology as in *Obelia* or corals or the modules may become separated from each other as in *Hydra*—but in both cases the processes that lead from the zygote to descendent zygotes* involve the iteration of basic modules of phenotypic expression. It is interesting that one of the classic studies of population dynamics, that of Gause on *Paramecium* and *Didynium* (Gause 1934), concerned the dynamics of clonal multiplication, not the dynamics of zygote to zygote multiplication.

Higher plants are all organized as iterations of a basic constructional module, the leaf with its axillary bud (or below ground as a unit of root growth, iterated by endogenous branches). The characteristics of the plant that are taxonomically useful are the features of the module—the shape and sizes of leaves, floral modules, etc.—not the shapes and form of the whole. A further level of modular construction can be recognized in the grouping of leaves and axillary buds into a higher level of repeated structure—e.g. long shoots and short shoots. Hallé & Oldeman (1970) have used the organization of such shoots (the blueprint) to characterize the architecture of tropical trees.

* We stress 'zygote to zygote' in the same way as Williams (1975) because it defines the beginning and the end of the life cycle as it is relevant to and subject to evolutionary change.

Oldeman (1974) has shown how a 'réitération' of this blueprint explains the apparent complexity of the majority of mature trees. In perennial herbs growing by stolons or rhizomes, characteristic shoot forms are reiterated to give a population of shoots (as in a tillering grass, or a clone of *Mercurialis perennis*).

There is a strong historic tendency among both zoologists and botanists to identify the unit of modular construction as 'the individual'. To most zoologists the hydranth of, for example, *Obelia* is the 'individual' and the individuals are assembled in colonies. To most botanists the rosette of the strawberry plant is the 'individual' and the concept of 'vegetative reproduction' emphasizes this. However, both the botanical concept of vegetative reproduction and the zoological concept of colony formation are really descriptions of a process of growth (not reproduction) by which a zygote gives rise to daughter zygotes. In multicellular organisms the process of reproduction can be distinguished from processes of growth (Bonner 1974). In *reproduction*, a new organization is initiated from a single cell. In *growth*, modular organisms simply iterate the basic structural unit from an already highly organized meristem *or* cellular structure.

In some modular organisms the products of growth fragment so that the genet (product of a single zygote) develops as physically separated parts as in *Lemna*, *Hydra* and strawberries. Fragmentation is just one variant in the pattern of growth by which a zygote of a modular organism is ultimately multiplied to give rise to daughter zygotes. Thus the study of the dynamics of population growth in duckweeds (*Lemna* species) is a study of the iteration of fronds produced by a fraction of the products of a single zygote (Clatworthy & Harper 1961). A population of duckweed fronds is made up of the parts of a single plant that continually falls to pieces and is formally equivalent to a population of *Hydra* (which is an *Obelia* that continually falls to pieces).

If we are to 'explain' the form and behaviour of organisms, the ultimate form of explanation has to be in evolutionary terms. 'Nothing in biology has meaning except in the light of evolution' (Dobzhansky 1973). The zygote is the point in the life cycle at which new genetic recombinants are produced and their relative fitness is expressed in daughter zygotes. The 'survival machines' of genes are the life cycles that link parental zygotes to progeny zygotes. The survival machine may be unitary (as in *Drosophila* or man) or modular (as in *Obelia* and buttercups). If the phenotype is modular, it may be expressed as a fragmentary system (*Hydra*, *Lemna*, strawberry runners) or maintain a colonial identity (*Obelia*, corals, oak trees, a bracken rhizome system). When fragmentation occurs it may be programmed (as in *Lemna*) or simply result from decay or damage to interconnections (*Trifolium repens* L., *Pteridium aquilinum* L.). The product of an individual zygote of *Hydra* may be represented by the whole population of free living polyps in a pond just as the product of an individual zygote of an oak is the tree with its whole population of shoots composing the canopy, or a zygote of *Lemna* may fill a pond with its separated fronds.

A whole set of demographic properties stems directly from modular patterns of growth. These appear to be common to plants and those animals that have modular growth patterns. The population dynamics of plants is not different from that of animals; rather, the population dynamics of organisms with modular construction is different from that of organisms of unitary construction. The differences are profound and have important consequences in population biology:

(i) The parts of modular organisms have their own birth and death rates; a genet has its own internal population dynamics *and*

(ii) the placement of modular units determines the form of the organism. Form is a consequence of the dynamics.

## THE POPULATION DYNAMICS OF MODULAR ORGANISMS

1 The populations of modular organisms need to be studied at two distinct levels—the number of zygotes that are represented ($N$) and the number of modules developed from each zygote ($\eta$). The fundamental equation of population biology,

$$N_{t+1} = N_t + \text{Births} - \text{Deaths} + \text{Immigrants} - \text{Emigrants} \qquad (2.1)$$

applies at both levels. This is easily illustrated by the growth of a grass, *Lolium perenne* L., sown from seed ($N$) and in which the plants grow by the multiplication of modular units, the tillers ($\eta$). In this example (Fig. 2.1) the number of zygotes represented in the population declines in a density-dependent manner while the number of tillers initially increases (the lower the density the faster the increase), reaches a peak and then declines. The early reflection of the effects of density appear in the *death* rate of genets ($N$) and in the *birth* rate of modules ($\eta$). The effect of these two processes together is that the population density of modules (tillers) rapidly becomes independent of the density of genets. Interpretation of the population dynamics of such a sward is meaningless without concern with the population processes at both the levels of $N$ and $\eta$ (Kays & Harper 1974). Ultimately, when such a grass sward comes to flower and produce seed, the seed progeny from the population will carry a biased balance of genes from the parental seeds—biased (a) by the elimination of some original genets and (b) by the differential production of tillers (and hence of flowers and seeds) by the surviving genets.

2 The product of a zygote of a modular organism has population properties of its own. Even an annual plant, because it is made up of iterated units of leaf

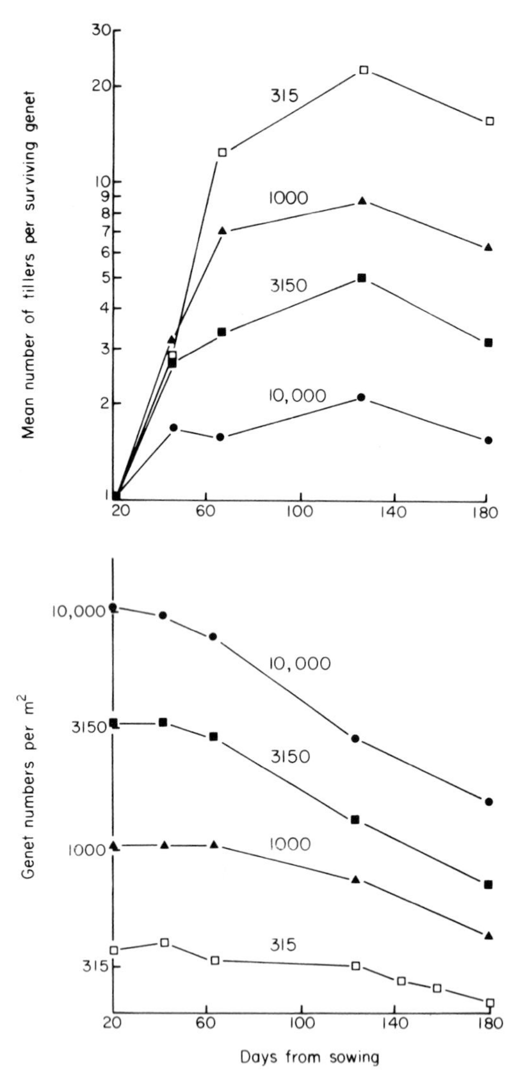


FIG. 2.1. The development of populations of *Lolium perenne* sown from seed at a range of densities. Calculated from data in Kays & Harper (1974).
(a) Changes in the average number of tiller modules ($\eta$) per surviving plant (genet).
(b) Changes in the number of surviving plants (genets).

and axillary bud and reiterated units of branching, is a population of parts with demographic properties. The annual *Linum usitatissimum* L. (flax or linseed) develops a population of leaves by iterative births of primordia at the stem apex. Some basal buds commonly grow out to form branches (reiterations of the form of the main shoot). Leaves have a limited length of life and the population of leaves on a plant has an age structure (Fig. 2.2). Demographic concepts such as expectation of life can realistically be applied to the modules of the shoot system (and presumably could be applied to the root as well). Figures 2.2 and 2.3 illustrate two applications of this type of analysis. Figure 2.2 shows the effects of light intensity on the changing age structure of leaves in the canopy of a population of plants of *Linum*. Leaves change their activity with age; in particular, their photosynthetic activity rises to a plateau and then declines. Hence the population of leaves that constitutes a canopy should have an activity determined by (i) the age structure of a canopy and (ii) the age-specific photosynthetic activity of the leaves. It should be possible to analyse the assimilation of a mass of vegetation as the product of (i) and (ii)—the equivalent of the classic $b_x l_x$ integration in population growth equations.

Figure 2.3 shows the effects of specific nutrients in the growth medium on the survivorship curves of the leaves of individual plants of *Linum* grown in water culture. In this case the population dynamics of the canopy reveals in a formal, quantitative manner, phenomena that are normally described only subjectively.

3 Modular organisms tend to grow exponentially at least in their early stages. The earliest modules developed from a zygote are usually vegetative units and later sexual modules are formed in a possibly infinitely extended phase of sexuality and the production of daughter zygotes. This is the pattern exemplified by the growth of a tree or the growth of a coral such as the Gorgonian coral illustrated in Fig. 2.4. Sometimes the vegetative phase, in which non-reproductive modules are produced, is ended abruptly by a lethal phase of sexual activity in which all modules are generative (e.g. the foxglove *Digitalis purpurea*, Harper 1977). More often zygotes are produced iteroparously through a prolonged reproductive life and the organisms grow by repeated production of a mixture of vegetative and generative modules. The pattern of growth of genets that continually increase their fecundity (though the expectation of life may continually fall) leads to some curious consequences. In higher plants and predictably in modular animals, the frequency distribution of fecundities of the genets becomes log-normal; a very few individuals within the population contribute overwhelmingly to the zygotes that are hazarded to the next generation (Harper 1977). This has effects on the potential rates of evolutionary change under selection pressure. Levins (1977) concludes that such a log-normal frequency distribution of fecundities will have the effect of greatly speeding up the fixation of genes in populations under selection pressure.

4 Modular organisms commonly show polymorphism in the form of the modules—particularly between 'vegetative' and 'generative' modules. At the very fundamental level at which classical morphologists describe plant form,

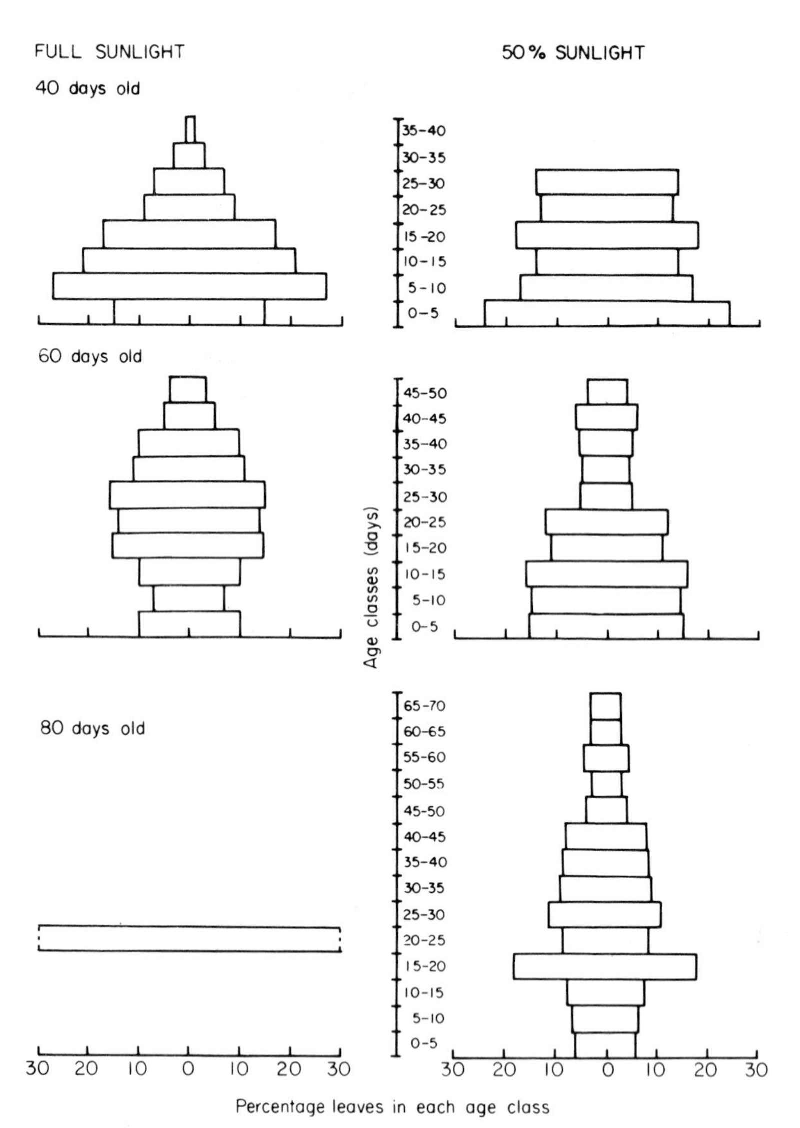


FIG. 2.2. The age structure of the population of leaves of *Linum usitatissimum* grown in full light and under plastic mesh that reduced light intensity by 50%. In full light the plants matured more rapidly so that 80-day-old plants had lost all leaves of more than 25 days old and had ceased to produce new leaves. In contrast, on the slower developing plants in the shade, leaves had a higher expectation of life and the birth of new leaves continued for longer, with the result that 80-day-old plants bore a wide age spectrum of leaves (calculated from data in Bazzaz & Harper 1977).

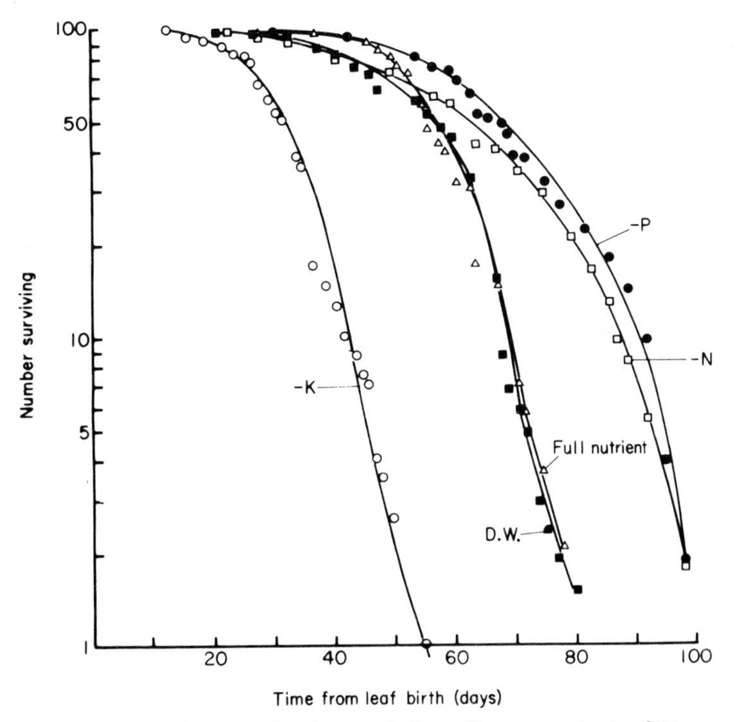


FIG. 2.3. Survivorship curves for the population of leaves on plants of *Linum usitatissimum* grown in sand culture and provided with (a) ■ deionised water, (b) △ full Long Ashton nutrient solution, (c) ○ as (b) but without potassium, (d) ● as (b) but without phosphorus and (e) □ without nitrogen.

the flowering plant has a polymorphism of modular construction involving stems bearing photosynthetic leaves, attractive flowers, stamens and carpels—homologous organs with different forms and functions. Comparable polymorphism in colonial hydroids is that of vegetative and generative polyps (in e.g. *Obelia*, *Podocoryne*, etc.). It is interesting to envisage natural selection acting on the balance of such somatic polymorphism, maximizing fitness by adjusting the proportion of modules allocated to conflicting demands for resources. A relevant form of analysis would be that of fitness sets as used by Wilson (1971) to analyse an analogous polymorphism—that of the balance between expenditure on different castes in a hymenopteran colony. In organisms with unitary construction somatic polymorphism has no real meaning, but in modular organisms it can reach very high complexity (in e.g. the Portuguese Man of War, *Physalia*). Another example of modular polymorphism is the development of short shoots and long shoots on the branches of trees, which makes a major contribution to their elusive quality of form.

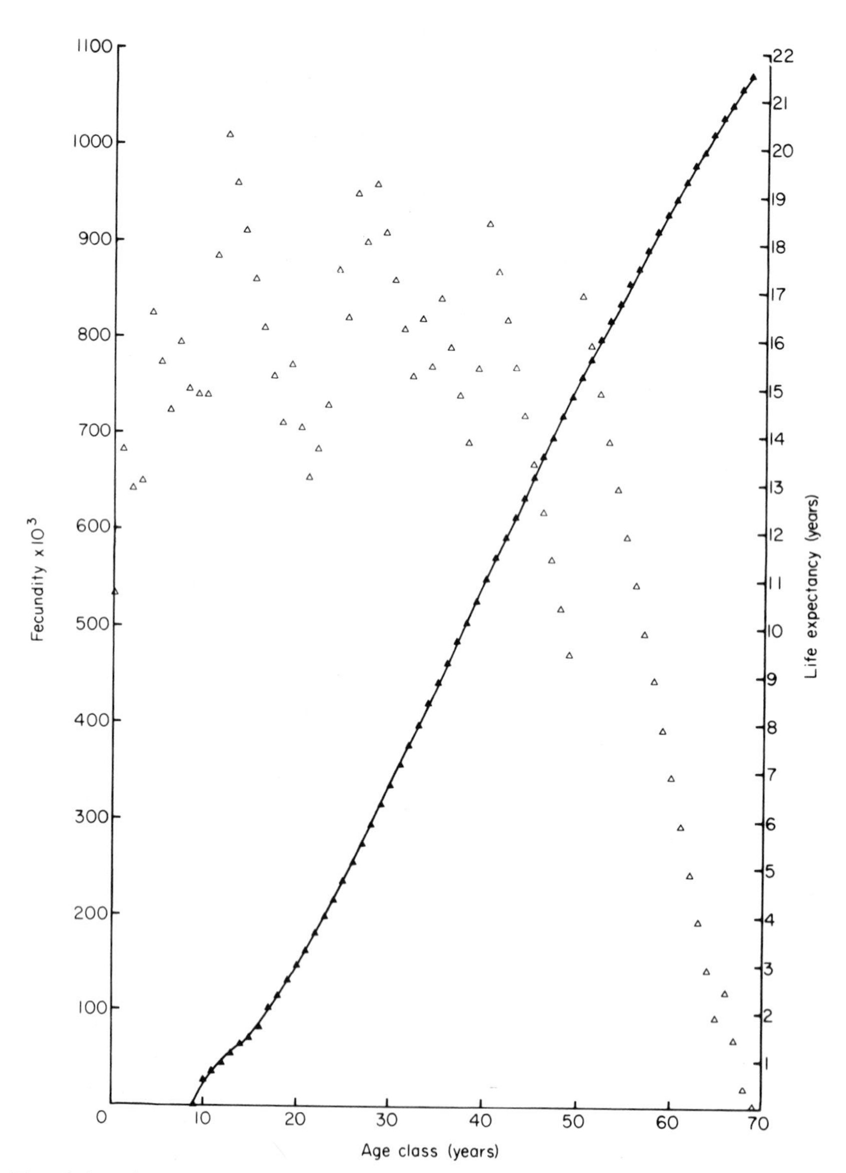


FIG. 2.4. The fecundity (▲) and life expectancy (△) of a Gorgonian coral (*Muricea californica*) drawn from data in Grigg (1977).

5 In unitary organisms (man, *Drosophila*) it is possible (and essential for predictive demography) to describe an age-specific fecundity schedule. Most of recent theorizing on the meaning of life cycle 'strategies' depends on the consequences of different fecundity schedules [growing directly from the work of Fisher (1929) and developed in the studies of Lamont Cole (Cole 1954), Lewontin (1965), etc.]. In most modular organisms age-specific fecundity schedules have almost no meaning because although a juvenile period can usually be recognized (when all the modules are vegetative, as in a young *Podocoryne*, or a young oak tree) the reproductive period can, at least in theory, continue indefinitely (see Fig. 2.4). Moreover, in many plants and modular animals new tissue continually grows away from and discards the ageing parts of the soma and there is no intrinsic cause of death. Single sporelings of *Pteridium aquilinum* are thought to have been continually expanding their territory for at least 1 400 years and a single clone has been recorded covering an area 474 × 292 m (Oinónen 1967). There is no reason to suppose that there is any inherent doom in the life of such organisms and one is led to question Hamilton's view that senescence is a necessary outcome of evolution (Hamilton 1966). Similarly, there is no apparent reason why a colony of bryozoans, e.g. *Alcyonidium*, should have any limit to the number of zooids produced per genet, except that its common habitat, a *Fucus* frond, has a limited life. Instead of the classic picture of organisms moving through phases of juvenility and reproductive activity to senescence and death, the genet expands indefinitely, steadily increasing the number of modules of which it is composed so that fecundity rises indefinitely. Moreover, a genet may grow to many modules, suffer from predation, competition or damage and then reiterate its form again from the few modules that remain. Such rejuvenation is not a feature associated with the natural life of unitary organisms. Such features have led some botanists to discard age structure as a useful element in population biology and to replace it by other concepts such as the 'life state' of organisms in a population (see discussion of the attitudes of the Russian school of Rabotnov *et al.* in Harper 1977).

6 Modular construction is (with a few exceptions) characteristic of anchored organisms, the rooted plants or attached seaweeds, or many of the fixed animals, such as corals or hydroids. For such organisms there is no escape from competitors or predators by running away. The consideration of the population biology of these organisms immediately reveals that the concept of density so widely used in the studies of population dynamics is an abstraction which is needed only by biologists who study motile organisms. Even in the case of motile organisms we know that the density of the population as a whole is not what is sensed by the reaction of any individual within it. Increasingly the behaviourist's emphasis on territory drives the interpretation of populational phenomena away from a concern with abstract 'density' to the behaviour of

individuals. In motile animals it is, however, usually difficult to isolate influences of competition at the level of interaction between individuals and it becomes necessary to use 'density' to summarize individual events at a statistical level. In the case of plants and modular fixed animals, density is an unnecessary concept because the spatial fixity of individuals defines which neighbours will influence them. Instead of density, the more precise measures of size and distance define the relationship between individuals.

The individual may react to the presence of its neighbours by a reduced birth rate of its parts and/or an increased death rate of its parts. Indeed one side of a tree or a Bryozoan may be dying under the influence of an aggressive neighbour while the other side of the same genet may be growing vigorously. The precision of the effects of neighbours has been demonstrated by Mack & Harper (1977), who have shown that 69% of the variance of size and fecundity among individuals of populations of dune annuals can be accounted for from a knowledge of the size, distance and species of the nearest neighbours. Stimson (1974) showed that the diameter of individual corals was inversely related to their distance apart. In populations of modular organisms the spatial distribution and the extension of individual genets is overwhelmingly important in the game of space capture, where space is the equivalent of resources (Ross & Harper 1972; Maguire & Porter 1977).

Among populations of modular organisms, for example plants growing in a pasture, trees in a canopy or corals on a shore, the shape and form of the developing genet has a special role in determining the nature and magnitude of the inter- and intra-specific contacts that are made. Within a plant community one can differentiate plants that expand their modules in a 'guerilla' strategy, e.g. extending stolons or branches into neighbouring vegetation and maximizing inter-specific contacts. In contrast, one can recognize 'phalanx' strategies in which the arrangement of modules of the genet ensures a tight packing so that the plant colonizes as a dense invasive front (Clegg 1978).

Community diversity takes on a new meaning for organisms of modular growth form. Yet again a concept that is widely used amongst students of motile animals seems irrelevant when considering communities of modular organisms. Species diversity as a characteristic of an area of land or of water is an abstraction which is often necessary where organisms are motile and where the individual experiences from inter- and intra-specific neighbours cannot be easily quantified. However, in a pasture or a coral community the diversity or monotony of the biotic environment can be measured as the experience of contacts made between individual modules. Turkington (1975) has recently analysed the contacts made by leaves within a permanent pasture in order to answer the question 'How does the individual organism sample the diversity of the community?' The grain of the biotic environment can then be measured as the cumulative experience of individuals instead of some statistical abstraction about communities or an area of land.

7 In classical predator–prey models it is assumed that a predator kills its prey. A predator–prey model involving modular organisms has, however, to take account of predators that may remove only parts of a genet. The activity of a grazing animal in a pasture involves the removal of modular units from plants. Most of the plants (genets) will not be killed but will reiterate. In the terminology used earlier one may say that the predator affects the numbers of modules ($\eta$) without necessarily affecting the number of genets ($N$) present in the population. A predator may affect individuals in the population differentially by taking more modules from some than from others and so affect their relative growth rates and their potential fecundity. This is apparently what happens when a predator such as a sheep selectively chooses to take clover leaves with one particular leaf mark from a polymorphic population of leaves (Cahn & Harper 1976).

8 Modular organisms that fragment may be dispersed as whole modules or groups of modules (e.g. fronds of *Lemna*, rosettes of *Pistia*, *Stratiotes* or individual hydroids of a genet of *Hydra* are dispersed as whole modules). More generally the dispersal phase of the life cycles of modular organisms is accomplished by very small propagules, usually zygotes or juveniles and usually produced in great abundance. There is an almost complete absence of any phenomenon that could be called search in the finding of new sites for establishment. This contrasts very strikingly with most motile animals in which it is adults that disperse and determine the habitats of juveniles.

## FORM AS A CONSEQUENCE OF THE POPULATION DYNAMICS OF MODULES

The size of a plant or a modular animal is a function of the number of the modular units produced and the number that have died. Thus, the development of a genet has a population dynamics in its own right. A bud may die, may remain dormant or may grow out to a shoot producing daughter buds. The form of the resultant genet is determined by these essentially demographic properties with the addition of geometric description of the placement of the new modules that are born and the old modules that die.

The description of a plant structure in terms of units of growth is not new (see Arber 1950). We have shown, however, that in certain circumstances the recognition that a plant is a population of parts with demographic properties begins to make it possible to monitor plant growth with greater sensitivity than in conventional growth analysis with its gross consideration of leaf area index or increase in dry weight, etc. Moreover, demographic analysis of plant growth can be done without destructive sampling.

In simple relatively unbranched annual plants such as *Linum usitassimum* it is convenient to analyse the demography of leaves as modules. In much-branched annuals and perennial plants the bud may often be the more appropriate unity. Every shoot on a plant was at one time merely a bud or, more correctly, a meristem. The population of meristems on a plant represents both the potential of that plant for further growth and reflects the history of that plant (every meristem has had ancestral meristems). By considering meristem potential, position and fate, we are able to combine an understanding of the organized structure of the plant with an appreciation of the demography of its parts. This means that it should become possible to model and predict the future performance and potential productivity of a plant and perhaps ultimately relate this to environmental influences on the birth and death rates of parts.

A study of the modular construction of plants has already provided the means of deciphering the architecture of tropical trees (Hallé & Oldeman 1970). Moreover, it has been possible to model the growth of rhizomatous perennials in order to compare morphological strategies of 'space' capture (Bell 1974, 1976, 1979; Bell & Tomlinson 1979).

A germinating seed commences life with a limited number of shoot meristems, that of the plumule and perhaps two more in the axils of the cotyledons in the case of a dicotyledon. However, each of these developing meristems may produce leaves, and each leaf generally subtends a new meristem.

Each member of this potentially expanding population is destined for one of four fates:

1 it may be aborted
2 it may be damaged and die
3 it may remain dormant
4 it may develop into a shoot.

If a shoot develops, the meristem may continue growth indefinitely and give rise to a monopodial branch, or it may cease growth and become one unit of a sympodial branch. Dormant meristems may eventually grow and then suffer one of fates (1), (2) or (4) above. When a meristem does develop it may do so rhythmically or continuously (Koriba 1958; Hallé & Oldeman 1970; Zimmermann & Brown 1971) and it will develop into one particular type of shoot or module, e.g. short shoot, long shoot, reproductive shoot. A plant may therefore be defined in terms of the types of module it bears, their location, time of growth and time of death.

The architecture of the plant represents 'successful' meristems and both the static and dynamic structure of the plant thus depends on its meristem 'bank', on meristem potential, position and fate. We have attempted to use these concepts in computer modelling of growth and form in several contrasting types of modular plant and animal. These are illustrated in the following examples.

### 1 Carex arenaria *L.* (*Sand sedge*)

*Carex arenaria* invades patches of bare sand at the base of mobile dunes by means of thin linear rhizomes. A study of the rhizome morphology shows that it is strictly sympodial and has a very conservative organization (Čelakovsky 1881; Noble, Bell & Harper in press.). Each unit of the sympodium consists of a horizontal region bearing scale leaves and roots, and a vertical region bearing foliage leaves and terminating in a potential inflorescence. When a sympodial unit becomes vertically orientated, the advance of the rhizome would cease except that a new sympodial unit invariably develops to replace it unless damaged.

The *Carex* system demonstrates the four fates of meristems outlined above, and it can be described in terms of a simple dimorphism of just two module types (see Fig. 2.5(a)). The plant consists of a sequence of sympodial units (Module Type 1) and 'tillers' (Module Type 2). The 'build-up' of the plant in terms of these modules is shown as a computer drawing in Fig. 2.5(b). A population of tillers is developing as Type 1 modules which bear more Type 1 modules, thus advancing the invasion of the community by the genet. At the same time Type 2 modules bear more Type 2 modules. The figure also shows one 'extra' Type 1 module that has initiated a new line of advance.

The size of a genet of *Carex arenaria* depends on the dormancy and potential of the one unaborted meristem that is found on the Type 2 module. There is evidence that dormancy of this bud is governed by nutrient status (Noble, Bell & Harper in press) and probably by water availability and the season. This dormant meristem may give rise to either a Type 1 or Type 2 module. Development to Type 2 will immediately lead to flowering and seed production; if the meristem forms a Type 1 module it reiterates the pattern of the parent shoot and initiates a new line of advance across the substrate.

A computer model of the expansion of a *Carex* rhizome system based on actual observations of module type, position and 'success rates' allows an assessment of the dynamics of these units both temporarily and spatially. Figure 2.6 is the plan view of such a model based on field data. Groups of tillers occur with one, two, three or four members, and some advancing Type 1 meristems have failed to grow (representing damaged meristems in the field) and this has led to the growth of some dormant Type 1 meristems.

This type of computerized graphic simulation provides data of position, age, density, type and numbers of the two types of modules in *Carex* and gives an indirect prediction of productivity in terms of numbers of leaves (based on average per tiller), and of seeds (based on seeds per inflorescence). This two-dimensional simulation can cope with most branching patterns using simple recursive probabilistic rules. We are developing a versatile three-dimensional version at the present time.

(a)

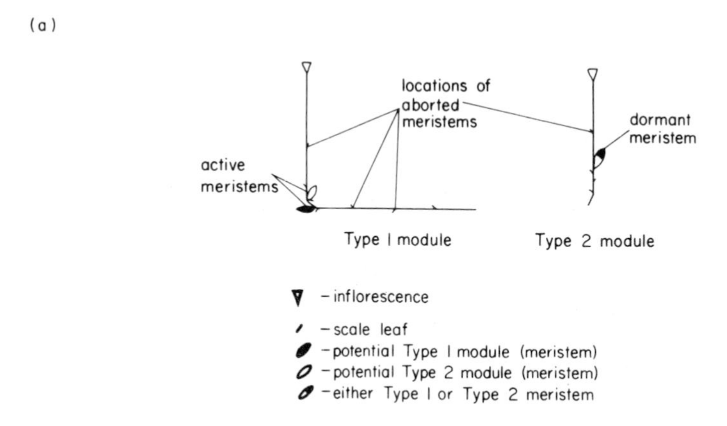


(b)

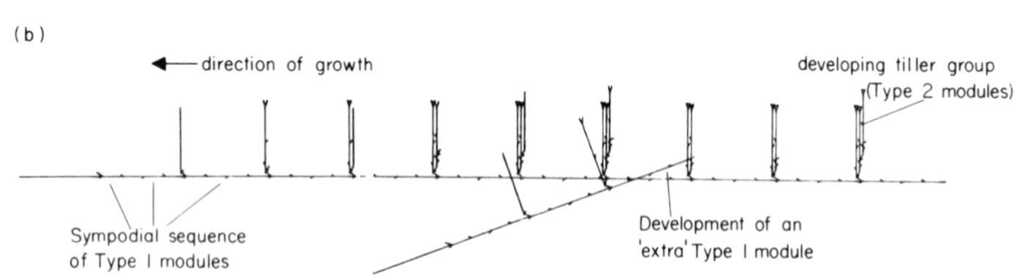


FIG. 2.5. The constructional organization of *Carex arenaria* in terms of modular units.
(a) Module types (profile view).
*Type 1 module:* The sympodial unit of which the *Carex* rhizome is built. It bears roots, foliage leaves and flowers in addition to one Type 1 meristem and one Type 2 meristem.
*Type 2 module:* A 'tiller' bearing foliage leaves and flowers plus one potentially dormant meristem which may be of either Type 1 or Type 2.
(b) A computer-generated sequence demonstrating the progressive accumulation of modules. In this instance dormant meristems on Type 2 modules have a 30% chance of remaining dormant, a 30% chance of developing as a Type 1 module and a 30% chance of developing as a Type 2 module. This diagram is one of a developmental series viewed on a graphics display screen.

## 2 Obelia *and* Alstonia

The procedure we have used to describe the growth of a clone of *Carex* can equally be applied to a modular animal such as a colony of *Obelia*. Figure 2.7(a)–(d) shows four replicate computer drawings of such colonies. We know of no published data of branch length, branch angle and hydranth location in *Obelia* and the rules of growth have therefore been synthesized from illustrations in Hincks (1868) and do not necessarily represent one particular species. These four figures are not simply copies of four *Obelia* colonies; they represent the application of simple rules of growth, based on probabilities of module type and birth, and so result in a series of colonies fundamentally all the same

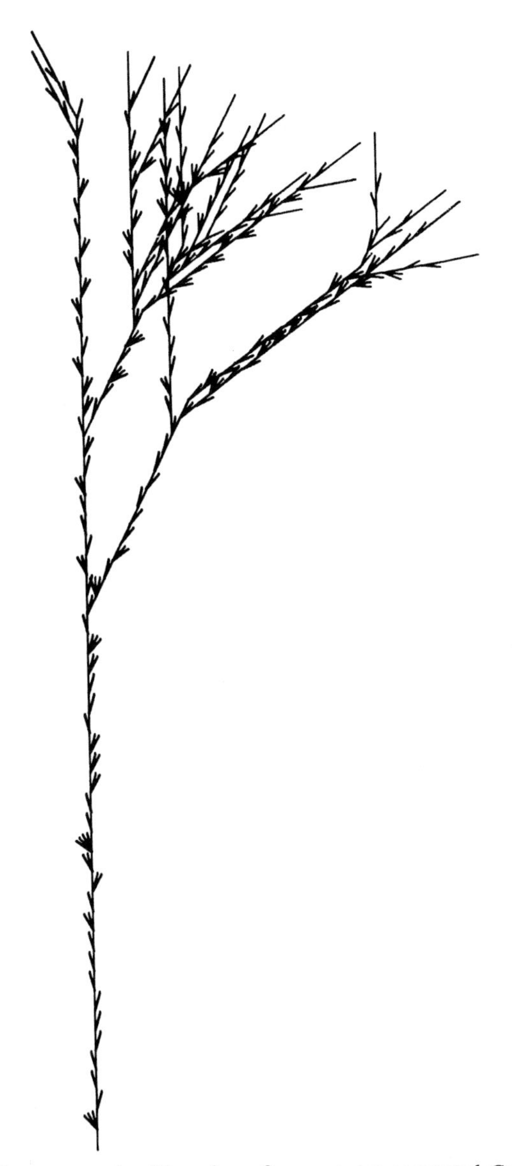

FIG. 2.6. *Carex arenaria*—Plan view of a computer-generated *Carex* rhizome system after 5 years' growth. The oldest part of the system is at the bottom of the illustration. The long lines represent linear sequences of Type 1 modules (see Fig. 2.5(a)); the short lines represent 'tillers' (Type 2 modules). These are displayed as rotated onto the horizontal plane. Each module may or may not grow depending on a programmed probability. These probabilities (divided into 'spring', 'summer', 'autumn' and 'winter') represent averages based on module numbers recorded in an excavated genet.

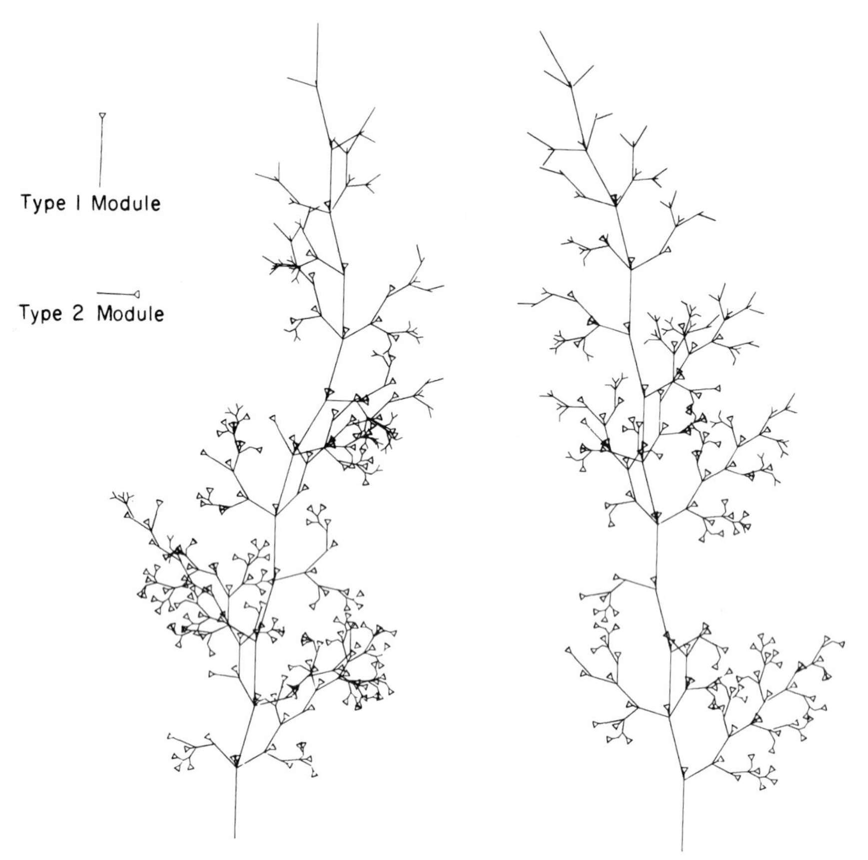


FIG. 2.7. *Obelia* sp.—A hydroid *Obelia* can be regarded as consisting of two module types (prior to the sexual phase). These are illustrated in the top left-hand corner. A triangle represents a hydranth. Module Type 1 is orthotropic (± vertical) and bears one Type 1 module and two Type 2 modules. Type 2 modules are plagiotropic (± horizontal) and bear a variable number of Type 2 modules.

The four illustrations are four computer-drawn replicates based on one set of growth rules. Type 1 modules always develop. Type 2 modules have a 75% success rate and are progressively shorter from one generation to the next.

although individually distinct. This is a familiar situation botanically. Any beech tree can be distinguished from a birch tree by shape alone, but equally, any two beech trees can be told apart by individual variations on the beech 'theme'.

*Obelia* does not have the precise meristem/leaf/internode organization of a plant, but its development is the equivalent of an elongated naked 'apical meristem' producing modular units at statistically defined locations. The structural demography of *Obelia* appears to follow precisely the same rules of growth

as any one of the numerous species of tree conforming to one of the architectural models of Hallé & Oldeman (1970) (the model of Prévost). There are basically two types of module, one of which is orientated orthotropically ('vertical'), the other plagiotropically ('horizontal'). Each module of the sympodial trunk ends in an aborted apex or inflorescence, or in the case of *Obelia*, in a hydranth. The branches are likewise composed of series of modules, each ending in an inflorescence (hydranth) and bearing similar modules.

Figure 2.8 is a diagrammatic simulation of a tree conforming to the model of Prévost (*Alstonia boonei* L.) using the same computer programme generated for the simulation of *Obelia* in Figure 2.7. Modular unit types, locations and lengths have not been altered, only their orientation and probabilities of growth have been changed to produce the strikingly different growth form.

These two simulated organisms have the same structural demography. Change in emphasis produces either a rigid stable tree with limited variability

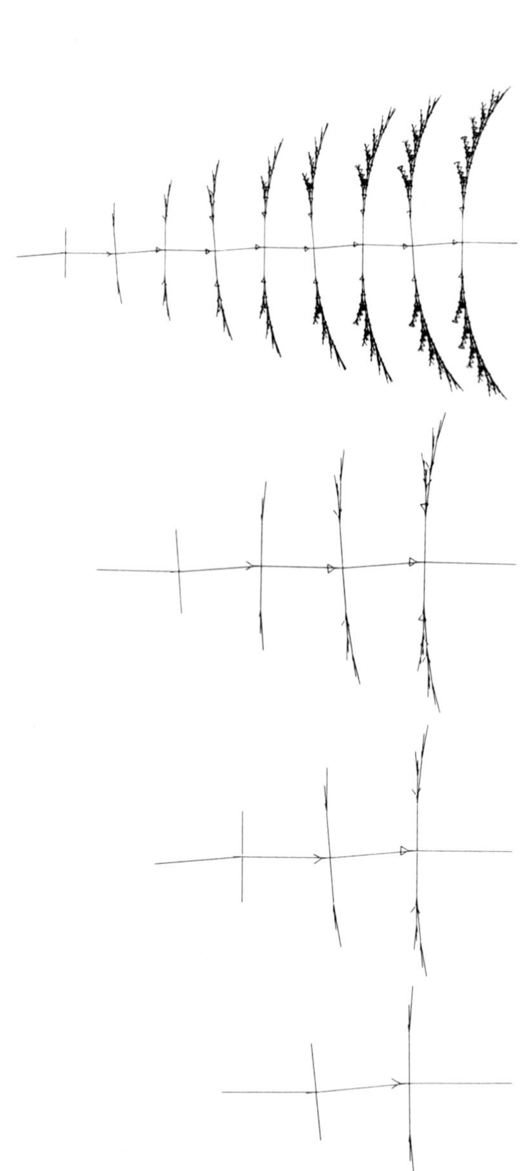

FIG. 2.8. *Alstonia boonei* (a tropical tree of family Apocynaceae)—Computer-generated developmental growth sequence for a tree (*Alstonia boonei*) conforming to the model of Prévost (Hallé & Oldeman 1970) shown at 3, 4, 5 and 10 years old. The trunk is formed of a series of sympodial units terminating in aborted apices. The branches consist of plagiotropic sequences of sympodial units terminating in flowers. This simulation is based on the *same* modular construction as the simulated *Obelia* replicates in Figure 2.7 and using the same programme: only angles and modular success probabilities have been altered. Triangles represent *either* aborted apices *or* inflorescences.

of angle and success of modules, or a physically flexible framework in the case of *Obelia*.

## 3 Acer pseudoplatanus *L.*

A number of branching systems found in plants are built up of more than two types of modules. We are amassing data which will allow us to begin to understand the structural demography of a number of perennial plants, including Sycamore (*Acer pseudoplatanus*). Figure 2.9 shows three replicates of computer drawings in two dimensions of Sycamore based on a pilot study. In this instance we are attempting to recreate the growth rules of a young mature flowering tree by a 'hit and miss' method. If a representation is produced that satisfies the eye that the drawing is indubitably sycamore, then the operating rules we have used may give us a clue to the plant's own 'rules'. This criterion is obviously not yet satisfied by Figure 2.9! We are developing programmes for three-dimensional descriptions which will allow the use of field data to refine the programme.

## 4 Podocoryne carnea

Stoloniferous hydroids such as *Podocoryne carnea* L. have many of the characteristics of a stoloniferous plant such as *Ranunculus repens* L., although in the hydroids converging stolons may anastomose and this does not occur between stolons of higher plants when they meet. Braverman & Schrandt (1966) attempted a computer simulation of the radial growth pattern of *Podocoryne* stolons. They decided upon a number of rules of growth principally related to the nutritional balance between stolon length and hydroid number. They assumed that stolon growth was linear and that the branching angles were of 90° (although this is not the natural situation) and they allowed for the anastomosis of convergent stolons; the graphic display of the stolon system and its supported hydranths was remarkably close to the form of natural colonies, despite the high artificiality of the 90° angle. We have attempted to model a similar system for a hypothetical stoloniferous plant in which no anastomosis occurs and in which there are just two module types—the stolon and the ramet or rooted individual (this type of organization occurs in a number of plants; see Jeannoda-Robinson 1977; Bell & Tomlinson 1979). The meristem success rates that have been used in this example (Fig. 2.10) might in reality reflect the availability of nutrients within the genet (as proposed for *Podocoryne*), or be an intrinsic feature of the morphology. Our model for a stoloniferous plant depends on quite different rules of growth from those applied by Braverman and Schrandt to *Podocoryne*. Nevertheless, the same pattern of colony development results from the two different programmes in contrast to the example of *Obelia* and *Alstonia*, where a single model fits two quite different constructional systems.

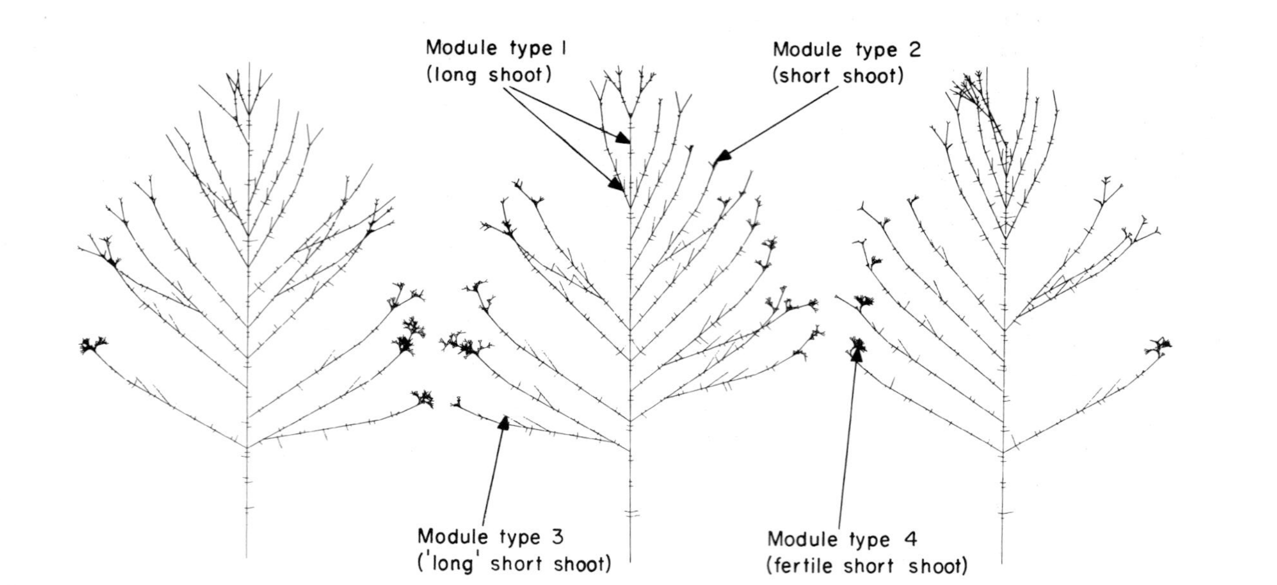


FIG. 2.9. *Acer pseudoplatanus*—Three replicates of a computer-generated sycamore 17 years old which has just begun to flower (triangle). This simulation is based on four module types.

Module Type 1—an orthotropic long shoot growing seasonally for a number of years before terminating in an inflorescence. Both the trunk and the main side branches are of this category.

Module Type 2—short shoots growing more or less indefinitely and bearing leaves.

Module Type 3—'long' short shoots. Lateral Type 1 modules bear extra Type 1 on their lower and 'long' short shoots on their upper surface.

Module Type 4—short shoots bearing flowers and leaves.

These module types are indicated on the diagram. Each simulation is based on the same set of rules.

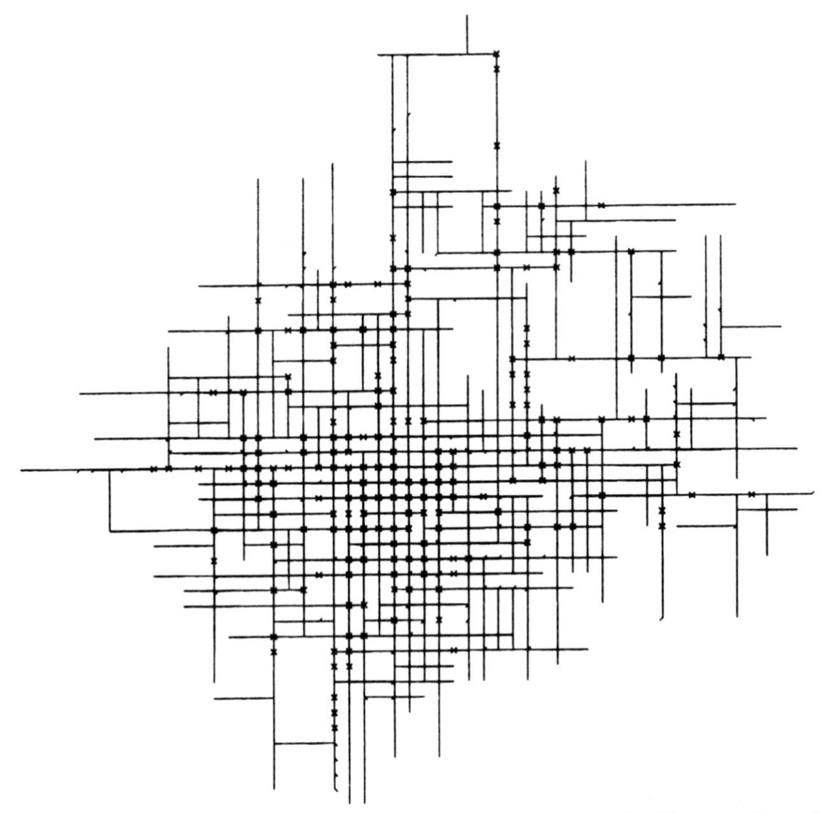

FIG. 2.10 *Podocoryne carnea*—Diagrammatic simulation of a stoloniferous colony of *Podocoryne carnea*. Angles have been arbitrarily set at 90°. Crosses mark hydroid locations with a probability of 25% success at each site. This diagram can be compared with Figures 10 and 15 in Braverman & Schrandt (1966).

## CONCLUSION

It has often been pointed out that plants are plastic in their development and contrast strongly with animals in this respect (e.g. Bradshaw 1965). It is the plasticity of plants that has so hindered the study of their population dynamics (Harper 1961). Nevertheless, we would stress that the problems are not unique to plants—animals with modular growth are also plastic. However, the plasticity of size in plants and modular animals does not extend to plasticity of the form of the individual modules. It is doubtful whether the phenotypic variation in the shape and size of a module of a tree or a coral is any greater than that of a rabbit, *Drosophila* or man.

The 'plasticity' of size and fecundity in modular organisms derives from the great variation in the number of modules per genet. We have shown that these variations, like those in more 'normal' populations are dependent on the birth and death rates of the modules and that the growth of a modular organism can be described as a demographic process.

The fact that the dynamics of a population of growth modules is represented by their fixed position in space means that even 'form' is the result of a demographic process and as such can be built into graphic computer models. We believe that such an interpretation offers quite new and promising opportunities for predictive plant ecology and perhaps also in the modelling of the growth of crops and the ecology of modular invertebrates.

## ACKNOWLEDGMENTS

The work reported in this paper was supported in part by a grant from the Natural Environment Research Council. We are grateful to Miss C. Sellek and also to the staff of the Computing Laboratory, U.C.N.W., Bangor, for invaluable assistance.

## REFERENCES

**Arber A. (1950)** *The Natural Philosophy of Plant Form.* Cambridge University Press, London.

**Bazzaz F.A. & Harper J.L. (1977)** Demographic analysis of the growth of *Linum usitatissimum. New Phytologist*, **78**, 193–208.

**Bell A.D. (1974)** Rhizome organization in relation to vegetative spread in *Medeola virginiana. Journal of the Arnold Arboretum, Harvard University*, **55**, 458–468.

**Bell A.D. (1976)** Computerized vegetative mobility in rhizomatous plants. *Automata, Languages, Development* (Ed. by A. Lindenmayer & G. Rozenberg) pp. 3–14. North-Holland Publishing Company, Amsterdam.

**Bell A.D. (1979)** The hexagonal branching pattern of *Alpinia speciosa* L. (Zingiberaceae). *Annals of Botany* (in press).

**Bell A.D. & Tomlinson P.B. (1979)** Adaptive architecture in rhizomatous plants. *Botanical Journal of the Linnaean Society* (in press).

**Bonner J.T. (1974)** *On Development*. Harvard University Press, Cambridge, Massachusetts.

**Bradshaw A.D. (1965)** Evolutionary significance of phenotype plasticity in plants. *Advances in Genetics*, **13**, 115–155.

**Braverman M.H. & Schrandt R.G. (1966)** Colony development of a polymorphic hydroid as a problem in pattern formation. *The Cnidaria and Their Evolution* (Ed. by W.J. Rees) pp. 169–198. Symposium of the Zoological Society of London, 16.

**Cahn M.A. & Harper J.L. (1976)** The biology of the leaf mark polymorphism in *Trifolium repens* L. II. Evidence for the selection of leaf marks by rumen fistulated sheep. *Heredity*, **37**, 327–333.

**Čelakovsky L. (1881)** Morphologische Beobachtungen. *Sitzungsberichte der Kaiserlichen Böhmischen Gesellschaft der Wissenschaften*, **1881**, 238–250.

**Clatworthy J.N. & Harper J.L. (1961)** The comparative biology of closely related species living in the same area. V. Inter- and intra-specific interference within cultures of *Lemna* spp. and *Salvinia natans. Journal of Experimental Botany*, **13**, 307–324.

**Clegg L. (1978)** The morphology of clonal growth and its relevance to the population dynamics of perennial plants. *Unpublished Ph.D. thesis, University of Wales.*

**Cole L.C. (1954)** The population consequences of life history phenomena. *Quarterly Review of Biology*, **29**, 103–137.

**Dobzhansky Th. (1973)** Nothing in biology makes sense except in the light of evolution. *American Biology Teacher*, March 1973.

**Fisher R.A. (1929, revised 1958)** *The Genetical Theory of Natural Selection* (revised edition). Dover Press, New York.

**Gause G.F. (1934)** *The Struggle for Existence.* Waverly Press, Baltimore, U.S.A.

**Grigg R.W. (1977)** Population dynamics of two Gorgonian corals. *Ecology*, **58**, 278–290.

**Hallé F. & Oldeman R.A.A. (1970)** Essai sur l'architecture et la dynamique de croissance des arbes tropicaux. *Collection de Monographies de Botanique et de Biologie Vegetale*, no. 6, Masson et Cie, Paris.

**Hamilton W.D. (1966)** The moulding of senescence by natural selection. *Journal of Theoretical Biology*, **12**, 12–45.

**Harper J.L. (1961)** Approaches to the study of plant competition. *Mechanisms of Biological Competition* (Ed. by F.L. Milthorpe) pp. 1–39. Symposium of the Society of Experimental Biology, 15.

**Harper J.L. (1977)** *The Population Biology of Plants.* Academic Press, London & New York.

**Hincks T. (1868)** *A History of the British Hydroid Zoophytes.* Vol. II. Van Voorst, London.

**Jeannoda-Robinson, V. (1977)** *Contribution a l'etude de l'architecture des herbes.* Unpublished Ph.D. thesis, Université des Sciences et Techniques du Languedoc Montpellier.

**Kays S. & Harper J.L. (1974)** The regulation of plant and tiller density in a grass sward. *Journal of Ecology*, **62**, 97–105.

**Koriba K. (1958)** On the periodicity of tree-growth in the tropics, with reference to the modes of branching, the leaf-fall, and the formation of the resting bud. *Gardens' Bulletin, Singapore*, **17**, 11–81.

**Levins D.A. (1977)** The genetic implication of different modes of reproduction in plants in relation to their environment. *A Synthesis of Demographic and Experimental Approaches to the Functioning of Plants.* International Symposium, Wageningen, Holland.

**Lewontin R.C. (1965)** Selection for colonizing ability. *The Genetics of Colonizing Species* (Ed. by H.G. Baker and G.L. Stebbins) pp. 77–91. Academic Press, New York & London.

**Mack R.N. & Harper J.L. (1977)** Interference in dune annuals: spatial pattern and neighbourhood effects. *Journal of Ecology*, **65**, 345–363.

**Maguire L.A. & Porter J.W. (1977)** A spatial model of growth and competition strategies in coral communities. *Ecological Modelling*, **3**, 249–271.

**Noble J.C., Bell A.D. & Harper J.L. (1979)** The population dynamics of plants with clonal growth. I. The morphology and structural demography of *Carex arenaria. Journal of Ecology*, **67**, (in press).

**Oinonen E. (1967)** The correlation between the size of Finnish bracken (*Pteridium aquilinum* (L.) Kuhn) clones and certain periods of site history. *Acta forestalia fennica*, **83**, 1–51.

**Oldeman R.A.A. (1974)** L'Architecture de la Forêt Guyanaise. *Memoires O.R.S.T.O.M. No. 73 Paris.*

**Ross M.A. & Harper J.L. (1972)** Occupation of biological space during seedling establishment. *Journal of Ecology*, **60**, 77–88.

**Stimson J. (1974)** An analysis of the pattern of dispersion of the hematypic coral *Pocillopora meadrina* var. *nobilis* Verill. *Ecology*, **55**, 445–449.

**Turkington R.A. (1975)** Relationships between neighbours among species of permanent grassland. *Unpublished Ph.D. thesis, University of Wales.*

*Evolution* (Ed. by E.R. Creed) pp. 20–50. Blackwell Scientific Publications, Oxford.

**Williams G.C. (1975)** *Sex and Evolution.* Princeton University Press, Princeton, New Jersey.

**Wilson E.O. (1971)** *The Insect Societies.* Belknap Press of Harvard University, Cambridge, Massachusetts.

**Zimmermann M.H. & Brown C.L. (1971)** *Trees, Structure and Function.* Springer, New York.

# 3. GENETICAL FACTORS IN ANIMAL POPULATION DYNAMICS

R. J. BERRY

*Royal Free Hospital School of Medicine, London WC1N 1BP**

There are only four factors which can change the size of a population: births (*B*), deaths (*D*), immigrants (*I*) and emigrants (*E*). Fifty years ago, Chapman (1928) urged ecologists to measure these factors, drawing an analogy with the usefulness of Ohm's Law in analysing electrical processes, since 'it seems evident that we have in nature a system in which the potential rate of reproduction of the animal is pitted against the resistance of the environment, and that the quantity of organisms which may be found is a result of the balance between the biotic potential, or the potential rate of reproduction, and the environmental resistance.' In crude terms, Chapman was right, and his exhortation has been well heeded. Indeed, there is no longer any need to encourage measurement: too often nowadays we are at risk of being submerged by a deluge of quantitative data on a wide variety of species. Unfortunately and ironically, this is tending towards an ecological straitjacket, as confining as taxonomic typology once was, and which will be relieved only when the intrinsic determinants as well as the extrinsic interactions of *B*, *D*, *I* and *E* are recognized. Put another way, we are in danger of being satisfied with such spurious achievements as accurate measures of the parameters in life-tables of particular species or the fitting of population numbers to simple theoretical models (such as those derived from the logistic equation and its derivatives); or of determining Volterra's 'coefficient of increase' or Gause's 'coefficient of competition' to several places of decimals, while failing to recognize that *B*, *D*, *I* and *E* are all subject to genetical variation and that a single population may contain different genetical types (or morphs) heterogeneous in both the absolute values of these factors and also their interactions.

One example of this will suffice. A vast amount of work has been done on the population dynamics of the flour beetle *Tribolium*, initially by Chapman (1928), but particularly over many years by Park (reviewed 1962). Traditionally these studies involved the determination of conventional growth and competition parameters, but increasingly it was realized that these parameters varied considerably between *Tribolium* strains, quite apart from differences between

* Present address: Department of Genetics & Biometry, University College, London WC1E 6BT.

species. For example, Sokal & Sonleitner (1968) showed that a black (*b*) mutant affected egg production, cannibalism and developmental rate, but that it remained segregating in laboratory stocks and the overall effect on fecundity changed markedly with genotype frequencies. Such results raise questions about the validity of earlier *Tribolium* experiments, which assumed that any particular strain had characteristic and stable properties, and throw doubt on the ideas of competition that have been based upon them (Sokoloff 1966; King & Dawson 1972).

Ecologists are, of course, aware that the population parameters they are seeking to measure are genetically determined and hence subject to inherited variation, but they have tended either to regard intra-population variation as negligible or, alternatively, to assume that selection will have maximized fitness. In both cases, the inference is that individuals are homozygous at all but a few gene-loci: in effect a 'population type' has replaced the species type of a century ago.

This simplicity is no longer acceptable following the demonstration by electrophoresis of proteins that a significant part of almost all genomes is heterozygous: on average 5 to 10% of vertebrate loci and 10 to 20% of invertebrate loci are present in the heterozygous state, and perhaps a third of all genes are segregating (Harris 1966; Lewontin & Hubby 1966; Selander & Kaufman 1973; Powell 1975) (Table 3.1). The absolute value of these estimates of

TABLE 3.1. Heterozygosity per individual (after Powell 1975; Selander 1976)

| | | Number of species or forms | Number of loci | % average heterozygosity |
|---|---|---|---|---|
| INSECTS | *Drosophila* | 28 | 24 | 15·0 ± 1·0 |
| | Other insects | 7 | 18 | 17·0 ± 2·7 |
| | Non-insect invertebrates | 13 | 15 | 10·2 ± 2·1 |
| MOLLUSCS | Land snails | 5 | 18 | 15·0 ± 3·6 |
| | Marine snails | 5 | 17 | 8·3 ± 3·2 |
| | INVERTEBRATES | | | 14·6 ± 0·9 |
| FISH | | 14 | 21 | 7·8 ± 1·2 |
| AMPHIBIANS | | 11 | 22 | 8·2 ± 0·8 |
| REPTILES | | 9 | 21 | 4·7 ± 0·8 |
| BIRDS | | 4 | 19 | 4·2 ± 2·1 |
| MAMMALS | Rodents | 26 | 26 | 5·4 ± 0·5 |
| | Large mammals | 4 | 40 | 3·7 ± 1·8 |
| | VERTEBRATES | | | 5·0 ± 0·4 |
| | PLANTS | | | 17·0 ± 3·1 |

variability may be too low or too high (Johnson 1977), but there can be no doubt that genetical variation rather than homogeneity is the normal state of animals.

Geneticists have similar hangovers from their history. Calculations on genetic load (Muller 1950) and the 'cost' of natural selection (Haldane 1957) convinced population geneticists that there was an upper limit to the amount of inherited variation which a population could carry before becoming extinct. The realization from electrophoretic work that this upper limit was almost universally over-stepped threw population geneticists into turmoil and led to the suggestion that most of the protein variation found in nature was neutral, i.e. had no significant effect on its carriers (King & Jukes 1969; Kimura & Ohta 1971). This meant that observed variation is biologically negligible, which is the same as the ecological assumption.

Furthermore, neither geneticists nor ecologists have really come to terms with the discoveries of E.B. Ford and his co-workers that selection intensities of 10% and more are common in natural situations. When this fact is combined with the observed large amounts of inherited variation, there is a strong possibility that populations can quickly respond to environmental pressures, even when such pressures vary in time or space.

## GENETICAL VARIATION IN THE DETERMINANTS OF POPULATION NUMBERS

In retrospect, it may have been a good thing that the rates of birth, death, immigration and emigration used to be regarded as species constants, because it has led to the development of reasonably sophisticated techniques for their measurement, and hence estimates of their variances. Nevertheless, there can be no dispute that they are all subject to intra-population variation.

Birch (1960), reviewing the situation twenty years ago, described the effect of crowding in producing genetical changes in two sub-species of the Queensland fruit fly *Dacus tryoni* and in the blow-fly *Lucilia cuprina* (Nicholson 1957; also Birch 1971). Ayala (1968) produced similar evidence from work on *Drosophila*, including his own studies on two Australian species, *D. serrata* and *D. birchi*. Parsons (1973) has summarized much further evidence from *Drosophila* species, and no attempt will be made to repeat this here. There are many further such examples, most of them from natural populations (Table 3.2).

An interesting point that emerges from Table 3.2 is that many of the 'classical' examples of natural selection can be regarded as affecting *B*, *D*, *I* and *E*, most of them (e.g. melanic frequencies in the Peppered Moth *Biston betularia* or morph distributions in *Cepaea nemoralis*) involving differential survival of different adult phenotypes.

The classification in Table 3.2 is artificial because the selection pressures

TABLE 3.2. Genetical influences on population number determinants

| Phenotype | Selective agent | References |
|---|---|---|
| 1. BIRTH | | |
| Aggressiveness in *Mus musculus* | Breeding success | Southwick 1955 |
| Transferrin polymorphism in several mammals | Infertility | Ashton & Dennis 1971 |
| Embryonic mortality from *t*-alleles | Distorted segregation ratios | Bennett 1975 |
| Nest selection in *Cuculus canorus* | Imprinting on host | Southern 1954 |
| Breeding season in melanic *Columba livia* | Steroid control of breeding season | Murton, Westwood & Thearle 1973; Murton, Thearle & Coombs 1974 |
| Colour phases in *Stercorarius parasiticus* | Food availability and pair formation | Berry & Davis 1970; O'Donald & Davis 1975 |
| Clutch size in birds | Food collecting ability | Lack 1968; Owen 1977 |
| Melanism in *Adalia bipunctata* | Rate of heat absorption | Muggleton, Lonsdale & Benham 1975 |
| 2. DEATH | | |
| Human haemoglobinopathies and other polymorphisms | Epidemic disease, especially malaria | Haldane 1949; Allison 1964; Vogel & Chakravarti 1966 |
| Energy mobilization in *Apodemus sylvaticus* | Food shortage? | Brown 1977 |
| Haemoglobin and blood efficiency in *Mus musculus* | Low (winter) temperature | Berry & Peters 1977 |
| Esterases and chick colour in *Lagopus scoticus* | Population density | Henderson 1977 |

| | | |
|---|---|---|
| Pesticide resistance in rodents, mosquitoes, aphids, pathogenic micro-organisms | Direct human offensive | Bishop & Hartley 1976; Cook & Wood 1976; Beranek & Oppenoorth 1977; Anderson 1968 |
| Melanic moths in industrial areas | Visual predation by birds | Kettlewell 1973; Lees & Creed 1975; Steward 1977 |
| *Cepaea nemoralis* colour and banding morphs | Visual predation; climatic tolerance | Jones, Leith & Rawlings 1977 |
| *Sphaeroma rugicauda* colour morphs | Temperature | Heath 1974 |
| *Spirorbis borealis* settling behaviour | Substrate and population density | Mackay & Doyle 1978 |
| Heavy metal tolerance in marine invertebrates and grasses | Toxic levels in the environment | Bryan 1973; Luoma 1977; Bradshaw 1976 |
| Stabilizing selection on morphological traits: *Passer domesticus*, *Nucella lapillus*, etc. | Climatic stress | Bumpus 1899; Berry & Crothers 1968 |
| 3. IMMIGRATION AND EMIGRATION | | |
| Persistence of small mammals in refuge habitats | Differential migration | Anderson 1970 |
| Population fluctuation in *Microtus* spp. and allozyme frequencies | Aggression and dispersal? | Semeonoff & Robertson 1968; Myers & Krebs 1971 |
| Spread of *Streptopelia decaocto* | Temperature tolerance?? | Mayr, quoted by Murton 1971 |
| Reduction in flight frequency in *Amathes glareosa* | Life on a small, windy island? | Kettlewell *et al.* 1969 |
| Outbreeding in angiosperms | Heterostyly, self-incompatibility, etc. | Darlington 1971 |
| Poor survival of metal-tolerant grasses in normal sward | Intolerance of crowding | Jain & Bradshaw 1966 |

that reveal inherited differences within populations do not act primarily on the parameters that ecologists chose to measure. This is clearest for stabilizing and endocyclic selections where variation eliminated during (usually) adult life is favoured at some other stage (Haldane 1959; Mather 1953, 1970). For example, the proportion of heterozygotes for the *d* and *s* alleles at the *Hbb* locus in house mice increases during the breeding season on the Welsh island of Skokholm, but decreases during the winter (Berry & Murphy 1970; Berry 1978a), i.e. a homozygote is at a relative disadvantage during the breeding season but is favoured during the harsh 'survival' period. In mice which breed all the year round on sub-Antarctic islands, the *d* allele is less common in older animals than younger ones; this implies that animals with the *s* allele are less successful at breeding than survival (Berry, Peters & Van Aarde 1978). Bumpus (1899) found that morphologically 'average' sparrows recovered better from storm exposure than more extreme ones; again the presumption is that the successful breeders are those which produce variable young. Exactly the same situation exists in the land snail *Clausilia laminata* (Weldon 1901) and the littoral whelk *Nucella lapillus* (Berry & Crothers 1968, 1970).

In other cases, inherited variants have been shown to be strongly correlated with environmental changes, but the physiological effects of the variants are unknown. Examples of this are the increase of B-chromosomes in the Mottled Grasshopper *Myrmeleotettix maculatus* in upland and eastern Britain with changes of aridity (Hewitt & Brown 1970), and of allozyme frequency changes in the snail *Theba pisana* in areas of varying temperature and humidity in Israel (Nevo & Bar 1976).

Genetical influences on emigration and immigration have been combined in Table 3.2. Inherited variance of these behaviours is less well known than for *B* and *D*, and movement *into* one group necessarily implies movement *out* of another.

### *Selection coefficients*

A common and entirely invalid assumption of population geneticists has been that the genetical pressures acting on a trait are effectively constant, i.e. that the fitness of a genotype is a characteristic of an allele rather than a highly variable product of interactions both within the genome and between genome and environment. The use of shorthand language to speak of the dominance or recessivity of an allele rather than as the property of a trait has encouraged this confusion. Geneticists, in fact, have misled themselves about the dynamic nature of genetical forces just as a past generation of ecologists did with ecological forces.

The simplest situation involving varying genetical pressures occurs where a species has to tolerate different environmental conditions. This is most easily quantifiable for littoral species exposed on sheltered and open beaches. Berry &

Crothers (1968) collected Dog-whelks *Nucella lapillus* from a range of exposures around the mouth and along the shores of Milford Haven. In all the more exposed populations there was a reduction in variance of a particular shell measurement between the youngest and oldest groups in the population, whilst in the most sheltered there was no significant change during life. Using an argument of Haldane (1959) that the amount of variance loss can be equated to the strength of natural selection, selection coefficients ranged from over 90% on a westward-facing headland to no selection at all acting on the trait measured on sheltered shores. The correlation between degree of exposure and strength of selection was 63%, or 76% if one anomalous sheltered population was omitted. An analogous situation in *Littorina* species has been investigated by Heller (1975) and Raffaelli & Hughes (1978).

Environmental gradients frequently produce changes in gene frequencies, which can be regarded as the result of different selection pressures at different places along the gradient (Haldane 1948; Endler 1973, 1977). A simple example of this is the correlation between melanic moth frequency and atmospheric pollution (Kettlewell 1973; Bishop *et al.* 1975), or between melanic morph frequencies in the ladybird *Adalia bipunctata* and local temperature (or sunshine) differences (Creed 1975; Muggleton, Lonsdale & Benham 1975). This argument can be used for environmental heterogeneity in either space or time (Haldane 1956; Anderson *et al.* 1975; Karlin & Richter-Dyn 1976).

A more complicated possibility of variability in selective values was considered by R.A. Fisher in 1930 in relation to its dependence on the frequency of a genotype in the population. Fisher was interested in Batesian mimicry among butterflies. If a distasteful model is common compared to its palatable mimic, predators will not often encounter the mimic and the latter will have a high chance of survival. However, the protective value of the relationship must decrease as the mimic becomes commoner, and predators fail to associate its appearance with unpalatability. Since the intensity of predation depends on the frequency of contact between prey and predator, selection upon Batesian predators will be both frequency- and density-dependent (Turner & Williamson 1968; Clarke 1972; Turner 1977). There is now a fair amount of experimental evidence that predators may select in a frequency-dependent way even when no mimicry is involved, perhaps due to predators forming 'searching images' of their prey (Tinbergen 1960; Harvey, Jordan & Allen 1974; Thompson & Vertinsky 1975).

Another facet of varying selection relates to parasites and pathogens, and was suggested by Haldane (1949, 1955). He argued that such organisms may act in the same was as mimetic butterflies, becoming adapted to the commonest biochemical or immunological variety of host. The effect of this is that there is a reduced defence against rarer varieties and they thus gain a selective advantage. Flor (1956) has shown a one-to-one relationship between the genes for rust-resistance in flax and the genes for virulence in the rust itself; and

Hatchett & Gallun (1970) have shown a similar interaction between wheat and an *animal* parasite (reviewed Day 1974). Clearly, a range of ecological situations may produce changes in selection coefficients (Murray 1972; Clarke 1975a; Berry 1977).

The question of variable selection pressures has been vividly summed up by Wallace under the guise of 'hard' and 'soft' selection (Wallace 1968, 1975). Wallace began from the unsatisfactoriness of the 'genetic load' concept (which implied that populations had an intolerable burden of detrimental alleles dragging them towards extinction, as Christian in Bunyan's *Pilgrim's Progress* was weighed down by his load of sin: Muller 1950; Wallace 1970). His terminology of 'hard' and 'soft' came from economics where 'soft currency' is usable within a particular country, but 'hard currency' maintains its value in all countries. In this sense, hard selection acts on alleles which produce their effects under all known conditions (for example, lethal genes are inevitably fatal to their carriers), whereas soft selection involves varying survival (or death) of individuals as conditions change. This means that the inferior fitness of certain genotypes is revealed by a direct comparison with others in the same population, not by mathematical calculation.

Hard selection can be predicted with no reference to individuals: the selection coefficient of a particular genotype is fixed. Consequently it is both density- and frequency-independent: it is a mathematical geneticist's delight, but an ecologist's nightmare.

In contrast, soft selection is both frequency- and density-dependent, and varies with population parameters. Since much genetical variation is cryptic (i.e. is not readily scored, unlike colour variants of Lepidoptera), the operation of soft selection can account for the existence of otherwise inexplicable differences between ostensible replicates. For example, Southwick (1955) set up six 'replicates' of house mouse populations beginning each with four pairs of apparently identical mice, and studied population growth. After two years, the numbers in each replicate had more or less stabilized—but they varied from 25 to 140 individuals in different cages. In each cage, the number of fights increased with population size, and growth continued until this measure of aggression reached approximately one fight per mouse per hour, at which point reproduction ceased. Presumably the populations were initially heterogeneous for factors affecting aggressiveness (see below).

Selection coefficients are generally acknowledged to vary, even if this knowledge is not acted upon. However, there are other variables which are not usually accepted as such.

Even such a basic population parameter as 'effective population size' is not a species constant, but depends on ecological conditions. For example, DeFries & McClearn (1972) have used the fine structure of house mouse populations as a model for understanding gene flow and microevolution in local populations. Summarizing laboratory experiments, computer simulations and extrapolation from particular natural situations, they concluded that mouse deme size is extremely small, such that the effective population size may be as low as four, meaning the consequent increase in inbreeding coefficient would be 12·5% every generation. Now there is no doubt that house mice do have a very small deme size in stable and crowded conditions, but such conditions are unlikely to persist for long in an opportunistic species: in a mouse population living independently of man on Skokholm, more than a quarter of the animals breed at a different site to the one where they were born (Berry & Jakobson 1974). Under these conditions the whole island population comes close to being a panmictic unit.

## THE FITNESS DOGMA

The fitness of an individual is the final outcome of all its developmental and physiological processes. If individuals are homozygous over most of their genome (as, we have seen, both geneticists and ecologists tacitly assume), it follows that fitness is a constant measure for particular environments. Unfortunately for convenience and calculation, genomes are highly heterogeneous and selection pressures can be very high, so that considerable fine genetical adjustment tends to take place in many populations (Berry 1971, 1978a; Bradshaw 1971, 1976). Consequently, the fitness of an individual or genotype has to be regarded as a variable in exactly the same way as, say, birth rate or gene frequency.

A further complication as far as fitness is concerned is that since many independent (albeit associated) factors interact to produce the quantity called fitness, strictly speaking it is an *epigenetic* and not a *genetic* trait. This is at once clear from Table 3.2, where a host of assorted characters are shown as being genetically variable and hence affecting selection on the basic population parameters. In a limited number of organisms (notably *Drosophila melanogaster* and *Mus musculus*), enough is known about the inheritance and interaction of the contributors to fitness to make their analysis possible. The complexity of the character we call fitness becomes evident when this is done. For example, there is in house mice a high correlation ($r = 0{\cdot}64 \pm 0{\cdot}07$) between maternal size and number of eggs ovulated, but this correlation does not exist in large free-living island populations: on Foula the correlation is $-0{\cdot}03 \pm 0{\cdot}27$, on the Isle of May it is $0{\cdot}02 \pm 0{\cdot}16$, and on Skokholm it is $0{\cdot}18 \pm 0{\cdot}18$. Clearly litter size in these island populations is affected by factors other than allometry (Lack 1948; Batten & Berry 1967).

In a few cases the contribution of different genes to fitness has been worked out. An excellent example of such complex trait analysis has been that of foulbrood disease resistance in honey bees (Rothenbuhler 1967). Two genes have been specifically identified as being involved in resistance: one determines 'larval cell uncapping' and the other 'larval removal'. The normal alleles at both these loci are necessary for a bee colony to be protected against foulbrood disease, although clearly other genes must be involved in both these complex behaviours—just as the interaction of at least seven loci are concerned with the determinants of blood clotting in mammals, which is a comparatively simple system.

The effect of gene action on behaviour affecting fitness is shown also by genes involved in tyrosine metabolism: black rats are less aggressive than agouti ones (King 1939); black rabbits are less timid than agouti ones, permitting longer feeding and hence survival in food-limited populations despite visual selection against them (Berry 1977); melanic cats are more successful in breeding under dense urban conditions than lighter animals (Clark 1975, 1976); dark phase Arctic Skua males pair more readily and therefore earlier than ones of pale or intermediate phase (Berry & Davis 1970); melanic Rock Doves breed all the year round, and hence increase in frequency when food is permanently available (Murton, Westwood & Thearle 1973; Murton, Thearle & Coombs 1974); human phenylketonurics are less pigmented than their normal sibs (Cowie & Penrose 1950). Since any random groupings of animals are likely to be genetically different from each other, it is not surprising that the apparently replicate lines set up by Southwick (1955) should differ in their levels of aggression (see above): laboratory strains of mice are markedly heterogeneous in their aggressive behaviour (McGill 1970).

It is worth noting that the recognition of inherited variance in aggression helps to reconcile apparently conflicting views about the causes of number fluctuations in microtine rodents. Chitty and Krebs have suggested that different growth and mortality rates at different stages of a 'cycle' in voles and lemmings may be the result of changes in the frequencies of alleles affecting aggressiveness, while Davis and Christian have argued that the physiological responses of individuals to crowding are sufficient to produce the observed demographic changes (Christian & Davis 1964; Chitty 1967; Krebs *et al.* 1973). Although changes in allele frequencies have been found during the progress of vole number fluctuations, it has not been possible to relate these to any change in 'fitness' traits. In fact, the genetical and physiological interpretations are not mutually exclusive, since the hormonal responses of stressed rodents described by Christian are inevitably the products of genes which are themselves subject to selection. When the genetical control of behaviour in rodents is better known, it could well prove that Christian and Davis have worked out part of the mechanism of the ecological responses studied by Chitty, Krebs, and their students.

### *Measures of variability*

Unfortunately there has been little effort by geneticists to dissect complex epigenetic traits like fitness. The classical divide between Mendelians and biometricians which enlivened and embittered the decade around 1900 still persists (Froggatt & Nevin 1971). The Mendelians are too often satisfied with evidence of selection as revealed by marker genes, while biometricians rarely pursue their analyses to the extent of identifying the action of individual genes. There are, of course, exceptions. The fine structure and biochemistry of the alcohol dehydrogenase (*Adh*) locus of *Drosophila melanogaster* has been intensively studied by Gibson (1972), Clarke (1975b) and others, while Vigue & Johnson (1973) have shown a marked latitudinal cline of two common alleles in the eastern U.S.A., with allele frequencies associated with a variety of characteristics of the habitat. (Unfortunately for a clear-cut interpretation, ADH is linked to a small inversion in *D. melanogaster* so that other factors beside selection on the locus itself may influence frequencies; furthermore, the enzyme catalyses a variety of other physiological processes besides ethanol breakdown, including the production of retinol from retinene which is critical to vision, and of glycerol from glyceraldehyde; Johnson 1976.)

The biometrical approach is more immediately attractive to ecologists, because characters of apparent adaptive value can be studied directly. For example, Mackay & Doyle (1978) compared the settling and gregarious behaviour of the larvae of *Spirorbis borealis* collected from three contrasting habitats, and found that the means of the traits they measured were interpretable in terms of adaptation to their habitat. Wool (1977) carried out a similar analysis on populations of gall-forming aphids. The problem in natural populations is to relate such phenotypic adaptation to gene action. Techniques for identifying the effect of particular chomosomal segments have been developed, but they are laborious and precise only for genetically well-known organisms (Thoday 1961; Kearsey & Kojima 1967; Swank & Bailey 1973).

However, the analysis of inherited variations has become confused by attempts to understand genetical heterogeneity revealed by electrophoresis. A few allozymic variants show correlations between their physiological properties and environmental conditions, and variation at these loci can be explained using classical selection criteria. For example, there is change in allele frequency with latitude in an esterase locus in the freshwater fish *Catostomus clarki* in U.S.A. Many of the populations sampled are isolated from each other, so it is clear that there is a selective influence varying with latitude. Part of the answer about this influence is that the allele commoner in the northern part of the area is more active at low temperatures than the variant more common in southern areas (Koehn & Rasmussen 1967; Koehn 1969). A similar situation occurs in the blenny *Anoplarchus purperescens*, where experimental alteration of temperature alters allelic frequencies at the lactic dehydrogenase locus (Johnson 1971). A different type of example occurs in organophosphorus-resistant peach potato aphids (*Myzus persicae*), where an electrophoretic band diagnostic of resistance in aphid fractions has been shown to be a hydrolase which breaks down the aphicide *in vitro* (Beranek & Oppenoorth 1977). For most species, however, allozyme frequencies are constant or varying in an unexplained or random way between populations (Frydenberg & Simonsen 1973; Hedrick, Ginevan & Ewing 1976; Berry & Peters 1977). The response of most workers to this mess of data is to speculate on the significance of overall genetical heterogeneity, measured by two easily determined values: the proportion of loci which are polymorphic (*P*) (defined as the commonest allele having a frequency of less than either 0·95 or 0·99); or average heterozygosity over all loci examined (*H*) (e.g. Nei 1972). This has elevated *P* and *H* into species characteristics, which they patently are not (Table 3.3).

For example, different enzymes have different variabilities in a population. Hence estimates of *H* and *P* depend on the number and nature of the loci which are scored. In two of the main laboratories in the U.S.A. carrying out electrophoretic studies of *Drosophila* species, Lewontin and his associates survey a high proportion of glucose metabolizing enzymes and derive *H* estimates of below 14%, whereas Ayala and his co-workers screen more non-specific enzymes and all but one of the *H* estimates from *their* laboratory are over 14% (Kojima, Gillespie & Tobari 1970; Selander 1976). In a study of our own on house mice from Peru, the mean heterozygosity after five loci had been scored was 16·4%; after ten it was 14·2%; 10·9% after twenty; and 8·2% after fifty (R.J. Berry & J. Peters, unpublished).

Notwithstanding, there are apparent patterns in the distribution of *P* and *H*, and considerable ingenuity has been devoted to speculation about their determinants.

TABLE 3.3. Intra-specific variation in mean heterozygosity

| | | %*H* |
|---|---|---|
| MAN (Nei & Roychoudhury 1974) | Caucasoid | 14·2 |
| | Negroid | 12·2 |
| | Mongoloid | 9·8 |
| HOUSE MOUSE (Berry & Peters 1977) | Faroe Is: Fugløy | 0 |
| | Sandøy | 7·8 |
| | Britain: Caithness | 11·4 |
| | Somerset | 4·7 |
| OLDFIELD MOUSE (Smith, Garten & Ramsey 1975) | Georgia | 3·9 |
| | South Carolina | 4·2 |
| | Florida | 9·6 |
| *Drosophila paulistorum* (Richmond 1972) | Andean | 18·9 |
| | Amazonian | 20·5 |
| | Central America | 13·4 |

1 *Neutrality*

The simplest hypothesis is to assume that allozymic variants are neutral in their effects. This means that variants will accumulate in direct proportion to the mutation rate: the older and larger a population or species, the more variation it will be expected to contain. The available information does not fit this idea at all. In the first place, the distribution of observed variants is bimodal: some variants are common and can be regarded as polymorphic, i.e. controlled by selective forces, whilst others are sufficiently rare to be interpreted as recent mutants (Harris, Hopkinson & Robson 1974; Berry & Peters 1977). Secondly, there does not seem to be an indefinite accumulation of variants, as would be expected in long-established and numerous populations (Fig. 3.1). For example, Ayala (1972) estimated that *Drosophila willistoni*, an abundant neotropical species, should have a heterozygosity of over 99%, whereas the observed $H$ is only about 18%. It is only possible to account for this on neutralist premises if the species has passed through a bottle-neck in numbers, when many of the variants would be lost (Smith 1970; Bonnell & Selander 1974; Nei, Maruyama & Chakraborty 1975). Gain in variants is much more difficult to explain on this hypothesis, yet some of the Hawaiian species of *Drosophila* are very young in evolutionary terms, but have heterozygosities of over 17% (Ayala 1975; Carson

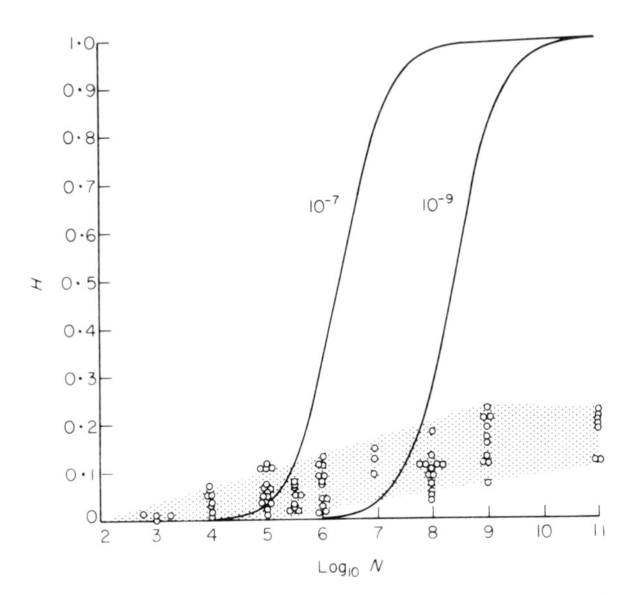


Fig. 3.1. Relationship between individual mean heterozygosity ($H$) and species size estimates. The dots are actual heterozygosity estimates; the two curves represent the expected accumulation of neutral variants assuming mutation rates per locus of $10^{-7}$ and $10^{-9}$ (after Soulé 1976).

& Kaneshiro 1976). Finally, a number of surveys have shown significantly different variances of different protein variant frequencies between populations, although all the variances should be equal if they are affected by nothing more than inbreeding (Nevo, Dessauer & Chuang 1975; Berry & Peters 1977). Lewontin & Krakauer (1973) have proposed a formal test for detecting selection from such evidence. It has been criticized on statistical grounds (Ewens & Feldman 1976), but at least these heterogeneities show that the influences acting on all loci are not equal.

The extreme neutralist position has now been modified to a version of the classical deleterious mutation hypothesis, which assumes that most new variants are detrimental to their carriers and will be eliminated by selection in time (Ohta 1974). This hypothesis was effectively destroyed by Fisher (1954).

In passing, it is worth recording that neutralists have tended to place considerable weight upon the accuracy of the 'protein clock' in producing identical rates of mutational substitutions in differing evolutionary lines, but this constancy has been shown by the large number of proteins now sequenced to be an artefact of ignorance produced by combining inadequate data (Sarich 1977).

## 2 *Niche width*

Selander & Kaufman (1973) summarized *H* data for twenty-four invertebrate and twenty-two vertebrate species. These showed that the mean *H* for invertebrates was 15·1%, but only 5·8% for vertebrates. They interpreted this as support for the hypothesis of adaptation to habitats of different 'grain' (Levene 1953; Levins 1968). The genetical version of this is that fine-grained species (i.e. those in which an individual can exploit all or nearly all of the patches in a heterogeneous environment) should have jack-of-all-trades alleles at most loci, whereas selection in course-grained species (in which individuals spend most of their lives in one of several possible niches) will favour different alleles in different patches.

Evidence collected to test this idea that genetical variation is correlated with environmental instability in time or space seemed at first to uphold it, particularly experiments in which laboratory cultures were subjected to either genetical or environmental perturbations (Powell 1971; Hedrick, Ginevan & Ewing 1976), but specific predictions have not been fulfilled. In particular, species from the supposedly constant deep sea environment have high values of *H*, as do species living in trophically stable tropical areas (Somero & Soulé 1974; Campbell, Valentine & Ayala 1975; Valentine 1976), while burrowing animals (in an allegedly 'monotonous subterranean niche': Nevo 1976) tend to have as high heterozygosities as their non-burrowing relatives (5·2 ± 0·6% and 5·9 ± 0·7% respectively: Selander 1976). Notwithstanding, there is clearly some correlation between heterozygosity and way of life.

3 *Resource availability*

Accepting the inadequacy of the basic niche-width variation hypothesis, Ayala & Valentine (1974) suggested that predictability or dependability of food resources is the key to heterozygosity. Their argument is that in situations where resource (mainly food) abundances are highly predictable, the best genetical strategy is to produce a variety of offspring able to exploit different foods, thus reducing sib and intra-population competition (Table 3.4). In unpredictable circumstances, a generalist and uniform genome is supposed to be better suited for coping with adversity. On this interpretation deep-sea invertebrates and tropical *Drosophila* species are expected to be composed of groups of trophic specialists capable of finding the foods for which they are genetically adapted. Valentine & Ayala (1976) cite differences in krill (*Euphausia*) species in support of their hypothesis: *E. superba* lives in the circumpolar area, is said to have a high trophic seasonality and possesses a heterozygosity of 5·7%; *E. distinguenda* is an eastern Pacific tropical species and *H* is 21·3%; while *E. mucronata* lives in the temperate transitional water of the Peru Current and has an intermediate *H* of 14·1%.

These three main hypotheses about the factors involved in determining genetical variability relate to population history, physical hazards and food availability; they encompass intra- and inter-specific interactions as well as straightforward autecological ideas. All these processes affect the portmanteau

TABLE 3.4. Faunal diversity (ranked 1–6) and heterozygosity in marine invertebrates (after Valentine 1971, 1976)

| | %*H* | Faunal diversity ranking |
|---|---|---|
| *Asterias vulgaris* | 1·1 | 2 |
| *Liothyrella notorcadensis* | 3·9 | 2 |
| *Homarus americanus* | 3·9 | 2 |
| *Asterias forbesi* | 2·1 | 2–3 |
| *Limulus polyphemus* | 5·7 | 2–3 |
| *Cancer magister* | 1·4 | 3 |
| *Crangon negricata* | 4·9 | 3 |
| *Upogebia pugettensis* | 6·5 | 3 |
| *Callianassa californiensis* | 8·2 | 3 |
| *Phoronopsis viridis* | 9·4 | 3 |
| *Crassostrea virginica* | 12·0 | 2–4 |
| Asteroids, four deep sea spp. | 16·4 | 5 |
| *Frieleia halli* | 16·9 | 5 |
| *Ophiomusium lymani* | 17·0 | 5 |
| *Tridacna maxima* | 21·6 | 5–6 |

'fitness' of a particular individual or population. The net result is that geneticists are increasingly invoking processes which are the stock-in-trade of ecologists, yet too often ignoring the complexities of those processes. There seems little hope of fully understanding the factors determining levels of variation until we can dissect meaningful from accidental variation, and unique from parallel adaptation. This cannot come from lumping all the components of estimated variation into single measures of polymorphism or heterozygosity, however convenient these may be (Soulé & Yang 1974; Smith, Garten & Ramsey 1975). Heterozygosity and fitness both need dissecting, and this requires the cooperation of ecologists and geneticists.

## GENETICAL AND ECOLOGICAL NICHES

Apart from the efforts to prove particular hypotheses about the causes of observed genetical variation summarized above, there have been a number of valiant attempts to relate amounts of genetical variation to ecological parameters in general. Many years ago J.B.S. Haldane suggested that heterozygotes might be more metabolically flexible than homozygotes, but biochemical tests of this have shown that hybrid enzyme polymers are no more efficient than non-hybrid ones (Berger 1974; Harris, Hopkinson & Edwards 1977). On another level, Bryant (1974) found that about 70% of geographical variation in heterozygosities in rodent and *Drosophila* species could be accounted for by crude climatic variability. It is unlikely that such statistical correlations will prove useful in the long run because individuals live or die, reproduce or not according to their personal adaptation in relevant circumstances. Any useful models will have to be multi-dimensional—and perhaps so multi-dimensional as to be unusable.

Southwood (1977) has traced the interactions of the factors defining the relationships between organism and habitat in both time and space, and derived a 'reproductive success matrix'. This is, of course, one way of beginning to dissect fitness. Southwood showed that such a matrix depends on five habitat and three organism characteristics and their variances:

For the habitat:
- (i) the length of the period which permits breeding ($F$)
- (ii) the length of the unfavourable period (permits existence) ($L$)
- (iii) the time during which the habitat remains suitable for breeding (which will be related to $F$, $L$ and also the organism's generation time)
- (iv) size of favourable and
- (v) unfavourable patches of habitat.

For the organism: (i) generation time
(ii) trivial range, over which the organism gathers its food
(iii) migratory range, over which it may move when not reproducing.

From the genetical point of view, we are concerned primarily with the characteristics of the organism, and it has to be recognized that:

1 All three of the factors identified by Southwood are subject to inherited variation, although the evidence for variation in development rate, fecundity, etc. is much greater than for genetical influences on range of movement. Notwithstanding

2 All three are correlated with size and therefore with homeostatic control (of both physiology and behaviour) (Southwood *et al.* 1974).

3 If these characters (or their correlates) are inherited additively, the variance of their means will be normally distributed, but such inheritance will only be additive at the grossest phenotypic level.

It is this last point where genetical considerations impinge most closely on population ecology. For example, the proportion of the mouse population on Skokholm which dies in different winters varies from 40 to 90%. In a mild winter the deaths are independent of genotype, but in cold winters animals carrying the *s* allele have a greater chance of surviving to breed when favourable weather returns. Survival depends on the properties of haemoglobins produced by the *d* and *s* alleles. The variance in death rate is continuous at the phenotypic level (i.e. an animal either dies or not) but is bi- (or tri-) modal at the genic level. Exactly the same considerations apply to all the examples in Table 3.2: survival in melanic moths or ladybirds, or breeding in melanic pigeons or skuas is different in different morphs—but this is recognized only because the respective morphs are visually distinct.

The same problem can be looked at from the ecological rather than the genetical point of view. It appears most clearly when considering the $r - K$ continuum. Krebs *et al.* (1973) have pointed out that the characteristics of individuals in the increasing and declining phases of microtine cycles are suggestive of a reversal of selection for $K$ to $r$ traits: dispersing animals are $r$-strategists, while those that remain behind are less influenced by increasing density and maximize their use of the habitat, i.e. they are $K$-strategists. Among Lepidoptera, various *Morpho* species are large, territorial and long-lived, i.e. they are towards the $K$-selected end of the spectrum. Young & Muyshondt (1972) have suggested that *M. polyphemus* and some related species have recently shifted towards an $r$-strategy following the disturbance of tropical forests by man and the extension of areas of secondary growth.

Lack (1968 and earlier) repeatedly justified his work about the evolution of clutch size in birds in the face of apparent exceptions on the grounds that comparatively recent changes in the environment will affect the ecology of a species first, and any genetical adjustment in its reproductive rate will follow comparatively slowly. For example, the breeding behaviour of the Great Tit (*Parus major*) supports Lack's arguments in its traditional deciduous wood habitat, but not in the coniferous forests which it has colonized in the recent past; Gannets *Sula bassana* and Fulmars *Fulmarus glacialis* lay only one egg despite being able to feed two young, but they only began increasing at a rapid rate in the North Atlantic within the past century. And so on; such examples as these are further evidence of inherited variation in the determinants of population numbers described above.

### *Problems for a deterministic theory*

We can summarize the difficulties of producing genetical predictions from ecological models under three headings:

1 *Inconstant selection.* This has been discussed already, but it is worth labouring because of the confusion produced by assuming constant selective values and the consequent elaboration of the genetic load idea. Brues (1969) has summarized the situation: 'Not a few of the difficulties of the load concept are due to the fact that it has attempted to wrap up in one measure both the mean and the variance of a population's viability, by a rule which implied that increased heterogeneity must be associated with decreased mean fitness. It would be advantageous to keep the mean and the variance conceptually distinct, recognizing that the *mean* fitness of a population involves biological and ecological factors of considerable complexity. The *variance* of fitness, with its specific effects on gene frequency through time, can be dealt with mathematically in an unambiguous way . . .'

A rather different but significant aspect of selection inconstancy is fitness difference acting under rarely experienced conditions—perhaps once a lifetime, perhaps once every ten or hundred generations. The results of selection (in terms of a particular allele frequency) will persist in a population or species, but will be unaffected by the environment most of the time, i.e. the alleles will appear 'neutral'. Neutrality in this sense will be spasmodic, and thus arguably spurious. This is the genetical equivalent of the ecological wisdom of Charles Elton (1927), that 'Animals . . . spend an unexpectedly large amount of their time doing nothing at all, or at any rate nothing in particular . . . [They] are not always struggling for existence.'

2 The genetical composition of any population is dependent on its history in two ways:

(a) Particular mutations occur at a finite and low rate. For example, pesticide resistance depends on appropriate variation arising (Cook & Wood 1976); the Rosy Minor moth *Miana literosa* was extinct in the Sheffield area for many years, and only managed to recolonize the city in the mid-1940's through a newly arisen mutant.

(b) Species rarely evolve in the progressive fashion suggested by the fossil record but undergo a nexus of local crisis, extinction and colonization. This produces very marked genetical effects, both in reducing variance by 'bottle-necking' and changing allele frequencies through the founder effect (Berry, Jakobson & Peters 1978). Consequently, it is hazardous to extrapolate the action of genes from survey data on present populations; information collected over several generations and a knowledge of the species biology are necessary (Berry 1978a).

3 Individuals can react to an environmental stress either phenotypically or genetically. Phenotypic adjustment is more likely in species capable of homeostatic response than in those not so buffered. This means that large homeotherms (i.e. *K*-strategists) are less likely to undergo genetic change than smaller poikilotherms, which in turn means that the examples of genetical change in this paper may be biased towards *r*-strategists. Notwithstanding, failure to find a genetical response predicted from a particular model of environmental stress need mean no more than that the stressed organism has reacted in a non-genetic way. (The switch whereby aphids, etc. may develop into markedly different morphs is another form of phenotypic response, but it is really *epi*genetic since it represents alternative pathways for a 'preprogrammed' genome. It is the extreme manifestation of a situation explored by Waddington 1953; Smith & Sondhi 1960; Milkman 1961.)

## EVOLUTIONARY GENETICS AND ECOLOGY

Evolutionary ecology is a fashionable interest. It was urged by Lack (1965), but became a major topic of theory and practice following the elegant work of (particularly) MacArthur on island biogeography (MacArthur & Wilson 1967; MacArthur 1972). Although the successes of the island biogeographers have been less than their claims (e.g. Lynch & Johnson 1974), the simplicity and elegance of the underlying postulates and models ensure that island ecology will always remain near the centre of ecological endeavour (Diamond & May 1976).

The three main factors involved in determining the content of island faunas and floras are:

1 Immigration, which is related to isolation from the source of colonizers.

2 Establishment, which is correlated with available habitat and hence with area.

3 Extinction, which depends on competition and the availability of resources.

Lack (1969) added a further point:

4 Ecological tolerance, so that a species capable of occupying a wide niche may exclude specialists.

Lack's contribution is both important and confusing: important because it moves species diversity from the mechanical consummation of numbers into the realms of ecological interaction (e.g. Johnson & Raven 1973), but confusing because niche width is a genetical property of a population, and hence variable. In the context of islands, it might be better to replace the idea of 'width' with 'opportunistic adjustment' (Berry 1978b).

This means that the topic of this essay is highly relevant to island biology theory. Indeed, the importance of genetical variation can perhaps be best investigated in island populations, both because of the experimental simplicity produced by the relative species, poverty on most islands, but also because of the genetical diversity in the terrestrial fauna of archipelagoes resulting from the repeated operation of the founder effect (Mayr 1954; Berry 1969, 1978b; Carson 1970).

The relatively few visual inherited polymorphisms in natural populations have been profitably exploited by geneticists (q.v. Ford 1975; Berry 1977). Intra-population genetical variation has been largely ignored by ecologists, probably because they have had no easy means of recognizing the effects of segregating genes. The availability of electrophoretic techniques has changed this, and population studies will increasingly be incomplete if genetical influences are ignored. To cite one personal example, the study of house mice on Skokholm was a pedestrian investigation of a particular local population until electrophoretic investigation of blood proteins showed that the population was a fascinating mosaic of different morphs, responding differently to a range of social and climatic conditions (Berry 1968, 1978a).

Evolutionary ecology is with us; evolutionary genetics is a discipline only on the verge of recognition. The important endeavours in the coming years for both geneticists and ecologists will be to dissect the contribution of different genes to births, deaths, immigration and emigration; in other words, to unravel the interplay of nature and nurture, of chance and purpose, of pattern and organization in the structure of individuals and communities. To be fair, this is only continuing the analysis which began with Adam's taxonomic exercise in the Garden of Eden; the difficulty has been that geneticists and ecologists were allotted rooms on different floors of the Tower of Babel and learnt to speak different languages.

## REFERENCES

**Allison A.C. (1964)** Polymorphism and natural selection in human populations. *Cold Spring Harbor Symposia on Quantitative Biology*, **29**, 137–149.

**Anderson E.S. (1968)** Drug resistance in *Salmonella typhimurium* and its implications. *British Medical Journal*, **iii**, 333–339.

**Anderson P.K. (1970)** Ecological structure and gene flow in small mammals. *Variation in Mammalian Populations* (Ed. by R.J. Berry & H.N. Southern) pp. 229–325. Academic Press, London.

**Anderson W., Dobzhansky T., Pavlovsky O., Powell J. & Yardley D. (1975)** Genetics of natural populations. XLII. Three decades of genetic change in *Drosophila pseudo-obscura*. *Evolution*, **29**, 24–36.

**Ashton G.C. & Dennis M.N. (1971)** Selection at the transferrin locus in mice. *Genetics*, **67**, 253–265.

**Ayala F.J. (1968)** Genotype, environment and population numbers. *Science*, **162**, 1453–1459.

**Ayala F.J. (1972)** Darwinian *versus* non-Darwinian evolution in natural populations of *Drosophila*. *Proceedings of Sixth Berkeley Symposium on Mathematical Statistics and Probability*, **5**, 211–236.

**Ayala F.J. (1975)** Genetic differentiation during the speciation process. *Evolutionary Biology*, **8** (Ed. by T. Dobzhansky, M.K. Hecht & W.C. Steere) pp. 1–78. Plenum, New York.

**Ayala F.J. & Valentine J.W. (1974)** Genetic variability in the cosmopolitan deep-water ophiuran *Ophiomusium lymani*. *Marine Biology*, **27**, 51–57.

**Batten C.A. & Berry R.J. (1967)** Prenatal mortality in wild-caught house mice. *Journal of Animal Ecology*, **36**, 453–463.

**Bennett D. (1975)** The *T*-locus of the mouse. *Cell*, **6**, 441–454.

**Beranek A.P. & Oppenoorth F.J. (1977)** Evidence that the elevated carboxylesterase (esterase 2) in OP resistant *M. persicae* (Sulz.) is identical with the organophosphate-hydrolyzing enzyme. *Pesticide Biochemistry & Physiology*, **7**, 16–20.

**Berger D. (1974)** Esterases of *Drosophila* II. Biochemical studies of esterase-5 in *D. pseudo-obscura*. *Genetics*, **78**, 1173–1183.

**Berry R.J. (1968)** The ecology of an island population of the house mouse. *Journal of Animal Ecology*, **37**, 445–470.

**Berry R.J. (1969)** History in the evolution of *Apodemus sylvaticus* (Mammalia) at one edge of its range. *Journal of Zoology*, **159**, 311–328.

**Berry R.J. (1971)** Conservation aspects of the genetical constitution of populations. *The Scientific Management of Animal and Plant Communities for Conservation* (Ed. by E. Duffey & A.S. Watt) pp. 177–206. Blackwell Scientific Publications, Oxford.

**Berry R.J. (1977)** *Inheritance and Natural History*. Collins New Naturalist, London.

**Berry R.J. (1978a)** Genetic variation in wild house mice: where natural selection and history meet. *American Scientist*, **66**, 52–60.

**Berry R.J. (1978b)** The Outer Hebrides: where genes and geography meet. *Proceedings of the Royal Society of Edinburgh*, *B*, **77** (in press).

**Berry R.J. & Crothers J.H. (1968)** Stabilizing selection in the Dog Whelk (*Nucella lapillus*). *Journal of Zoology*, **155**, 5–17.

**Berry R.J. & Crothers J.H. (1970)** Genotypic stability and physiological tolerance in the Dog Whelk (*Nucella lapillus* L.). *Journal of Zoology*, **162**, 293–302.

**Berry R.J. & Davis P.E. (1970)** Polymorphism and behaviour in the Arctic Skua (*Stercorarius parasiticus* (L.)). *Proceedings of the Royal Society of London*, *B*, **175**, 255–267.

**Berry R.J. & Jakobson M.E. (1974)** Vagility in an island population of the house mouse. *Journal of Zoology*, **173**, 341–354.

**Berry R.J., Jakobson M.E. & Peters J. (1978)** The house mice of the Faroe Islands: a study in microdifferentiation. *Journal of Zoology*. **185**, 73–92.

**Berry R.J. & Murphy H.M. (1970)** Biochemical genetics of an island population of the house mouse. *Proceedings of the Royal Society of London, B*, **176**, 87–103.

**Berry R.J. & Peters J. (1977)** Heterogeneous heterozygosities in *Mus musculus* populations. *Proceedings of the Royal Society of London, B*, **197**, 485–503.

**Berry R.J., Peters J. & Van Aarde R.J. (1978)** Sub-Antarctic house mice: colonization, survival and selection. *Journal of Zoology*, **184**, 127–141.

**Birch L.C. (1960)** The genetic factor in population ecology. *American Naturalist*, **94**, 5–24.

**Birch L.C. (1971)** The role of environmental heterogeneity and genetical heterogeneity in determining distribution and abundance. *Dynamics of Populations* (Ed. by P.J. Den Boer & G.R. Gradwell) pp. 109–126. Centre for Agricultural Publishing and Documentation, Wageningen.

**Bishop J.A., Cook L.M., Muggleton J. & Seaward M.R.D. (1975)** Moths, lichens and air pollution along a transect from Manchester to North Wales. *Journal of Applied Ecology*, **12**, 83–98.

**Bishop J.A. & Hartley D.J. (1976)** The size and age structure of rural populations of *Rattus norvegicus* containing individuals resistant to the anticoagulant poison warfarin. *Journal of Animal Ecology*, **45**, 623–646.

**Bonnell M.L. & Selander R.K. (1974)** Elephant seals: genetic variation and near extinction. *Science*, **184**, 908–909.

**Bradshaw A.D. (1971)** Plant evolution in extreme environments. *Ecological Genetics and Evolution* (Ed. by E.R. Creed) pp. 20–50. Blackwell Scientific Publications, Oxford.

**Bradshaw A.D. (1976)** Pollution and evolution. *Effects of Air Pollutants on Plants* (Ed. by T.A. Mansfield) pp. 135–159. *Society of Experimental Biology Seminar Series.* University Press, Cambridge.

**Brown A.J.L. (1977)** Physiological correlates of an enzyme polymorphism. *Nature, London*, **269**, 803–804.

**Brues A.M. (1969)** Genetic load and its varieties. *Science*, **164**, 1130–1136.

**Bryan G.W. (1973)** Adaptation of the polychaete *Nereis diversicolor* to estuarine sediments containing high concentrations of zinc and cadmium. *Journal of the Marine Biological Association*, **53**, 839–857.

**Bryant E.H. (1974)** On the adaptive significance of enzyme polymorphisms in relation to environmental variability. *American Naturalist*, **108**, 1–19.

**Bumpus H.C. (1899)** The elimination of the unfit as illustrated by the introduced sparrow. *Biological Lectures, Woods Hole for 1898*, 209–226.

**Campbell C.A., Valentine J.W. & Ayala F.J. (1975)** High genetic variability in a population of *Tridacna maxima* from the Great Barrier Reef. *Marine Biology*, **33**, 341–345.

**Carson H.L. (1970)** Chromosome tracers of the origin of species. *Science*, **168**, 1414–1418.

**Carson H.L. & Kaneshiro K.Y. (1976)** *Drosophila* of Hawaii: systematics and ecological genetics. *Annual Review of Ecology and Systematics*, **7**, 311–345.

**Chapman R.N. (1928)** The quantitative analysis of environmental factors. *Ecology*, **9**, 111–122.

**Chitty D. (1967)** The natural selection of self-regulatory behaviour in animal populations. *Proceedings of the Ecological Society of Australia*, **2**, 51–78.

**Christian J.J. & Davis D.E. (1964)** Endocrines, behavior and population. *Science*, **146**, 1550–1560.

**Clark J.M. (1975)** The effects of selection and human preference on coat colour gene frequencies in urban cats. *Heredity*, **35**, 195–210.

**Clark J.M. (1976)** Variation in coat colour gene frequencies and selection in the cats of Scotland. *Genetica*, **46**, 401–412.

**Clarke B.C. (1972)** Density-dependent selection. *American Naturalist*, **106**, 1–13.
**Clarke B.C. (1975a)** The causes of biological diversity. *Scientific American*, **233**, 50–60.
**Clarke B.C. (1975b)** The contribution of ecological genetics to evolutionary theory: detecting the direct effects of natural selection on particular polymorphic loci. *Genetics*, **79**, 101–113.
**Cook L.M. & Wood R.J. (1976)** Genetic effects of pollutants. *Biologist*, **23**, 129–139.
**Cowie V. & Penrose L.S. (1950)** Dilution of hair colour in phenylketonuria. *Annals of Eugenics*, **15**, 297–301.
**Creed E.R. (1975)** Melanism in the two spot ladybird: the nature and intensity of selection. *Proceedings of the Royal Society of London, B*, **190**, 135–148.
**Darlington C.D. (1971)** The evolution of polymorphic systems. *Ecological Genetics and Evolution* (Ed. by E.R. Creed) pp. 1–19. Blackwell Scientific Publications, Oxford.
**Day P.R. (1974)** *Genetics of Host-parasite Interaction.* Freeman, San Francisco.
**DeFries J.C. & McClearn G.E. (1972)** Behavioral genetics and the fine structure of mouse populations: a study in microevolution. *Evolutionary Biology*, **5** (Ed. by Th. Dobzhansky, M.K. Hecht & W.C. Steere) pp. 279–291. Appleton-Century-Crofts, New York.
**Diamond J.M. & May R.M. (1976)** Island biogeography and the design of nature reserves. *Theoretical Ecology* (Ed. by R.M. May) pp. 163–186. Blackwell Scientific Publications, Oxford.
**Elton C. (1927)** *Animal Ecology.* Sidgwick & Jackson, London.
**Endler J.A. (1973)** Gene flow and population differentiation. *Science*, **179**, 243–250.
**Endler J.A. (1977)** *Geographic Variation, Speciation, and Clines.* University Press, Princeton.
**Ewens W.J. & Feldman M.W. (1976)** The theoretical assessment of selective neutrality. *Population Genetics and Evolution* (Ed. by S. Karlin & E. Nevo) pp. 303–337. Academic Press, New York.
**Fisher R.A. (1954)** Retrospect of the criticisms of the theory of natural selection. *Evolution as a Process* (Ed. by J.S. Huxley, A.C. Hardy & E.B. Ford) pp. 84–98. Allen & Unwin, London.
**Flor H.H. (1956)** The complementary genic systems in flax and flax rust. *Advances in Genetics*, **8**, 29–54.
**Ford E.B. (1975)** *Ecological Genetics*, 4th ed. Chapman & Hall, London.
**Froggatt P. & Nevin N.C. (1971)** The 'Law of Ancestral Heredity' and the Mendelian–Ancestrian controversy in England, 1889–1906. *Journal of Medical Genetics*, **8**, 1–36.
**Frydenberg O. & Simonsen V. (1973)** Genetics of *Zoarces* populations. V. Amount of protein polymorphism and degree of genic heterozygosity. *Hereditas*, **75**, 221–232.
**Gibson J.B. (1972)** Differences in the number of molecules produced by two allelic electrophoretic enzyme variants in *Drosophila melanogaster. Experientia*, **28**, 975–976.
**Haldane J.B.S. (1948)** The theory of a cline. *Journal of Genetics*, **48**, 277–284.
**Haldane J.B.S. (1949)** Disease and evolution. *Ricerca Scientia, Supplement*, **19**, 68–76.
**Haldane J.B.S. (1955)** Population genetics. *New Biology*, **18**, 34–51.
**Haldane J.B.S. (1956)** The theory of selection for melanism in Lepidoptera. *Proceedings of the Royal Society of London, B*, **145**, 303–306.
**Haldane J.B.S. (1957)** The cost of natural selection. *Journal of Genetics*, **55**, 511–524.
**Haldane J.B.S. (1959)** Natural selection. *Darwin's Biological Work* (Ed. by P.R. Bell) pp. 101–149. University Press, Cambridge.
**Harris H. (1966)** Enzyme polymorphisms in man. *Proceedings of the Royal Society of London, B*, **164**, 298–310.

**Harris H., Hopkinson D.A. & Edwards Y.H. (1977)** Polymorphism and the sub-unit structure of enzymes: a contribution to the neutralist–selectionist controversy. *Proceedings of the National Academy of Sciences of the United States of America*, **74**, 698–701.

**Harris H., Hopkinson D.A. & Robson E.B. (1974)** The incidence of rare alleles determining electrophoretic variants: data on 43 enzyme loci in man. *Annals of Human Genetics*, **37**, 237–253.

**Harvey P.H., Jordan C.A. & Allen J.A. (1974)** Selection behaviour of wild blackbirds at high prey densities. *Heredity*, **32**, 401–404.

**Hatchett J.H. & Gallun R.L. (1970)** Genetics of the ability of the hessian fly, *Mayetiola destructor*, to survive on wheats having different genes for resistance. *Annals of the Entomological Society of America*, **63**, 1400–1407.

**Heath D.J. (1974)** Seasonal changes in frequency of the 'yellow' morph of the isopod, *Sphaeroma rugicauda. Heredity*, **32**, 299–307.

**Hedrick P.W., Ginevan M.E. & Ewing E.P. (1976)** Genetic polymorphism in heterogeneous environments. *Annual Review of Ecology and Systematics*, **7**, 1–32.

**Heller J. (1975)** Visual selection of shell colour in two littoral prosobranchs. *Zoological Journal of the Linnean Society*, **56**, 153–170.

**Henderson B.A. (1977)** The genetics and demography of a high and low density of red grouse *Lagopus l. scoticus. Journal of Animal Ecology*, **46**, 581–592.

**Hewitt G.M. & Brown F.M. (1970)** The *B*-chromosome system of *Myrmeleotettix maculatus*. V. A steep cline in East Anglia. *Heredity*, **25**, 363–371.

**Jain S.K. & Bradshaw A.D. (1966)** Evolutionary divergence among adjacent plant populations. I. The evidence and its theoretical analysis. *Heredity*, **21**, 407–441.

**Johnson G.B. (1976)** Genetic polymorphism and enzyme function. *Molecular Evolution* (Ed. by F.J. Ayala) pp. 46–59. Sinauer, Sunderland, Massachusetts.

**Johnson G.B. (1977)** Assessing electrophoretic similarity. *Annual Review of Ecology and Systematics*, **8**, 309–328.

**Johnson M.P. & Raven P.H. (1973)** Species number and endemism: the Galapagos revisited. *Science*, **179**, 893–895.

**Johnson M.S. (1971)** Adaptive lactate dehydrogenase variation in the crested blenny, *Anoplarchus. Heredity*, **7**, 205–226.

**Jones J.S., Leith B.H. & Rawlings P. (1977)** Polymorphism in *Cepaea*: a problem with too many solutions? *Annual Review of Ecology and Systematics*, **8**, 109–143.

**Karlin S. & Richter-Dyn N. (1976)** Some theoretical analyses of migration selection interaction in a cline: a generalized two range environment. *Population Genetics and Ecology* (Ed. by S. Karlin & E. Nevo) pp. 659–706. Academic Press, New York.

**Kearsey M.J. & Kojima K. (1967)** The genetic architecture of body weight and egg hatchability in *Drosophila melanogaster. Genetics*, **56**, 23–37.

**Kettlewell H.B.D. (1973)** *The Evolution of Melanism.* Clarendon, Oxford.

**Kettlewell H.B.D., Berry R.J., Cadbury C.J. & Phillips G.C. (1969)** Differences in behaviour, dominance and survival within a cline. *Heredity*, **24**, 15–25.

**Kimura M. & Ohta T. (1971)** Protein polymorphism as a phase of molecular evolution. *Nature*, **229**, 467–469.

**King C.E. & Dawson P.S. (1972)** Population biology and the *Tribolium* model. *Evolutionary Biology*, **5** (Ed. by T. Dobzhansky, M.K. Hecht & W.C. Steere) pp. 133–227. Appleton-Century-Crofts, New York.

**King H.D. (1939)** Life processes in gray Norway rats during fourteen years in captivity. *American Anatomical Memoir*, No. 17.

**King J.L. & Jukes T.H. (1969)** Non-Darwinian evolution. *Science*, **164**, 788–798.

**Koehn R.K. (1969)** Esterase heterogeneity: dynamics of a polymorphism. *Science*, **163**, 943–944.

**Koehn R.K. & Rasmussen D.I. (1967)** Polymorphic and monomorphic serum esterase heterogeneity in catostomid fish populations. *Biochemical Genetics*, **1**, 131–144.

**Kojima K., Gillespie J. & Tobari Y.N. (1970)** A profile of *Drosophila* species' enzymes assayed by electrophoresis. I. Number of alleles, heterozygosities, and linkage disequilibrium in glucose-metabolizing systems and some other enzymes. *Biochemical Genetics*, **4**, 627–637.

**Krebs C.J., Gaines M.S., Keller B.L., Myers J.H. & Tamarin R.H. (1973)** Population cycles in small rodents. *Science*, **179**, 35–41.

**Lack D. (1948)** The significance of litter size in mice. *Journal of Animal Ecology*, **17**, 45–50.

**Lack D. (1965)** Evolutionary ecology. *Journal of Animal Ecology*, **34**, 223–231.

**Lack D. (1968)** *Ecological Adaptations for Breeding in Birds*. Methuen, London.

**Lack D. (1969)** The numbers of bird species on islands. *Bird Study*, **16**, 193–209.

**Lees D.R. & Creed E.R. (1975)** Industrial melanism in *Biston betularia*: the role of selective predation. *Journal of Animal Ecology*, **44**, 67–83.

**Levene H. (1953)** Genetic equilibrium when more than one ecological niche is available. *American Naturalist*, **87**, 311–313.

**Levins R. (1968)** *Evolution in Changing Environments*. University Press, Princeton.

**Lewontin R.C. & Hubby J.L. (1966)** A molecular approach to the study of genic heterozygosity in natural populations. II. Amount of variation and degree of heterozygosity in natural populations of *Drosophila pseudoobscura*. *Genetics*, **54**, 595–609.

**Lewontin R.C. & Krakauer J. (1973)** Distribution of gene frequency as a test of the theory of the selective neutrality of polymorphisms. *Genetics*, **74**, 175–195.

**Luoma S.N. (1977)** Detection of trace contaminant effects in aquatic ecosystems. *Journal of the Fisheries Research Board of Canada*, **34**, 436–439.

**Lynch J.F. & Johnson N.K. (1974)** Turnover and equilibria in insular avifaunas, with special reference to the California Channel Islands. *Condor*, **76**, 370–384.

**MacArthur R.H. (1972)** *Geographical Ecology*. Harper & Row, New York.

**MacArthur R.H. & Wilson E.O. (1967)** *The Theory of Island Biogeography*. University Press, Princeton.

**McGill T.E. (1970)** Genetic analysis of male sexual behavior. *Contributions to Behavior—Genetic Analysis. The Mouse as a Prototype* (Ed. by G. Lindzey & D.D. Thiessen) pp. 57–88. Appleton-Century-Crofts, New York.

**Mackay T.F.C. & Doyle R.W. (1978)** An ecological genetic analysis of the settling behaviour of a marine polychaete: 1. Probability of settlement and gregarious behaviour. *Heredity*, **40**, 1–12.

**Mather K. (1953)** The genetical structure of populations. *Symposia of the Society for Experimental Biology*, **7**, 76–95.

**Mather K. (1970)** The nature and significance of variation in wild populations. *Variation in Mammalian Populations* (Ed. by R.J. Berry & H.N. Southern) pp. 27–39. Academic Press, London.

**Mayr E. (1954)** Change of genetic environment and evolution. *Evolution as a Process* (Ed. by J. Huxley, A.C. Hardy & E.B. Ford) pp. 157–180. Allen & Unwin, London.

**Milkman R.D. (1961)** The genetic basis of natural variation. III. Developmental lability and evolutionary potential. *Genetics*, **46**, 25–38.

**Muggleton J., Lonsdale D. & Benham B.R. (1975)** Melanism in *Adalia bipunctata* L. (Col., Coccinellidae) and its relationship to atmospheric pollution. *Journal of Applied Ecology*, **12**, 451–464.

**Muller H.J. (1950)** Our load of mutations. *American Journal of Human Genetics*, **2**, 111–176.

**Murray J.J. (1972)** *Genetic Diversity and Natural Selection.* Oliver & Boyd, Edinburgh.

**Murton R.K. (1971)** *Man and Birds.* Collins New Naturalist, London.

**Murton R.K., Thearle R.J.P. & Coombs C.F.B. (1974)** Ecological studies of the feral pigeon *Columba livia* var. *Journal of Applied Ecology*, **11**, 841–854.

**Murton R.K., Westwood N.J. & Thearle R.J.P. (1973)** Polymorphism and the evolution of a continuous breeding season in the pigeon, *Columba livia. Journal of Reproduction and Fertility*, supplement, **19**, 563–577.

**Myers J.H. & Krebs C.J. (1971)** Genetic, behavioral, and reproductive attributes of dispersing field voles *Microtus pennsylvanicus* and *Microtus ochrogaster. Ecological Monographs*, **41**, 34–78.

**Nei M. (1972)** Genetic distance between populations. *American Naturalist*, **106**, 283–292.

**Nei M., Maruyama T. & Chakraborty R. (1975)** The bottleneck effect and genetic variability in populations. *Evolution*, **29**, 1–10.

**Nei M. & Roychoudhury A.K. (1974)** Genic variation within and between the three major races of man, Caucasoids, Negroids, and Mongoloids. *American Journal of Human Genetics*, **26**, 421–443.

**Nevo E. (1976)** Genetic variation in constant environments. *Experientia*, **32**, 858.

**Nevo E. & Bar Z. (1976)** Natural selection of genetic polymorphisms along climatic gradients. *Population Genetics and Ecology* (Ed. by S. Karlin & E. Nevo) pp. 159–184. Academic Press, New York.

**Nevo E., Dessauer H.C. & Chuang K-C. (1975)** Genetic variation as a test of natural selection. *Proceedings of the National Academy of Sciences of the United States of America*, **72**, 2145–2149.

**Nicholson A.J. (1957)** The self-adjustment of populations to change. *Cold Spring Harbor Symposia on Quantitative Biology*, **22**, 153–173.

**O'Donald P. & Davis J.W.F. (1975)** Demography and selection in a population of Arctic Skuas. *Heredity*, **35**, 75–83.

**Ohta T. (1974)** Mutational pressure as the main cause of molecular evolution and polymorphism. *Nature, London*, **252**, 351–354.

**Owen D.F. (1977)** Latitudinal gradients in clutch size: an extension of David Lack's theory. *Evolutionary Ecology* (Ed. by B. Stonehouse & C.M. Perrins) pp. 171–179. Macmillan, London.

**Park T. (1962)** Beetles, competition and populations. *Science*, **138**, 1369–1375.

**Parsons P.A. (1973)** *Behavioural and Ecological Genetics. A Study in Drosophila.* Clarendon, Oxford.

**Powell J.R. (1971)** Genetic polymorphism in varied environments. *Science*, **174**, 1035–1036.

**Powell J.R. (1975)** Protein variation in natural populations of animals. *Evolutionary Biology*, **8** (Ed. by T. Dobzhansky, M.K. Hecht & W.C. Steere) pp. 79–119. Appleton-Century-Crofts, New York.

**Raffaelli D.G. & Hughes R.N. (1978)** The effects of crevice size and availability on populations of *Littorina rudis* and *Littorina neritoides. Journal of Animal Ecology*, **47**, 71–83.

**Richmond R.C. (1972)** Enzyme variability in the *Drosophila willistoni* group. III. Amounts of variability in the superspecies, *D. paulistorum. Genetics*, **70**, 87–112.

**Rothenbuhler W.C. (1967)** Genetic and evolutionary considerations of social behavior of honeybees and some related insects. *Behavior—Genetic Analysis* (Ed. by J. Hirsch) pp. 61–111. McGraw-Hill, New York.

**Sarich V.M. (1977)** Rates, sample sizes, and the neutrality hypothesis in evolutionary studies. *Nature, London*, **265**, 24–28.

**Selander R.K. (1976)** Genic variation in natural populations. *Molecular Evolution* (Ed. by F.J. Ayala) pp. 21–45. Sinauer, Sunderland, Massachusetts.

**Selander R.K. & Kaufman D.W. (1973)** Genic variability and strategies of adaptation in animals. *Proceedings of the National Academy of Sciences of the United States of America*, **70**, 1875–1877.

**Semeonoff R. & Robertson F.W. (1968)** A biochemical and ecological study of plasma esterase polymorphism in natural populations of the field vole, *Microtus agrestis* L. *Biochemical Genetics*, **1**, 205–227.

**Smith J.M. (1970)** The causes of polymorphism. *Variation in Mammalian Populations* (Ed. by R.J. Berry & H.N. Southern) pp. 371–383. Academic Press, London.

**Smith J.M. & Sondhi K.C. (1960)** The genetics of a pattern. *Genetics*, **45**, 1039–1050.

**Smith M.H., Garten C.T. & Ramsey P.R. (1975)** Genic heterozygosity and population dynamics in small mammals. *Isozymes. Vol. 4. Genetics and Evolution* (Ed. by C.L. Markert) pp. 85–102. Academic Press, New York & London.

**Sokal R.R. & Sonleitner F.J. (1968)** The ecology of selection in hybrid populations of *Tribolium castaneum*. *Ecological Monographs*, **38**, 345–379.

**Sokoloff A. (1966)** The genetics of *Tribolium* and related species. *Advances in Genetics*, supplement 1.

**Somero G.N. & Soulé M. (1974)** Genetic variation in marine fishes as a test of the niche-variation hypothesis. *Nature, London*, **249**, 670–672.

**Soulé M. (1976)** Allozyme variation: its determinants in space and time. *Molecular Evolution* (Ed. by F.J. Ayala) pp. 60–77. Sinauer, Sunderland, Massachusetts.

**Soulé M. & Yang S.Y. (1974)** Genetic variation in side-blotched lizards on islands in the Gulf of California. *Evolution*, **27**, 593–600.

**Southern H.N. (1954)** Mimicry in cuckoo's eggs. *Evolution as a Process* (Ed. by J.S. Huxley, A.C. Hardy & E.B. Ford) pp. 219–232. Allen & Unwin, London.

**Southwick C.H. (1955)** Regulatory mechanisms of house mouse populations: social behaviour affecting litter survival. *Ecology*, **36**, 627–634.

**Southwood T.R.E. (1977)** Habitat, the templet for ecological strategies? *Journal of Animal Ecology*, **46**, 337–365.

**Southwood T.R.E., May R.M., Hassell M.P. & Conway G.R. (1974)** Ecological strategies and population parameters. *American Naturalist*, **108**, 791–804.

**Steward R.C. (1977)** Melanism and selective predation in three species of moths. *Journal of Animal Ecology*, **46**, 483–496.

**Swank R.T. & Bailey D.W. (1973)** Recombinant inbred lines: value in the genetic analysis of biochemical variants. *Science*, **181**, 1249–1252.

**Thoday J.M. (1961)** The location of polygenes. *Nature, London*, **191**, 368–370.

**Thompson W.A. & Vertinsky I. (1975)** Bird flocking revisited: the case with polymorphic prey. *Journal of Animal Ecology*, **44**, 755–765.

**Tinbergen L. (1960)** Natural control of insects in pine-woods. 1. Factors influencing intensity of predation by song birds. *Archives neerlandaises de zoologie*, **13**, 265–336.

**Turner J.R.G. (1977)** Butterfly mimicry: the genetical evolution of an adaptation. *Evolutionary Biology*, **10** (Ed. by M.K. Hecht, W.C. Steere & B. Wallace) pp. 163–206. Plenum, New York.

**Turner J.R.G. & Williamson M.H. (1968)** Population size, natural selection and the genetic load. *Nature, London*, **218**, 700.

**Valentine J.W. (1971)** Resource supply and species diversity patterns. *Lethaia*, **4**, 51–61.

**Valentine J.W. (1976)** Genetic strategies of adaptation. *Molecular Biology* (Ed. by F.J. Ayala) pp. 78–94. Sinauer, Sunderland, Massachusetts.

**Valentine J.W. & Ayala F.J. (1976)** Genetic variability in krill. *Proceedings of the National Academy of Sciences of the United States of America*, **73**, 658–660.

**Vigue C.L. & Johnson F.M. (1973)** Isozyme variability in species of the genus *Drosophila*. VI. Frequency–property–environment relationships of allelic alcohol dehydrogenases in *D. melanogaster*. *Biochemical Genetics*, **9**, 213–227.

**Vogel F. & Chakravarti M. (1966)** ABO blood groups and smallpox in a rural population of West Bengal and Bihar (India). *Humangenetik*, **3**, 166–180.

**Waddington C.H. (1953)** Genetic assimilation of an acquired character. *Evolution*, **7**, 118–126.

**Wallace B. (1968)** Polymorphism, population size, and genetic load. *Population Biology and Evolution* (Ed. by R.C. Lewontin) pp. 87–108. University Press, Syracuse.

**Wallace B. (1970)** *Genetic Load.* Prentice-Hall, Englewood Cliffs.

**Wallace B. (1975)** Hard and soft selection revisited. *Evolution*, **29**, 465–473.

**Weldon W.F.R. (1901)** A first study of natural selection in *Clausilia laminata* (Montagu). *Biometrika*, **1**, 109–124.

**Wool D. (1977)** Genetic and environmental components of morphological variation in gall-forming aphids (Homoptera, Aphididae, Fordinae) in relation to climate. *Journal of Animal Ecology*, **46**, 875–889.

**Young A.M. & Muyshondt A. (1972)** Biology of *Morpho polyphemus* (Lepidoptera: Morphidae) in El Salvador. *Journal of New York Entomological Society*, **80**, 18–42.

# 4. ECOLOGICAL DETERMINANTS IN THE EVOLUTION OF LIFE HISTORIES

RICHARD LAW

*Department of Botany,*
*University of Sheffield, Sheffield S10 2TN*

## INTRODUCTION

In the evolution of life histories, there is an intriguing area of convergence between interests of population ecologists and geneticists. The reason for this is simply that life histories—sets of age-specific rates of reproduction and risks of death—are central to the ideas of both. To ecologists, their average values are essential to predict future numbers in populations. To geneticists, the extent to which they differ from genotype to genotype allows prediction of the relative contribution of their genes to future generations. The study of life-history evolution attempts to draw together these two lines of thought. We suppose that rates of reproduction and death are genetically variable characters in populations, and we consider the consequences of this on future states of populations when the environment changes in ecologically realistic ways.

In this paper, I consider some consequences of genetic variation in life histories which might be of interest to ecologists. To do this, I draw on a small part of a large literature on life-history evolution. (Stearns (1976, 1977) gives a comprehensive review.) I start by filling in some background—two prerequisites for life histories to evolve. This leads us to the problem of how to find the particular set of rates of reproduction and risks of death which maximize the contribution of progeny to future generations. Armed with a solution to this, it is possible to make some predictions about the kinds of life histories which should be selected under different environmental conditions. I confine my attention to differences in the availability of resources and intensity of predation, illustrating the effects where possible with work on the grass *Poa annua*. As far as possible, I am motivated to identify theories which can be tested on natural or experimental populations, without too much difficulty. I am therefore ultimately interested in changes taking place *within* populations, as it is here that we must look for the forces which drive life-history evolution.

## CONSTRAINTS IN DESIGN OF ORGANISMS

Our first prerequisite for evolution of life histories is best introduced by defining some characteristics of an organism in which the problems of maximizing numbers of progeny have all been solved. This organism starts to produce progeny almost immediately after its own birth, producing very large numbers at frequent intervals as it grows older. It experiences no mortality, and its capacity for dispersal and finding mates knows no bounds. Such an organism can appropriately be called a 'Darwinian demon'.

We all know that this is an absurd proposition. But the reason for its absurdity sheds some light on one of the central issues of life-history evolution. This is that there are constraints in the amount of resources which an individual can accumulate over any interval of time. For organisms living in environments with only a limited supply of resources, this is self-evident; but it is also true even if the supply of resources is unlimited, since an organism's machinery for assimilation must have a limited capacity. Once assimilated, these hard-won resources must be divided among all the activities of the organism, so if more go to one, less go to others.

One of the first people to appreciate the evolutionary implications of this was Fisher (1958, p. 47). He drew attention to the division of resources between gonads and soma. As more resources are allocated to reproduction, less remain available for non-reproductive activities. Both contribute to the number of offspring which an individual ultimately leaves—reproduction directly and non-reproductive activities indirectly, by increasing the chance of still being alive to reproduce in the future. Fisher suggested that it might prove most illuminating to study the balance achieved between these activities in different environments.

Fisher's idea has generated much interest, and has become embodied in the term *reproductive effort* (Williams 1966a,b; Tinkle 1969; Gadgil & Bossert 1970; Gadgil & Solbrig 1972; Schaffer 1974a,b; Taylor *et al.* 1974; Hirschfield & Tinkle 1975; Charlesworth & Léon 1976; Michod 1978). In the broadest sense, this is the proportion of resources which are used in reproductive activities over a convenient interval of time. The resource considered is generally energy, but this is over-restricting, as others might be more limiting, such as certain minerals in plants (Harper 1977, p. 656). Various attempts have been made to measure reproductive effort in natural populations (Harper & Ogden 1970; Gadgil & Solbrig 1972; Abrahamson & Gadgil 1973; Goodman 1974; Menge 1974; Ogden 1974; Schaffer & Elson 1975; Tinkle & Hadley 1975; Werner & Platt 1976; Hickman 1977; Swingland 1977), but they have often been beset with problems; for although reproductive effort is simple to define in theory, it can be very difficult to measure in practice (Hirschfield & Tinkle 1975).

It is important to keep the role of reproductive effort in perspective. It identifies only two of a wide range of sinks into which resources can be channelled. For example, a constant amount of resource to reproduction can be divided among a small number of large progeny or a large number of small ones (Williams 1966b, p. 167; Smith & Fretwell 1974). A constant amount of resource to non-reproductive activities can be channelled into the production of new tissue, the maintenance of tissue already in existence, into the capture of more resources, the defence against predators and pathogens, or into dispersal to new places, to name but a few possibilities. A life history represents a point of balance between the conflicting resource requirements of all these activities insofar as they influence rates of reproduction and risks of death.

However, it would be extremely difficult to consider all these sinks simultaneously. At best, we can only study simple combinations of two or three, while holding all others constant. I adopt such an approach here, concentrating on the division of resources between reproduction, maintenance of existing tissue and growth of new tissue. This approach has been used on a number of occasions, because these sinks have particularly direct relationships with rates of reproduction and risks of death (Gadgil & Bossert 1970; Schaffer 1974a; Charlesworth & Léon 1976; Michod 1978). More resources to reproduction can be expected to give greater rates of reproduction. More resources to maintenance can be expected to give reduced risks of death. More resources to growth can be expected to give greater body sizes. Large body size may in turn eventually give greater rates of reproduction where reproductive rates are functions of body size (Bagenal 1967; Harper & White 1974). In addition, they may reduce risks of death if a large body size enables a disproportionate share of resources to be obtained (Harper 1977, p. 187). (It is important to distinguish here between the short- and long-term effects of allocating resources to growth. In the short term, diverting resources away from reproduction and maintenance may cause a reduction in rates of reproduction and survival.) In adopting this approach, we obtain a much simplified picture of the range of alternative life histories but one which is amenable to rigorous analysis.

In principle, it is straightforward to test the theory that reproduction, maintenance and growth are alternative sinks for a limited supply of resources. Organisms with high rates of reproduction should have lower rates of growth and/or greater risks of death. I have attempted to test this with the grass *Poa annua* (Law 1979a) by comparing families which differ in their rates of reproduction. Consider first the relationship between numbers of inflorescences produced in the first year of growth and plant size in the second. Figure 4.1 shows clear indications that families with large numbers of inflorescences in the first year tend to be smaller in the second. The effect of this smaller size on the rate of reproduction in the second year can also be seen (Fig. 4.2). Families with large numbers of inflorescences in the first year tend to have low numbers in the second. It is less easy to see any relationship between rates or reproduction and

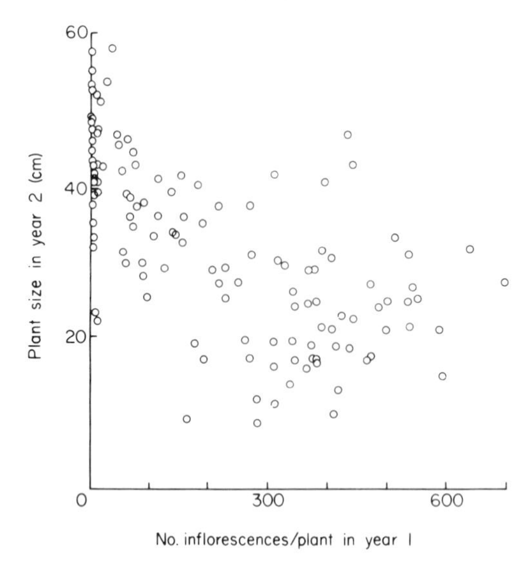


FIG. 4.1. Numbers of inflorescences per plant in year 1 and plant size in year 2 in *Poa annua*. Each point gives numbers averaged over several progeny of a single mother. Probability of achieving this relationship by chance less than 0·001 by Spearman's rank correlation coefficient ($r_s$). (From Law 1979a.)

risks of death, but if we consider only the numbers of inflorescences produced in the very early stages of reproduction, a significant relationship can be observed, those with large numbers being less likely to survive (Fig. 4.3).

There are various other sources of information which suggest the existence of a relationship between reproduction, maintenance and growth. Reduced rates of growth following years of high reproduction have been demonstrated clearly in several species of trees (Holmsgaard 1956; Hustich 1956; Eklund 1957; Fielding 1960; Eis, Garman & Ebell 1965). A relationship between growth and reproduction has also been observed in tobacco plants (Namkoong & Matzinger 1975), in barnacles and bivalves (Barnes 1962; Seed & Brown 1978) and in a number of species of fish (Iles 1974; Tyler & Dunn 1976; Wootton 1977). Between reproduction and risks of mortality the evidence is rather less convincing. There are several cases of high rates of reproduction being associated with short lives (Böcher 1949; Langer 1956; Böcher & Larsen 1958; Langer, Ryle & Jewiss 1964; Murdoch 1966; Tinkle 1969; Calow & Woolhead 1977). There are, however, at least two cases of high rates of reproduction being associated with long lives (Sarukhán 1974), although it is possible that these demonstrate only that both reproduction and survival increase with body size. Information on relationships between growth and risks of death are more difficult to find. Lints & Soliman (1977) found a negative correlation in *Drosophila melanogaster* and a positive one in *Tribolium castaneum*.

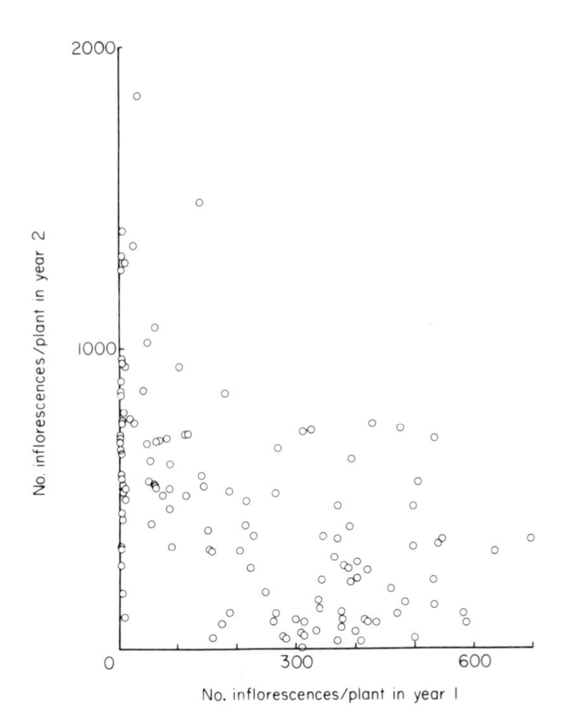


FIG. 4.2. Numbers of inflorescences per plant in *Poa annua* during first and second year's growth. Each point gives numbers averaged over several progeny of a single mother ($r_s < 0{\cdot}001$). (From Law 1979a.)

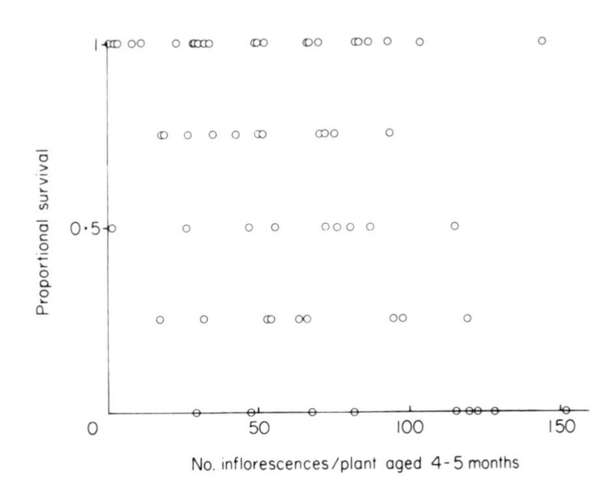


FIG. 4.3. Numbers of inflorescences per plant aged 4–5 months and chance of surviving subsequently. Each point gives numbers of inflorescences averaged over several progeny of a single mother and the proportion of the progeny surviving to end of experiment. There are nine points at (0,1), not shown individually ($r_s < 0{\cdot}01$). (From Law 1979a.)

While the evidence is far from conclusive, such information that we do have generally suggests that allocation of resources to one activity entails the loss of resources to others. We are thus led to envisage a life history as a point of balance between the conflicting requirements of different activities of an organism. This is of central importance to the evolution of life histories. It means that there can be no unique solution for maximizing numbers of progeny in all environments—there can be no 'Darwinian demon'. Different environments push the point of balance in different directions, depending on the relative importance of reproduction, maintenance and growth. The aim of studies of life-history evolution is to determine the direction in which the point of balance is pushed under different environmental conditions.

## GENETIC VARIATION IN LIFE HISTORIES

Constraints in design of organisms do not on their own lead to the evolution of life histories. Our second prerequisite is that the point of balance achieved in allocation of resources between reproduction, growth and maintenance should be under genetic control, and should vary from genotype to genotype within populations. Only then does genetic change in rates of reproduction and risks of death, of the kind considered in this paper, become possible.

Let us look first at some evidence for genetic variation in life histories of *Poa annua*. Much of the scatter of points in Figures 4.1, 4.2 and 4.3 is in fact attributable to genetic differences between families (Law, Bradshaw & Putwain 1977). Some variation arises from differences between populations and is not therefore available for selection. Even after variation between populations has been removed, however, the remaining variation between families within populations is still highly significant (Table 4.1). We can reasonably argue, then, that these populations contain appropriate genetic variation in life histories.

*Poa annua* is not alone in possessing genetic variation in life histories. Genetic variation in fecundity has been observed in a wide range of organisms (Falconer 1955; King & Henderson 1954; Dickerson 1955; Robertson 1957; Nordskog, Festing & Verghese 1967; Hulata, Moav & Wohlfarth 1974; Land, Russell & Donald 1974; Perrins & Jones 1974; Solbrig & Simpson 1974). The period of development before first reproduction is also known to be genetically variable (Böcher 1949; Dickerson 1955; Robertson 1957; Böcher & Larsen 1958; Dawson 1965; Hulata, Moav & Wohlfarth 1974; Hamrick & Allard 1975; McLaren 1976). There are certain instances of genetic variation in viability (Böcher 1949; Dickerson 1955; Böcher & Larsen 1958). In addition, genetic variation has been observed in size and rates of growth (Whatley 1942; Bates & Henson 1955; Robertson 1957; Bradshaw 1959; Hulata, Moav & Wohlfarth 1974; Hamrick & Allard 1975; McLaren 1976). It is also worth

TABLE 4.1. Results of analysis of variance between families within populations. $F$ ratio is the ratio of the between family within population mean square to the within family mean square. $p$ is the probability of null hypothesis: no difference between families within populations, with $n_1$ numerator and $n_2$ denominator degrees of freedom. Inflorescence data transformed to square root $(x + 0{\cdot}5)$ (from Law, Bradshaw & Putwain 1977)

| Character | $n_1$ | $n_2$ | $F$ ratio | $p$ |
|---|---|---|---|---|
| Pre-reproductive period | 112 | 414 | 5·6288 | <0·001 |
| Plant diameter at age 7 months | 112 | 416 | 3·2173 | <0·001 |
| Plant diameter at age 17 months | 101 | 316 | 2·6489 | <0·001 |
| Number of inflorescences of 4-month-old plants | 112 | 416 | 5·6239 | <0·001 |
| Number of inflorescences of 5-month-old plants | 112 | 416 | 6·4676 | <0·001 |
| Number of inflorescences of 6-month-old plants | 112 | 416 | 5·2215 | <0·001 |
| Number of inflorescences of 7-month-old plants | 112 | 415 | 4·5498 | <0·001 |
| Number of inflorescences of 8-month-old plants | 112 | 409 | 4·3119 | <0·001 |
| Number of inflorescences of 9-month-old plants | 112 | 406 | 3·5429 | <0·001 |
| Number of inflorescences of 10-month-old plants | 112 | 396 | 2·8622 | <0·001 |
| Number of inflorescences of 11-month-old plants | 111 | 377 | 2·5123 | <0·001 |
| Number of inflorescences of 12-month-old plants | 110 | 370 | 2·1743 | <0·001 |
| Number of inflorescences of 13-month-old plants | 106 | 351 | 1·9918 | <0·001 |
| Number of inflorescences of 14-month-old plants | 104 | 342 | 2·8940 | <0·001 |
| Number of inflorescences of 15-month-old plants | 104 | 339 | 2·8695 | <0·001 |
| Number of inflorescences of 16-month-old plants | 103 | 331 | 2·7812 | <0·001 |
| Number of inflorescences of 17-month-old plants | 102 | 316 | 1·8634 | <0·001 |

noting that there are various examples of successful selection for these characters (Falconer 1953, 1955; Dawson 1965; Englert & Bell 1970; Sokal 1970; Mertz 1975; Namkoong & Matzinger 1975; Moav & Wohlfarth 1976). There can be little doubt, then, that many populations contain genotypes with different rates of reproduction, growth and survival (although the extent to which the differences arise from resource allocation to reproduction, maintenance and growth remains to be established).

## OPTIMAL LIFE HISTORIES

We should now be able to envisage a population as a set of individuals which differ genetically from one another in their rates of reproduction and risks of death. The set of genotypes reflects a range of historical factors, such as the selective forces imposed by past environments, migration, mutation and random genetic drift, together with the breeding structure of the population. Their life histories reflect the constraints inherent in division of resources between reproduction, maintenance and growth. The problem which now faces us is: which genotype makes the greatest contribution to future generations?

Which set of genes balances resource allocation between reproduction, maintenance and growth in such a way that the total lifetime's production of progeny is maximized?

A precise answer to this question would require detailed information on genetic variation in life histories—information which is not available. In its absence, a rather less rigorous approach has been adopted (Schaffer 1974a; Charlesworth & Léon 1976; Michod 1978). We make some general assumptions about the effect on survival and growth of diverting resources to reproduction. We suppose that all points of compromise in resource allocation exist (from no resources to all resources allocated to reproduction). We then find the point of compromise which maximizes the contribution to future generations (the *optimal life history*). The limitations to this should be self-evident. There is no certainty that a genotype with the optimal life history exists; selection can only operate on variation which is actually present. Some caution is therefore advisable when adopting this approach.

In a formal sense, the optimal life history can be defined as the one which maximizes the rate of increase ($\lambda$) in the equation

$$1 = \sum_{j=0}^{m} \lambda^{-(j+1)} L_j b_j \tag{4.1}$$

where

$$L_j = \prod_{k=0}^{j-1} p_k g_k \qquad \text{for } j > 0 \qquad (L_0 = 1)$$

(Schaffer 1974a; Charlesworth & Léon 1976). In this equation, $m$ is the oldest age of reproduction and $b_j$, $g_j$ and $p_j$ are the rates of reproduction per unit size, of growth per unit size and of survival respectively in the $j$th age class. By expressing the demographic equation in this form, we allow reproduction to be a function of size. $\lambda$ is then a function of $b_j$, $g_j$ and $p_j$ over all $j$, where $p_j$ and $g_j$ are functions of $b_j$ set by constraints and design, $p_j$ and $g_j$ being taken as inseparable.

There is an alternative way of defining the optimal life history which emphasizes the importance of resource partitioning in each age class. Consider the contribution which an individual aged $i$ makes to future generations. It contains two components, one being immediate reproduction, the other being reproduction deferred to the future. Formally, this is a modified form of Fisher's (1958) reproductive value ($V_i$) (Schaffer 1974a):

$$V_i = (\underbrace{b_i}_{\text{reproduction at age } i} + \underbrace{p_i g_i V_{i+1}}_{\text{reproduction deferred to the future}})/\lambda \tag{4.2}$$

where $V_{i+1}$ is a measure of the present value of future progeny. By adjusting resource allocation so that the reproductive value in each age class is maximized, we obtain exactly the same optimal life history as before (Taylor *et al.* 1974).

This analysis leads us to a central division in life histories between organisms which reproduce only once and those which reproduce repeatedly (Schaffer 1974a; Charlesworth & Léon 1976). The difference arises from the relationship between reproduction, growth and survival, expressed formally as $p_i g_i(b_i)$, a relationship which reflects the constraints controlling design (Fig. 4.4). If $p_i g_i(b_i)$ is convex (i.e. $d^2 p_i g_i / db_i^2$ is positive) in all age classes, the optimum is semelparous (monocarpic), reproduction being confined to a single lethal burst. If it is concave (i.e. $d^2 p_i g_i / db_i^2$ is negative), the optimum can be iteroparous (polycarpic) with intermediate rates of reproduction, allowing further survival and growth to reproduce when older. In this case, the optimal balance between reproduction and growth and survival in each age class is defined as the point at which the gradient of $p_i g_i(b_i)$ is $-1/V_{i+1}$ (differentiating eqn (4.2) with respect to $b_i$). Notice that the function might alternatively be both convex and concave in different parts, in which case there could be more than one local optimum.

Despite the importance of the form of relationship between reproduction, growth and survival, we know very little about it in practice (but see Schaffer & Schaffer 1977). In *Poa annua* I have built a model based on some general features of its biology in an attempt to discover the form which it might take (Law 1979a). I suppose that an individual is built of modular units (tillers), and that they constitute a limited resource. A tiller can bear an inflorescence, but if it does so it dies. If it does not bear one, it may survive to produce more tillers. In many circumstances, this gives a relationship with a region of convexity and one of concavity (e.g. Fig. 4.5). There are then two local optima, one with a low

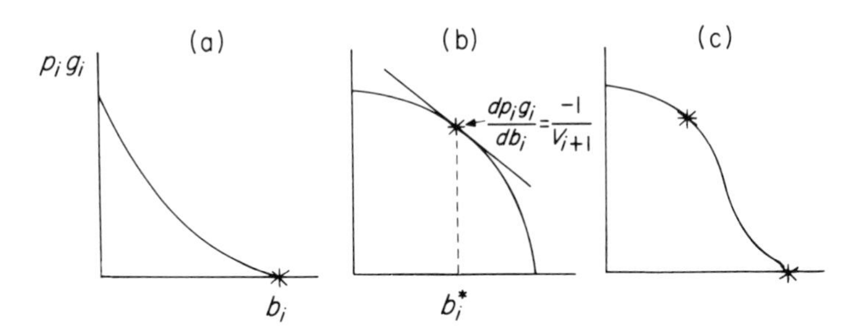


FIG. 4.4. Three forms of the function $p_i g_i(b_i)$ arising from different kinds of constraints between reproduction, maintenance and growth. * is the point on the curve giving values of $b_i$, $p_i$ and $g_i$, maximizing the contribution of age class $i$ to future generations.
(a) $d^2 p_i g_i / db_i^2$ positive;
(b) $d^2 p_i g_i / db_i^2$ negative;
(c) $d^2 p_i g_i / db_i^2$ positive and negative.

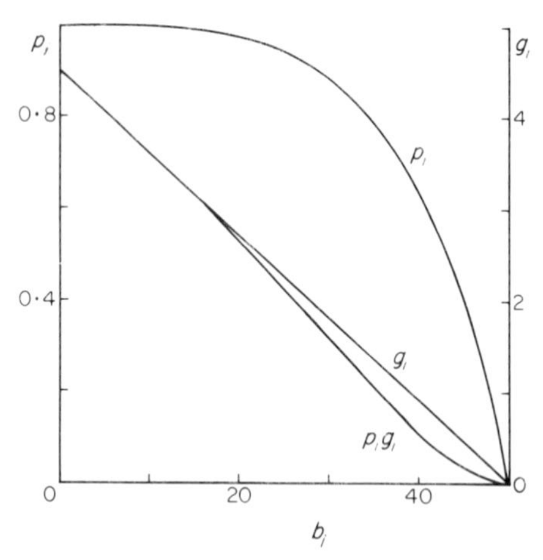


Fig. 4.5. A possible form of the relationship between reproduction, maintenance and growth in *Poa annua*. Based on the functions:

$$p_i = 1 - (0{\cdot}1 + 0{\cdot}018b_i)^5$$
$$g_i = 4{\cdot}5 - 0{\cdot}09b_i$$

(For details of model see Law 1979a.)

rate of reproduction and the other with a lethal rate. However, it is not entirely clear how the prediction relates to observations of natural populations.

The constraints operating between reproduction, growth and maintenance therefore play a central role in determining the optimal life history. They provide a set of alternative life histories from which the optimum is chosen, and can be expected to differ from species to species. But superimposed on this the optimal life history is a function of the prevailing environmental conditions. Different environments can be expected to tilt the balance in different directions, depending on the extent to which they favour reproduction, growth or maintenance. It is to this problem that I want to turn in the next sections, using the analytical methods just discussed for finding optimal life histories. (For clarity of presentation, I confine my arguments to iteroparous optima; analogous arguments can also be developed for semelparous ones.)

## DETERMINISTIC CHANGE IN RESOURCE AVAILABILITY

One of the most important environmental variables in the life of an organism is its supply of resources. The supply may be changed by biotic factors, such as the number of other individuals with which the resources must be shared, or it may be changed by abiotic factors, such as random fluctuations in resources from year to year. Here I consider only one kind of deterministic change—a gradual reduction in resources which might occur, for example, as numbers of individuals increase during colonization. Our problem is to find how the optimal balance of resources between reproductive and non-reproductive activities changes as the supply becomes more and more restricted.

Interest in this problem stems largely from work of MacArthur & Wilson (1967) on the changing forces of natural selection during colonization. They identified two extremes of a continuum of selective forces. During early stages of colonization when resources are plentiful, they argued that genotypes most productive in converting resources into progeny should be most successful ($r$-selection). As the supply becomes more restricted, they argued that genotypes more efficient in converting them into progeny should become increasingly advantageous ($K$-selection). (The terms $r$ and $K$ are taken from the logistic equation of population growth, (Pearl & Reed 1920.) Their ideas have since provoked a number of theoretical developments (Hairston, Tinkle & Wilbur 1970; Pianka 1970, 1972; King & Anderson 1971; Roughgarden 1971; Charlesworth 1971; Gadgil & Solbrig 1972).

Some consequences of a declining supply of resources on the optimal life history can be predicted from eqn (4.2). As the supply declines, there are less resources available to reproductive or non-reproductive activities, or both, at some stage in life. In theory, this can have two effects on eqn (4.2). Firstly, it should lead to a reduction in the rate of increase ($\lambda$), since reproduction, growth or survival must decrease at some stage in life. As a result, the present value of future progeny ($V_{i+1}$) increases, placing an increased emphasis on non-reproductive activities at age $i$. Secondly, if reproduction, growth or survival are altered at ages greater than $i$, there will be a tendency for the present value of future progeny to decrease, placing a reduced emphasis on non-reproductive activities at age $i$. Thus the direction in which resource allocation is tilted depends on the balance between these opposing forces. The problem is greatly simplified by making two assumptions: (a) that risks of mortality increase disproportionally among young individuals ($< i$) as resource input declines, and (b) that rates of reproduction decrease proportionally in all age classes. In such circumstances, forces causing the present value of future progeny to increase should outweigh those causing it to decrease. The balance is then tilted away from reproduction at age $i$ as the supply of resources declines.

I have tested these predictions, comparing populations of *Poa annua* from environments with contrasting supplies of resources (Law, Bradshaw & Putwain 1977). I assumed relatively large amounts of resources to be available in environments which existed only transiently so that population density remained low (e.g. derelict sites) and in more permanent environments where plants were subject to other forces keeping densities low (for example, tracks and car parks). Populations from such environments can be called 'opportunist'. Conversely, I assumed relatively small amounts of resources to be available where a large number of plants had to share a restricted supply (e.g. pastures). Samples from these environments showed some quite striking genetic differences (Fig. 4.6). Opportunist plants tended to be shorter lived, with more rapid development to reproduction and larger numbers of inflorescences in the first year of growth. Pasture ones produced many more inflorescences in the second year of growth.

Evidence from other sources also supports the predictions. There are some particularly clear differences between populations of dandelions (*Taraxacum officinale*) from sites with differing degrees of disturbance (Gadgil & Solbrig 1972; Solbrig & Simpson 1974, 1977). More circumstantial evidence is available from other species (Böcher 1949; McNaughton 1975; Sterk 1975; Vasek & Clovis 1976) and from comparisons between species (Baker 1965; Abrahamson & Gadgil 1973; Gaines *et al.* 1974; Sarukhán & Gadgil 1974). However, there are other instances in which the predicted differences do not occur (Hickman 1975, 1977; Swingland 1977). It should be pointed out, though, that the circumstantial evidence is generally deficient as a test of the theory, as it does not unambiguously demonstrate the existence of genetic differences between populations.

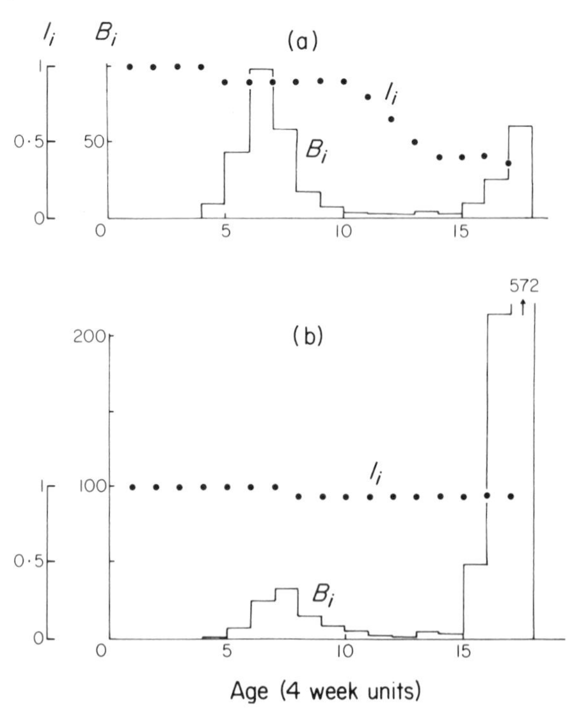


FIG. 4.6. Rates of reproduction and survival in samples of *Poa annua* from (a) an opportunist and (b) a pasture environment, grown from seed under similar experimental conditions. $B_i$ is the number of inflorescences produced over monthly intervals. $l_i$ is the proportion of plants surviving from age zero to age $i$. (Populations *a* and *o* from Law, Bradshaw & Putwain 1977.)

However, it is important to appreciate that a declining supply of resources does not *inevitably* tilt the optimal life history away from reproduction. There are two reasons for this. Firstly, our arguments have been dependent on mortality being concentrated on young individuals ($< i$) and rates of reproduction declining proportionally in all age classes as resource availability declines. If this is not so, i.e. older individuals are more vulnerable than younger ones, there could be age classes in which the present value of future progeny ($V_{i+1}$) decreases, tilting the balance in favour of increased reproduction. The second reason arises from constraints in design. We have already seen that an optimal life history depends on internal constraints as much as the external environment. Now the functional form of the constraints could itself be changed by a decreasing supply of resources. For example, growth and survival could be particularly sensitive to decreasing resource availability, whereas reproduction might be relatively insensitive. In such circumstances, the tendency for present value of future progeny to increase could be more than counterbalanced by the reduction in growth and survival ($p_i$, $g_i$). The optimal allocation of resources to reproduction would then increase as the supply of resources declined (Fig. 4.7).

Thus, we should not be too surprised to find cases in which a declining supply of resources tilts the optimal life history towards reproduction. Our evidence is that such circumstances do not occur very often, which is probably a reflection of the sensitivity of the present value of future progeny ($V_{i+1}$) to the rate of increase ($\lambda$). However, it would be interesting to study systems in which,

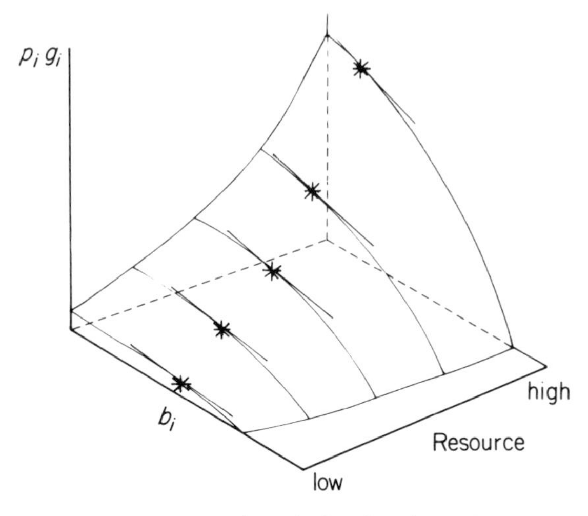


FIG. 4.7. Reproduction, growth and survival as functions of resource supply. $p_i g_i$ is much more sensitive to declining availability of resources than $b_i$. In these circumstances the optimal rate of reproduction (denoted by *) increases as the supply of resources decreases.

for example, old individuals were more vulnerable than younger ones, to see if they run counter to our currently held views of *r*- and *K*-selection.

## AGE-SPECIFIC PREDATION

A second environmental variable which may alter the optimal balance between reproductive and non-reproductive activities is age-specific predation. Predators are almost always selective in their choice of prey, and one criterion in their choice is likely to be age. Even if they have no direct knowledge of prey age, they are likely to make their choice on the basis of characters associated with age, such as size. Is it then possible to predict how age-specific predation will alter the optimal balance between reproductive and non-reproductive activities? (My arguments here are directed at animal prey populations rather than plants.)

It may not be immediately clear why age-specific predation should have any effect at all on an optimal life history. So let us begin with a simple example. Consider a prey population with some individuals aged, say, ten years. Suppose that they differ from one another genetically in the proportion of resources allocated to reproduction at this age. In the absence of predation, we suppose that one of them makes the greatest overall contribution to future generations by allocating some resources to reproduction and some to growth and maintenance, so that it can reproduce when older. Now suppose a predator appears and kills all ten-year-old individuals after they have reproduced. The genotype which was previously optimal is no longer so, because the resources it saved for reproduction when older are wasted. Instead a genotype which diverts all resources to reproduction at age ten has the greatest advantage.

Several papers have pointed to the importance of environmental sources of mortality as determinants of optimal life histories (Gadgil & Bossert 1970; Schaffer 1974a; Taylor *et al.* 1974; Perrins 1977; Michod 1978; Law 1979b). As before, some outcomes can be predicted using eqn (4.2), by determining the effect of predation on the present value of future progeny ($V_{i+1}$). Predation, like declining resource supply, can influence the present value of future progeny in two ways. Firstly, it must always reduce the rate of increase ($\lambda$), tending to make their value greater. But, secondly, if individuals older than $i$ are killed, the direct effect of reduced survival is to make their value smaller. It is the balance between these opposing forces which causes the present value of future progeny to increase or decrease, favouring more or less reproduction at age $i$.

Two predictions can be made without difficulty (Law 1979b). Firstly, if predation is confined to individuals younger than $i$, the optimal rate of reproduction at age $i$ decreases. This follows from the fact that the present value of future progeny ($V_{i+1}$) here is only influenced by the rate of increase ($\lambda$). Secondly, if predation is confined to individuals older than $i$, the optimal rate of reproduction at age $i$ increases. This is because the direct effect of reduced survival always outweighs that of reduced rates of increase. However, if predation acts on individuals both younger and older than $i$, it is more difficult to make predictions, since either reduced or increased reproduction may be optimal. The effects of some patterns of age-specific predation are illustrated in Figure. 4.8.

However, it is as well to bear in mind that this analysis may ignore important features of some organisms. I have assumed that predation exerts a direct influence only on the chance of survival of prey, not on their rates of growth and reproduction. This makes it appropriate to animal populations where predation generally leads to death, but less so to plants where predation often leads only to loss of parts of individuals (Harper 1977, p. 385). I have also assumed that no amount of resource allocation to avoiding predation has any effect on the chance of predation, thereby ignoring evolution of defence mechanisms. In addition, I have ignored the possibility that predation could leave more resources to surviving prey, a problem which has been considered in depth by Michod (1978).

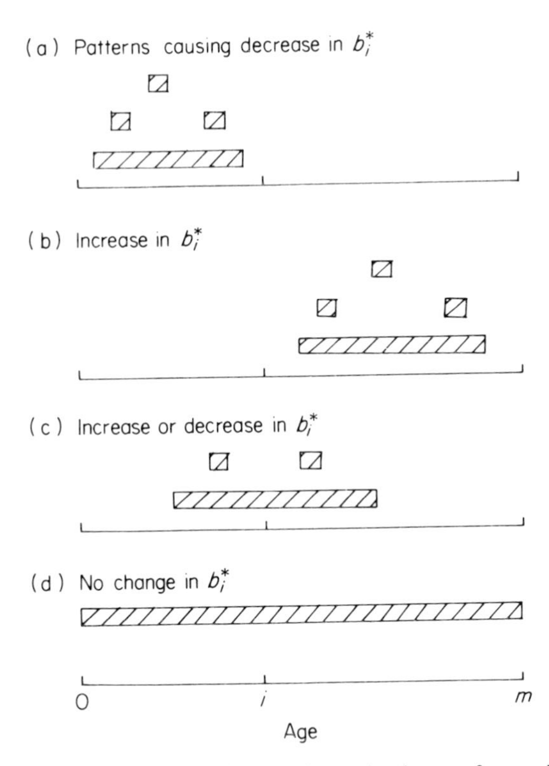


FIG. 4.8. Effect of age-specific predation on the optimal rate of reproduction per unit size ($b_i^*$). Predated age classes are shown as hatched areas, each line being an alternative pattern. (Simplified from Law 1979b.)

There has as yet been little attempt to test these predictions about genetic change under predation on experimental or natural populations. I know of only two experiments, in fact carried out to investigate the evolution of senescence, but none the less relevant to our problem. These were experiments on flour beetles (*Tribolium castaneum*), in which some populations were subjected to intense mortality almost immediately after the start of reproductive life and others were allowed to live out their full natural life span. After several generations of selection, samples from each population were compared under standard conditions. Sokal (1970) found that those from predated populations had lower rates of survival than those from unpredated ones. Mertz (1975) found that those from predated populations had greater rates of reproduction early in adult life. Both experiments are compatible with the theory that predation on adults selects genotypes with greater resource allocation to reproduction.

## RATE OF GENETIC CHANGE OF LIFE HISTORIES

We have seen how different environments can select genotypes which differ in their allocation of resources to reproduction, maintenance and growth. This may seem to be of little importance to a population ecologist, whose interests lie primarily in short-term flux in populations. It seems to be widely believed by ecologists that genetic change is sufficiently slow for it to be ignored on an ecological time scale. But how far are we justified in making such an assumption when considering rates of reproduction and risks of death?

Studies of optimal life histories are of no help in answering this question. They point only to the direction of the new optimum, not how long it takes to reach it. In fact, the rate of genetic change is controlled by several factors. It depends on the degree of environmental change, the extent to which selection pressures are altered. It depends on the extent of genetic variation—if genotypes differ only slightly in their life histories, we can expect genetic change to be very slow. It depends on the genetic basis of components of life histories—if variation is non-additive, no amount of selection is going to bring about genetic change (Falconer 1960). It depends on the relationship between different components of life histories—a gene which influences more than one component may experience conflicting selection pressures. In all these areas we are seriously lacking in understanding. Until more is known about them, we cannot make predictions about rates of change.

Lacking theoretical predictions about rates of genetic change, is there any evidence from experimental or natural populations? I have attempted to put an upper limit on the rate of change in a field experiment on *Poa annua* (Law 1975). The experiment consisted of a set of populations, some of which were maintained under conditions relatively rich in resources (by keeping densities low) and others which were maintained under relatively limited conditions (by keeping densities high). All the populations began with the same stock of seed, synthesized from a wide range of natural populations to maximize the input of genetic variation. After twenty-one months, samples were taken from each, and the early part of their life histories analysed to test for signs of divergence. The samples showed some small but significant differences in reproductive characters early in life. One of these, the period of time from germination to emergence of the first inflorescence, is illustrated in Figure 4.9. Their means were 21·0 weeks for plants from low-density populations and 23·9 for those from high-density ones. An analogous experiment on *Taraxacum officinale* (Solbrig & Simpson 1977) showed marked changes in frequency of clones over a period of four years. This experiment is particularly informative, since only two genotypes were present—much less genetic variation than would usually be present in natural populations. Finally, we have already seen that short-term genetic change can take place in *Tribolium castaneum* when subjected to different regimes of mortality (Mertz 1975). These experiments therefore show genetic change taking place over short time spans, well within the limits of interest to ecologists.

It would be unwise to draw any firm conclusions with such limited evidence. Nevertheless, the results do point to the *possibility* that short-term genetic change in rates of reproduction and death can take place as an environment changes. If this was found to be true more generally, it would be of considerable

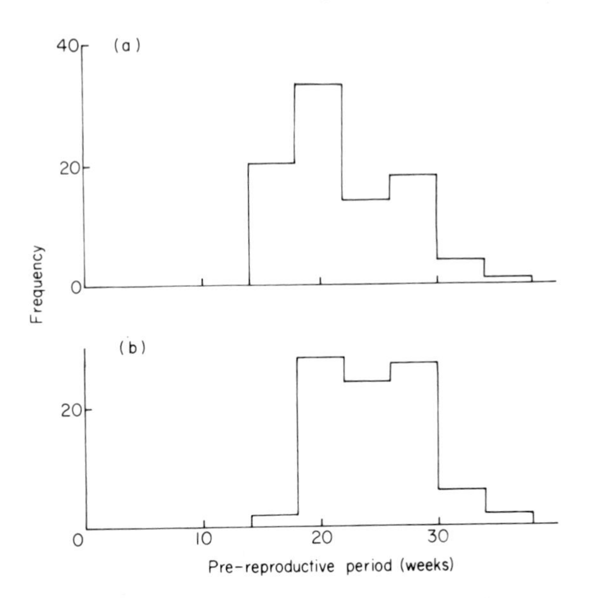


FIG. 4.9. Frequency distributions of pre-reproductive periods in *Poa annua* after selection at (a) low and (b) high density for 21 months. Measurements made on plants grown from seed under similar experimental conditions. Pre-reproductive periods measured as time from germination to emergence of first inflorescence. (From Law 1975.)

importance to ecologists. It would mean that over and above immediate effects of environmental change (e.g. density dependence), the environment would be acting as a selective force, favouring genotypes with different rates of reproduction and death. Consequently, to predict future states of populations, even over short periods of time, we would have to take into account genetic changes in rates of reproduction and risks of death.

To resolve the issue we clearly need much more information about rates of genetic change of life histories. Unfortunately, experiments are not usually designed in a way which provides this evidence. The dilemma is well illustrated with a population of pike (*Esox lucius*) in Lake Windermere. Since 1944 this population has been subjected to heavy adult predation and, at first sight, might appear to be suitable for testing rates of genetic change. Indeed, egg production has been estimated each year since 1963, and has risen in three-year-olds from about 50 000 to 80 000 per female (Bagenal pers. com.). At present, however, there is no way of discovering whether this represents an improvement in the environment favouring greater rates of reproduction irrespective of genotype, whether it is the result of genetic change due to age-specific predation or whether it is due to some combination of both. If we are to distinguish between alternatives like this, samples need to be compared under standard environmental conditions.

It is of more than academic interest to determine rates of genetic change in life histories. Natural populations which we exploit commercially experience strong age-specific mortality. As predators, we may be bringing about genetic change in rates of reproduction, growth and survival in our prey. The sheer intensity of our predatory activities should lend urgency to study of this problem. Furthermore, our management models for harvest optimization generally advocate a high degree of age specificity in predation, while at the same time assuming genetic uniformity (e.g. MacArthur 1960; Rorres 1976). It is arguable that we might do better to develop management regimes which select genotypes with desirable life histories as well as maximizing short-term yields.

## DISCUSSION

I have been trying to develop a view of populations as units in which life histories can evolve under pressures exerted by prevailing environments. The argument rests on two conditions. Firstly, we require that there should be constraints in design of organisms so that their life histories reflect points of balance in resource allocation among their activities. Secondly, we require that these points of balance should be genetically determined. There are good grounds to suppose that both these conditions are satisfied in many populations, so we can reasonably expect genetic change in life histories to be taking place. To predict the kind of genetic change is rather harder. At present it requires some general assumptions about operation of the constraints, and the presupposition that all points of compromise are represented by different genotypes. We can then find the optimal life history and establish the directions in which it is pushed by environmental forces.

However, it is essential to remember that the way in which an optimal life history changes is as much a function of the internal constraints as of the external environment. To some extent, I have avoided this issue by confining my attention to iteroparous optima; rather different changes in life histories would be expected under constraints leading to semelparous optima. However, the problem runs deeper than this. Environmental changes may bring about changes in the nature of the constraints themselves. One case of this arose as resource availability declined (Fig. 4.7). This is likely to create serious difficulties in making general predictions about the ways in which environments mould life histories. Much may depend on a detailed knowledge of the biology of the organism being studied.

Several other restrictions to my analysis should be recorded. Firstly, in considering only two kinds of deterministic change in environments, we have seen no more than the tip of the iceberg. There are many other kinds, for example fluctuating environments, both periodic and stochastic, could yield interesting results. Secondly, I have only discussed the consequences of resource division between reproduction, growth and maintenance. Again this identifies only three (admittedly important) of many activities which could influence life histories. Thirdly, there are inadequacies in optimal history models. The most that can be said for them is that they do give analytical solutions, and that the use of more sophisticated models at present would only further enlarge the gap between available data and theory.

Nevertheless, our main conclusions are unaffected by limitations of the theoretical models. Life histories—rates of reproduction and risks of death—are evolving under forces imposed by prevailing environments. It is possible that they may be doing so at an appreciable pace, well within time scales of ecological studies. If this is so, ecologists will need to consider selective forces exerted by environments on rates of reproduction and risks of death when making predictions about future states of populations.

## ACKNOWLEDGMENTS

Much of this work was carried out during the tenure of a Research Studentship and Research Fellowship of the Natural Environmental Research Council. I would like to thank A.D. Bradshaw, B. Charlesworth, J.P. Grime, J.L. Harper, J. Maynard Smith and P.D. Putwain for helpful discussions at many stages during its preparation.

## REFERENCES


**Abrahamson W.G. & Gadgil M. (1973)** Growth form and reproductive effort in Goldenrods (*Solidago compositae*). *American Naturalist*, **107**, 651–661.

**Bagenal T.B. (1967)** A short review of fish fecundity. *The Biological Basis of Freshwater Fish Production* (Ed. by S.D. Gerking) pp. 89–111. Blackwell Scientific Publications, Oxford.

**Baker H.G. (1965)** Characteristics and modes of origin of weeds. *The Genetics of Colonising Species* (Ed. by H.G. Baker & G.L. Stebbins) pp. 77–94. Academic Press, New York.

**Barnes H. (1962)** So-called anecdysis in *Balanus balanoides* and the effect of breeding upon the growth of the calcareous shell of some common barnacles. *Limnology and Oceanography*, **7**, 462–473.

**Bates R.P. & Henson P.R. (1955)** Studies of inheritance in *Lespedeza cuneata* Don. *Agronomy Journal*, **47**, 503–507.

**Böcher T.W. (1949)** Racial divergences in *Prunella vulgaris* in relation to habitat and climate. *New Phytologist*, **48**, 295–314.

**Böcher T.W. & Larsen K. (1958)** Geographical distribution of initiation of flowering, growth habit, and other characters in *Holcus lanatus* L. *Botaniska Notiser*, **111**, 289–300.

**Bradshaw A.D. (1959)** Population differentiation in *Agrostis tenuis* Sibth. I. Morphological differentiation. *New Phytologist*, **58**, 208–227.

**Calow P. & Woolhead A.S. (1977)** The relationship between ration reproduction effort and age-specific mortality in the evolution of life-history strategies—some observations on freshwater triclads. *Journal of Animal Ecology*, **46**, 765–781.

**Charlesworth B. (1971)** Selection in density-regulated populations. *Ecology*, **52**, 469–474.

**Charlesworth B. & Léon J.A. (1976)** The relation of reproductive effort to age. *American Naturalist*, **110**, 449–459.

**Dawson P.S. (1965)** Estimation of components of phenotypic variance for developmental rate in *Tribolium*. *Heredity*, **20**, 403–417.

**Dickerson G.E. (1955)** Genetic slippage in response to selection for multiple objectives. *Cold Spring Harbor Symposium on Quantitative Biology*, **20**, 213–224.

**Eis S., Garman E.H. & Ebell L.F. (1965)** Relation between cone production and diameter increment of Douglas fir (*Pseudotsuga menziesii* (Mirb.) Franco), grand fir (*Abies grandis* (Dougl.) Lindl.) and western white pine (*Pinus monticola* Dougl.). *Canadian Journal of Botany*, **43**, 1553–1559.

**Eklund B. (1957)** Annual ring variation of spruce in central Norrland and its relation to climate. *Meddelelser fra Statens Skogsforskningsinst*, **47**, 1–63.

**Englert D.C. & Bell A.E. (1970)** Selection for time of pupation in *Tribolium castaneum*. *Genetics*, **64**, 541–552.

**Falconer D.S. (1953)** Selection for large and small size in mice. *Journal of Genetics*, **51**, 470–501.

**Falconer D.S. (1955)** Patterns of response in selection experiments with mice. *Cold Spring Harbor Symposium on Quantitative Biology*, **20**, 178–196.

**Falconer D.S. (1960)** *Introduction to Quantitative Genetics*. Oliver & Boyd, Edinburgh.

**Fielding J.M. (1960)** Branching and flowering characteristics of Monterey pine. *Forestry and Timber Bureau Australia*, **Bulletin 37.**

**Fisher R.A. (1958)** *The Genetical Theory of Natural Selection*, 2nd edition. Dover, New York.

**Gadgil M. & Bossert W.H. (1970)** Life historical consequences of natural selection. *American Naturalist*, **104**, 1–24.

**Gadgil M. & Solbrig O.T. (1972)** The concept of *r*- and *K*-selection: evidence from wild flowers and some theoretical considerations. *American Naturalist*, **106**, 14–31.

**Gaines M.S., Vogt K.J., Hamrick J.L. & Caldwell J. (1974)** Reproductive strategies and growth patterns in sunflowers (*Helianthus*). *American Naturalist*, **108**, 889–894.

**Goodman D. (1974)** Natural selection and a cost ceiling for reproductive effort. *American Naturalist*, **108**, 247–268.

**Hairston N.G., Tinkle D.W. & Wilbur H.M. (1970)** Natural selection and the parameters of population growth. *Journal of Wildlife Management*, **34**, 681–690.

**Hamrick J.L. & Allard R.W. (1975)** Correlations between quantitative characters and enzyme genotypes in *Avena barbata*. *Evolution*, **29**, 438–442.

**Harper J.L. (1977)** *Population Biology of Plants*. Academic Press, London.

**Harper J.L. & Ogden J. (1970)** The reproductive strategy of higher plants. I. The concept of strategy with special reference to *Senecio vulgaris* L. *Journal of Ecology*, **58**, 681–698.

**Harper J.L. & White J. (1974)** The demography of plants. *Annual Review of Ecology and Systematics*, **5**, 419–463.

**Hickman J.C. (1975)** Environmental unpredictability and plastic energy allocation strategies in the annual *Polygonum cascadense* (*Polygonaceae*). *Journal of Ecology*, **63**, 689–701.

**Hickman J.C. (1977)** Energy allocation and niche differentiation in four co-existing annual species of *Polygonum* in western North America. *Journal of Ecology*, **65**, 317–326.

**Hirschfield M.F. & Tinkle D.W. (1975)** Natural selection and the evolution of reproductive effort. *Proceedings of the National Academy of Sciences of the United States of America*, **72**, 2227–2231.

**Holmsgaard E. (1956)** Effect of seed bearing on the increment of European beech (*Fagus sylvatica* L.) and Norway spruce (*Picea abies* (L.) Karst). *12th Congress of the International Union Forest Research Organisation.*

**Hulata G., Moav R. & Wohlfarth G. (1974)** The relationship of gonad and egg size to weight and age in the European and Chinese races of the common crop *Cyprinus carpio* L. *Journal of Fish Biology*, **6**, 745–758.

**Hustich I. (1956)** Notes on the growth of Scotch pine in Utsjoki in northernmost Finland. *Acta Botanica Fennica*, **56**, 3–13.

**Iles T.D. (1974)** The tactics and strategy of growth in fishes. *Sea Fisheries Research* (Ed. by F.R. Harden Jones) pp. 331–345. Elek Science, London.

**King C.E. & Anderson W.W. (1971)** Age-specific selection. II. The interaction between *r* and *K* during population growth. *American Naturalist*, **105**, 137–156.

**King S.C. & Henderson C.R. (1954)** Heritability studies of egg production in the domestic fowl. *Poultry Science*, **33**, 155–169.

**Land R.B., Russell W.S. & Donald H.P. (1974)** The litter size and fertility of Finnish Landrace and Tasmanian Merino sheep and their reciprocal crosses. *Animal Production*, **18**, 265–271.

**Langer R.H.M. (1956)** Growth and nutrition of timothy (*Phleum pratense*). I. The life history of individual tillers. *Annals of Applied Biology*, **44**, 166–187.

**Langer R.H.M., Ryle S.M. & Jewiss O.R. (1964)** The changing plant and tiller populations of timothy and meadow fescue swards. I. Plant survival and the pattern of tillering. *Journal of Applied Ecology*, **1**, 197–208.

**Law R. (1975)** Colonisation and the evolution of life histories in *Poa annua*. *Unpublished Ph.D. thesis, University of Liverpool.*

**Law R. (1979a)** The cost of reproduction in annual meadow grass. *American Naturalist*, **113**, 3–16.

**Law R. (1979b)** Optimal life histories under age-specific predation. *American Naturalist* (in press).

**Law R., Bradshaw A.D. & Putwain P.D. (1977)** Life history variation in *Poa annua*. *Evolution*, **31**, 233–246.

**Lints F.A. & Soliman M.H. (1977)** Growth rate and longevity in *Drosophila melanogaster* and *Tribolium castaneum*. *Nature, London*, **266**, 624–625.

**MacArthur R.H. (1960)** On the relation between reproductive value and optimal predation. *Proceedings of the National Academy of Sciences of the United States of America*, **46**, 143–145.

**MacArthur R.H. & Wilson E.O. (1967)** *The Theory of Island Biogeography*. Princeton University Press, Princeton.

**McLaren I.A. (1976)** Inheritance of demographic and production parameters in the marine copepod *Eurytemora herdmani*. *Biological Bulletin*, **151**, 200–213.

**McNaughton S.J. (1975)** *r*- and *K*-selection in *Typha*. *American Naturalist*, **109**, 251–261.

**Menge B.A. (1974)** Effect of wave action and competition on breeding and reproduction effort in the sea aster *Leptasterias hexactis*. *Ecology*, **55**, 84–93.

**Mertz D.B. (1975)** Senescent decline in flour beetle strains selected for early adult fitness. *Physiological Zoology*, **48**, 1–23.

**Michod R. (1978)** Life histories, mortality factors and reproductive value. *American Naturalist* (in press).

**Moav R. & Wohlfarth G. (1976)** Two-way selection for growth rate in the common carp (*Cyprinus carpio* L.). *Genetics*, **82**, 83–101.

**Murdoch W.W. (1966)** Population stability and life history phenomena. *American Naturalist*, **100**, 5–11.

**Namkoong G. & Matzinger D.F. (1975)** Selection for annual growth curves in *Nicotiana tabacum* L. *Genetics*, **81**, 377–386.

**Nordskog A.W., Festing M. & Verghese M.W. (1967)** Selection for egg production and correlated responses in the fowl. *Genetics*, **55**, 179–191.

**Ogden J. (1974)** The reproductive strategy of higher plants. II. The reproductive strategy of *Tussilago farfara* L. *Journal of Ecology*, **62**, 291–324.

**Pearl R. & Reed L.J. (1920)** On the rate of growth of the population of the United States since 1790 and its mathematical representation. *Proceedings of the National Academy of Sciences of the United States of America*, **6**, 275–288.

**Perrins C.M. (1977)** The role of predation in the evolution of clutch size. *Evolutionary Ecology* (Ed. by B. Stonehouse & C.M. Perrins) pp. 181–191. Macmillan, London.

**Perrins C.M. & Jones P.J. (1974)** The inheritance of clutch size in the great tit (*Parus major* L.). *Condor*, **76**, 225–229.

**Pianka E.R. (1970)** On *r*- and *K*-selection. *American Naturalist*, **104**, 592–597.

**Pianka E.R. (1972)** *r* and *K* selection or *b* and *d* selection? *American Naturalist*, **106**, 581–588.

**Robertson F.W. (1957)** Studies in quantitative inheritance. XI. Genetic and environmental correlation between body size and egg production in *Drosophila melanogaster*. *Journal of Genetics*, **55**, 428–443.

**Rorres C. (1976)** Optimal sustainable yield of a renewable resource. *Biometrics*, **32**, 945–948.

**Roughgarden J. (1971)** Density-dependent natural selection. *Ecology*, **52**, 453–468.

**Sarukhán J. (1974)** Studies on plant demography: *Ranunculus repens* L., *R. bulbosus* L. and *R. acris* L. II. Reproductive strategies and seed population dynamics. *Journal of Ecology*, **62**, 151–177.

**Sarukhán J. & Gadgil M. (1974)** Studies on plant demography: *Ranunculus repens* L., *R. bulbosus* L. and *Ranunculus acris* L. III. A mathematical model incorporating multiple modes of reproduction. *Journal of Ecology*, **62**, 921–936.

**Schaffer W.M. (1974a)** Selection for life histories: the effects of age structure. *Ecology*, **55**, 291–303.

**Schaffer W.M. (1974b)** Optimal reproductive effort in fluctuating environments. *American Naturalist*, **108**, 783–790.

**Schaffer W.M. & Elson P.F. (1975)** The adaptive significance of variations in life history among local populations of Atlantic salmon in North America. *Ecology*, **56**, 577–590.

**Schaffer W.M. & Schaffer M.V. (1977)** The adaptive significance of variations in reproductive habit in the *Agavaceae*. *Evolutionary Ecology* (Ed. by B. Stonehouse & C. Perrins) pp. 261–276. Macmillan, London.

**Seed R. & Brown R.A. (1978)** Growth as a strategy for survival in two marine bivalves, *Cerastoderma edule* and *Modiolus modiolus*. *Journal of Animal Ecology*, **47**, 283–292.

**Smith C.C. & Fretwell S.D. (1974)** The optimal balance between size and number of offspring. *American Naturalist*, **108**, 499–506.

**Sokal R.R. (1970)** Senescence and genetic load: evidence from *Tribolium*. *Science*, **167**, 1733–1734.

**Solbrig O.T. & Simpson B.B. (1974)** Components of regulation of a population of dandelions in Michigan. *Journal of Ecology*, **62**, 473–486.

**Solbrig O.T. & Simpson B.B. (1977)** A garden experiment on competition between biotypes of the common dandelion (*Taraxacum officinale*). *Journal of Ecology*, **65**, 427–430.

**Stearns S.C. (1976)** Life-history tactics: a review of the ideas. *Quarterly Review of Biology*, **51**, 3–47.

**Stearns S.C. (1977)** The evolution of life history traits: a critique of the theory and a review of the data. *Annual Review of Ecology and Systematics*, **8**, 145–171.

**Sterk A.A. (1975)** Demographic studies of *Anthyllis vulneraria* L. in the Netherlands. *Acta Botanica Neerlandica*, **24**, 315–337.

**Swingland I.R. (1977)** Reproductive effort and life history strategy of the Aldabran giant tortoise. *Nature, London*, **269**, 402–404.

**Taylor H.M., Gourley R.S., Lawrence C.E. & Kaplan R.S. (1974)** Natural selection of life history attributes: an analytical approach. *Theoretical Population Biology*, **5**, 104–122.

**Tinkle D.W. (1969)** The concept of reproductive effort and its relation to the evolution of life histories in lizards. *American Naturalist*, **103**, 501–516.

**Tinkle D.W. & Hadley N.F. (1975)** Lizard reproductive effort: caloric estimates and comments on its evolution. *Ecology*, **56**, 427–434.

**Tyler A.V. & Dunn R.S. (1976)** Ration, growth and measures of somatic and organ condition in relation to meal frequency in winter flounder *Pseudopleuronectes americanus* with hypothesis regarding population homeostasis. *Journal of the Fisheries Research Board of Canada*, **33**, 63–75.

**Vasek F.C. & Clovis J.F. (1976)** Growth forms in *Arctostaphylos glauca*. *American Journal of Botany*, **63**, 189–195.

**Werner P.A. & Platt W.J. (1976)** Ecological relationships of co-occurring Goldenrods (*Solidago: Compositae*). *American Naturalist*, **110**, 959–971.

**Whatley J.A. (1942)** Influence of heredity and other factors on 180-day weight in Poland China swine. *Journal of Agricultural Research*, **65**, 249–264.

**Williams G.C. (1966a)** Natural selection, the costs of reproduction, and a refinement to Lack's principle. *American Naturalist*, **100**, 687–690.

**Williams G.C. (1966b)** *Adaptation and Natural Selection*. Princeton University Press, Princeton.

**Wootton R.J. (1977)** Effect of food limitation during the breeding season on the size, body components and egg production of female sticklebacks (*Gasterosteus aculeatus*). *Journal of Animal Ecology*, **46**, 823–834.

# 5. SYCAMORE APHID NUMBERS: THE ROLE OF WEATHER, HOST AND APHID

A. F. G. DIXON

*School of Biological Sciences, University of East Anglia, Norwich NR4 7TJ*

## INTRODUCTION

At a previous symposium of this Society I discussed the effect of food quality and availability of space on sycamore aphid populations. This revealed the importance of seasonal changes in growth and development of sycamore in determining the reproductive activity of its aphid, *Drepanosiphum platanoidis* Schr. In most summers, when leaves are mature, this aphid does not reproduce but aestivates as an adult. This contrasts with the autumn and spring which are favourable periods for aphid development, certain autumns being better for reproduction than others (Dixon 1970a). An attempt to understand what regulates the numbers of sycamore aphid has shown that this could result in part from an insect–host interaction.

The aim of this paper is to review our understanding of the sycamore aphid/sycamore system.

## SCOPE OF THE STUDY

My field study has been on up to eight sycamore trees over 15 years in the west of Scotland. The trees are not all in the same locality but the maximum distance apart is 35 kilometres, with the largest tree being 20 m high and bearing 116 000 leaves and as many as $2\frac{1}{2}$ million aphids at one time. However, sycamore occurs throughout the United Kingdom and therefore, in terms of the dynamics of the sycamore aphid, my study covers a minute fraction of the population and range of a highly mobile insect.

Fortunately, the Rothamsted Insect Survey (R.I.S.) (Taylor 1974, 1977) over a period of 10 years has been recording aerial densities of the sycamore aphid at several sites throughout the U.K. Dr. L.R. Taylor has very generously consented to us using these data and Mercer (unpublished) has established that catches of sycamore aphids at Dundee, 106 kilometres away on the east coast of Scotland, are closely correlated with what happened in my study area. Therefore, flight activity recorded by the R.I.S. does reflect the number of aphids

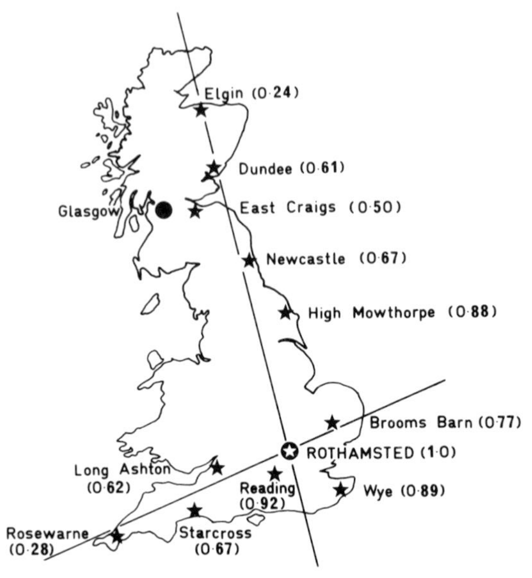


FIG. 5.1. The location of some of the Rothamsted Insect Survey suction traps in the United Kingdom and the correlation coefficients between the number of sycamore aphids caught by these traps and the Rothamsted trap.

present on sycamore. In addition, although the number caught by each trap varies greatly from year to year, the catches are correlated with one another, which indicates that good and bad years for sycamore aphid flights are generally so throughout the U.K. However, the farther apart the traps the weaker the correlation, possibly reflecting regional differences in weather from year to year (Fig. 5.1).

In summary, the R.I.S. enables us to extrapolate the understanding obtained from an intensive study of the aphids on eight trees to sycamore aphid populations throughout the U.K.

## RELATIONSHIP BETWEEN SPRING AND AUTUMN NUMBERS

In common with most aphid species, sycamore aphid populations consist of pathogenetically reproducing clones each initiated by a fundatrix (stem mother) and terminated by death or the development of males and oviparae (egg-laying forms) at the end of the season. A clone is comparable to an individual in a non-pathogenetic insect. Janzen (1977) refers to such clones as evolutionary units. Each morph in a clone has a particular role to perform and is part of a sequence which ends with the egg-laying form. The most successful genotype is that which optimizes the combined strategies of all the morphs in the sequence (Dixon 1977).

Thus, in the analysis of changes in sycamore aphid numbers a meaningful first step is to determine the relationship between the number of aphids present in spring after egg hatch and those present in autumn prior to egg laying. Over the period 1961–69 it was noticed that when there were many aphids on sycamore leaves in spring the number in autumn is invariably low; and, conversely, when there were few aphids in spring there were large numbers in autumn (Figs 5.2 & 5.3, Dixon 1970b). With two exceptions the size of the autumn population each year was related to the spring population size. The peak number achieved in the autumn of 1972 was greatly in excess of any prediction based on the pattern of observations since 1961. Although sampling in 1960 was relatively infrequent and involved counting aphids on only a few leaves, the autumn peak number for 1960 was also considerably greater than that predicted.

The weather in the autumns of both 1960 and 1972 was unusual. In particular, wind speeds were well below average in both years with August 1960 the stillest for the period 1960–77 and September 1972 the driest since 1894. High wind dislodges large numbers of aphids and as a consequence is an important mortality factor (Dixon & McKay 1970). Therefore, in still autumns when wind-induced mortality is low, aphids are likely to become extremely abundant.

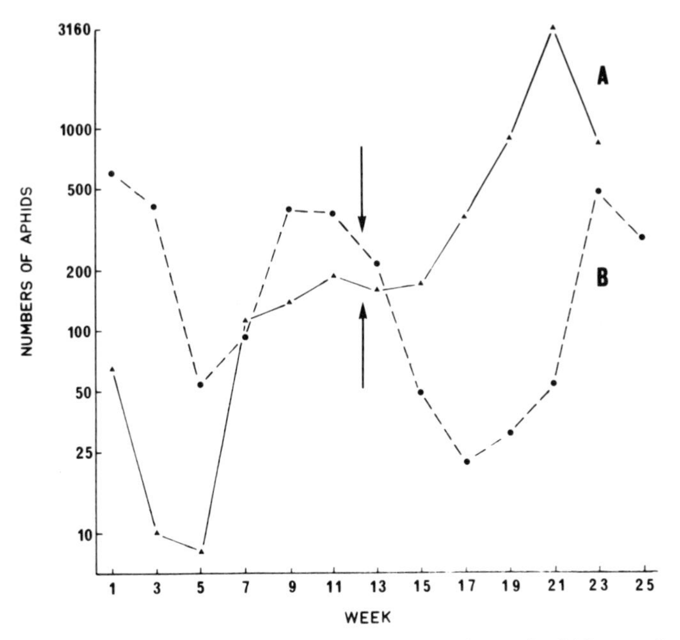


FIG. 5.2. The population trends in years when the numbers of aphids were low (A, 1973) and high (B, 1968) in spring (⇅ indicates the onset of autumn).

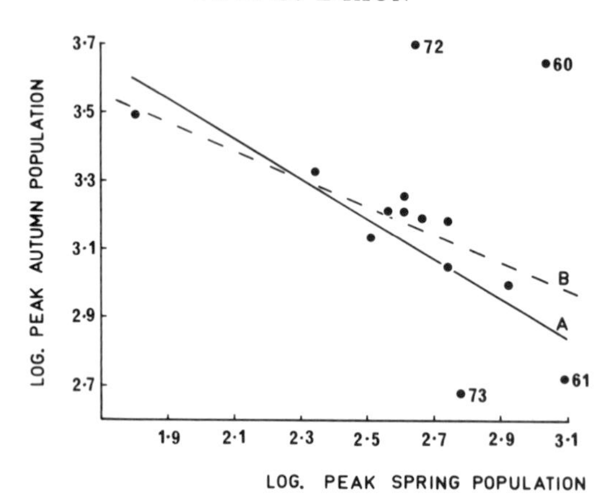


FIG. 5.3. Relationship between spring and autumn peak numbers of aphids for each year from 1960–73. (Regression line A, $y = 4{\cdot}65 - 0{\cdot}59x$, $r = 0{\cdot}81$, $p < 0{\cdot}01$, does not include 1960 and 1972; B, $y = 4{\cdot}25 - 0{\cdot}41x$, $r = 0{\cdot}91$, $p < 0{\cdot}01$, does not include 1960, 1961, 1972 and 1973.)

The effect of wind and fecundity on the rate of increase in aphid numbers in autumn is revealed by determining the relationship between the increase in the populations each week, average fecundity per adult per week ($F$) and average wind speed over that period ($W$). The relationship derived from pooling the results for seven autumns for one tree is:

$$\text{Log}\left(\frac{N_{t+1}}{N_t}\right) = 0{\cdot}107 + 0{\cdot}051F - 0{\cdot}029W \qquad (5.1)$$

($R = 0{\cdot}686$, $n = 57$, both regression coefficients are significant at $p < 0{\cdot}01$).

The effect of wind can also be expressed in terms of the difference between the logarithm of the number of aphids expected and the logarithm of the number observed each week, i.e. an estimated mortality attributable to wind ($k_w$). The potential recruitment to the population each week is the fecundity ($F$) times the average number of adults present. The relationship is:

$$k_w = 0{\cdot}891 \log W - 0{\cdot}526 \qquad (5.2)$$

($r = 0{\cdot}484$, $n = 57$, regression coefficient $p < 0{\cdot}001$)

which indicates that at wind speeds above 3 knots wind-induced mortality can be high; for example, at 12 knots it is 65%.

Fecundity is determined by keeping aphids in clip cages which affords the aphids shelter from wind. Therefore, wind could act by disturbing aphids and so preventing them from realizing their potential fecundity. However, in high winds large numbers of aphids are dislodged from trees which indicates that high wind can cause the death of aphids (Dixon & McKay 1970).

The relative contribution of changes in wind and fecundity from week to week and between years on the peak autumnal numbers achieved is shown by expressing the changes in numbers from week to week as follows:

$$\text{Log}\, N_{t+1} = 0{\cdot}152 + 0{\cdot}825\, \text{Log}\, N_t + 0{\cdot}25\, \text{Log}\, R - 0{\cdot}0322W \quad (5.3)$$

where $R$ is the average fecundity per adult per week ($F$) multiplied by the average number of adults present.

(Multiple correlation coefficient = 0·95, $n$ = 57, all regression coefficients are significant at $p < 0{\cdot}01$.)

Although no particular significance can be attributed to the high value of the correlation coefficient, eqn (5.3) is useful in revealing the effect of wind speed and of poor and good recruitment on peak numbers achieved. Both wind speed and recruitment have a marked effect on the peak numbers (Fig. 5.4). When wind speed is kept constant, differences in fecundity observed in the autumns of 1968 and 1973 can account for a 20-fold difference in the peak numbers achieved; likewise, a drop in average wind speed from 9 to 6 knots can also result in a 15-fold increase in the peak numbers achieved.

The numbers in the autumns of the years 1961 and 1973 were markedly lower than predicted (Fig. 5.3). These were years when there were very high numbers of aphids on the buds prior to bud-burst. Although many of the aphids disappeared before bud-burst, large numbers were successful in colonizing leaves (Dixon 1976).

Over 8 years' records of aphids caged on the leaves of their host trees showed that spring mortality is higher in years when aphids are numerous on both buds

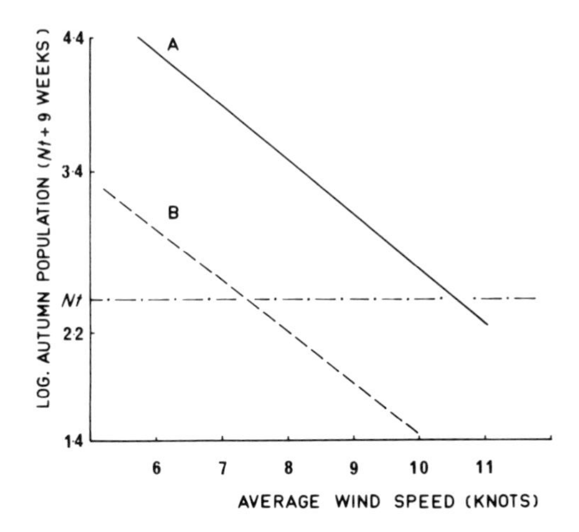


FIG. 5.4. Relationship between average wind speed in autumn, good (A, 1968) and poor (B, 1973) autumnal reproductive rates and peak populations achieved in autumn. Populations at the beginning of each autumn ($N_t$) is 300 aphids. The numbers plotted are those present 9 weeks later.

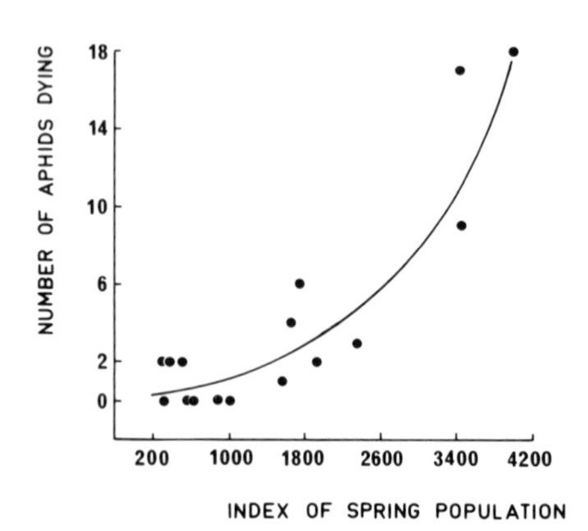


FIG. 5.5. Relationship between the number of caged adult aphids dying in the first 3 weeks of spring and the degree of crowding they experienced during development. In those years when many aphids died the maximum daily mortality was 40%.

and leaves than when the aphids are scarce (Fig. 5.5). In addition, aphids are smaller in springs when aphids are numerous than when they are scarce (Fig. 5.6) and small aphids are less fecund than large aphids (Dixon 1970a, 1975a). In the field the spring reproductive rate is also related to aphid weight but not to temperature. Migratory activity is also more marked in years when aphids are numerous in spring (Dixon 1969). Aphids isolated in cages on heavily infested trees initially produce as many offspring as those on uninfested trees, but in the second ten days of adult life they produce half as many ($F = 11{\cdot}84$,

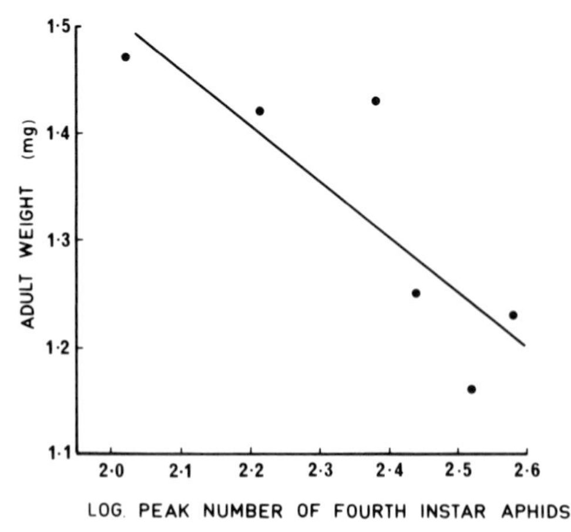


FIG. 5.6. Relationship between the weight of first generation adults and the crowding they experience during development—measured as the peak number of fourth instar aphids ($y = 2{\cdot}56 - 0{\cdot}52x$, $r = 0{\cdot}85$, $p < 0{\cdot}02$).

d.f. = 1/47, $p < 0{\cdot}01$). This possibly results from the early cessation of growth shown by infested trees. Therefore, a number of density-dependent processes occur in spring which result in the differences in the magnitude of the decline in the numbers of aphids in the first generation. Nevertheless, the number of second generation aphids present at the onset of autumn is generally higher in years when aphids are abundant in spring than when they are scarce, even in 1973 (Fig. 5.2).

In summary, this long-term study of the sycamore aphid reveals an inverse relationship between numbers of aphids present in spring and those present in autumn, with the two years 1960 and 1972 experiencing the highest numbers, well in excess of those predicted, which can be attributed to exceptionally still autumns in those years. Thus the regulatory mechanism(s) overcompensate when the population is deflected from equilibrium (Dixon 1970b), and the major component of this response acts in autumn. Weather, particularly wind, acts as a disturbing factor.

## NUMBERS IN AUTUMN

Both windiness and the reproductive rate of the aphid determine the numbers of aphids that will be achieved in autumn (Fig. 5.4). The reproductive rate in early autumn varies greatly from year to year (Dixon 1975b) and is not attributable to differences in temperature but is inversely related to the number of aphids present earlier in the year (Fig. 5.7). Both the length of time spent in

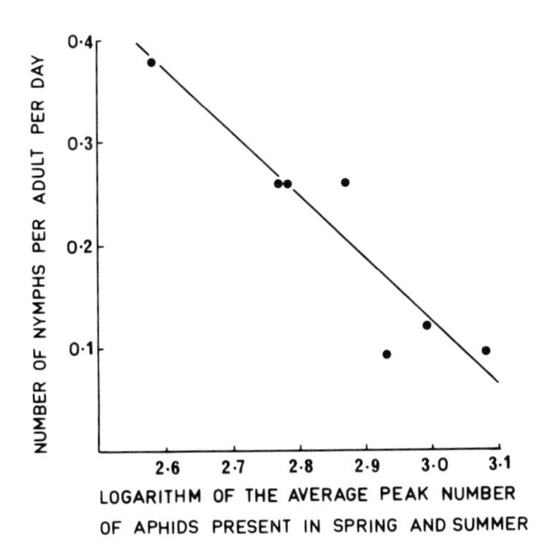


FIG. 5.7. Relationship between the number of offspring produced per adult per day from 15 July to 15 August and the logarithm of the average peak number of aphids present in spring and autumn ($y = 1{\cdot}98 - 0{\cdot}62x$, $r = 0{\cdot}93$, $p < 0{\cdot}01$).

aestivation and the number of offspring produced per adult per day after emergence from aestivation determine the variation in the autumnal reproductive rate from year to year.

### (a) *Aestivation*

As early as 1896 Mordwilko noticed that from June or early July until the end of July or August in the Botanic Gardens, Warsaw, although adult sycamore aphids were abundant on sycamore, nymphs were absent. He related this to a possible unsuitability of the phloem sap of sycamore from June to August (Mordwilko 1908). This was the first description of the aestivation of second generation adults which occurs in most years.

Even when second generation aphids are reared on opening buds and unfurling leaves, the conditions normally experienced by first generation aphids, they are large and have a well-developed fat body and poorly-developed gonads as if about to enter aestivation, i.e. they anticipate the onset of harsh conditions, associated with the cessation of growth of sycamore (Dixon 1975a). Second generation aphids are also different from those of the first generation in other respects: they have fewer ovarioles (ten rather than twelve) (Wellings unpublished), longer appendages and a greater number of rhinaria on their antennae (Dixon 1974), a higher wing beat frequency and they fly longer and lower (Mercer unpublished).

Variations in autumnal reproductive rates from year to year are associated with changes in the numbers of aphids present in spring and summer (Fig. 5.7), i.e. aphids of the first and second generations. Whether second generation aphids aestivate and for how long depends in part on their size and the degree of crowding they (Dixon 1975a) and their mothers (first generation) have experienced (Chambers unpublished, Fig. 5.8). Small second generation aphids show a longer pre-reproductive delay than do large aphids (Dixon 1970a). However, the experiment carried out by Chambers shows that the pre-reproductive delay can be further lengthened, without any extra reduction in adult weight, by crowding first generation aphids during their development. The mechanism by which the density experience of first generation aphids is transferred to second generation individuals is not through the plant since the two generations were reared on different groups of plants.

High population density in summer is a consistent feature of sycamore aphid populations and at this time of the year the aphids are neatly spaced out. Although they repel one another by kicking, they also attract one another and settle so that they touch their neighbours with their antennae. This is described as 'spaced-out gregariousness' by Kennedy & Crawley (1967). Although gregarious, the aphid is more widely spaced at low than at high population densities (Dixon & Logan 1972). By using leaf cages the number of similar-sized

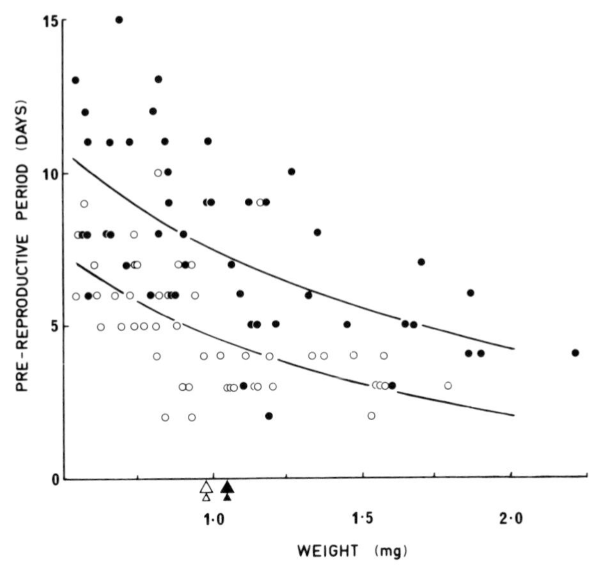


FIG. 5.8. Relationship between the pre-reproductive period and a second generation aphid's teneral weight (● both first and second generation individuals crowded during development, ○ first generation individuals isolated and second generation individuals crowded during development; △ ▲, the average teneral weight of the second generation aphids, are not significantly different). (From Chambers unpublished.)

adult aphids in a group has been varied, or the aphids have been kept apart by dividers, while keeping the area of leaf available to each aphid constant. This has revealed that on mature leaves aphids in large groups give birth to fewer offspring than aphids kept in small groups or in isolation (Dixon 1970a). Many of the individuals in the large groups lack well-developed embryos in their ovarioles (Dixon 1975a). Therefore, the low reproductive rate is a consequence of many of the aphids remaining in aestivation.

The small size of second generation adults is a consequence not only of the poor nutrition they experience as the leaves mature, but also of high temperatures and overcrowding. Compared with aphids reared on nutritious leaves, those reared on poor leaves grow and develop more slowly but, as the rate of development is less retarded than the rate of growth, the aphids are smaller. Aphids reared at high temperatures grow faster than those reared at low temperatures but, as high temperatures have an even greater effect in speeding up metamorphosis, aphids reared at high temperatures are small. Crowding acts in the same way as poor nutrition by reducing the growth rate and prolonging the time taken to reach maturity, although not proportionately, and the consequence is a small aphid (Chambers unpublished). Therefore the poor nutrition, overcrowding and high temperatures experienced by second generation aphids in their nymphal stages all contribute to the small size of these aphids and the length of aestivation.

Occasionally, in summer, the leaves of twigs deep within the canopy of a tree senesce while the leaves of adjacent twigs are still mature. Such senescent leaves offer a rich food supply and are attractive to aphids which then come out of aestivation and reproduce, despite the fact that the local adult population density is high (Dixon 1966). As food quality affects the length of time spent in aestivation, by reducing the quality of their food supply, high numbers of aphids could prolong the time spent in aestivation.

Aestivation in the field can be prolonged, lasting as long as 8 weeks. However, in laboratory populations kept on sycamore saplings, even at high densities and temperatures of 20°C, second generation individuals, although small and crowded, remain in aestivation for only three weeks (Chambers unpublished). This is puzzling but may result from differences between saplings and mature trees, the absence of wind effect in laboratory populations and/or the lack of cumulative effects through the plant from year to year, i.e. trees that experience high numbers in spring would also be heavily infested the previous autumn (p. 117). Second generation aphids kept on saplings in the laboratory and exposed to the buffeting action of wind do not spend longer in aestivation than control aphids (Wellings unpublished).

In summers when numbers are low, aphids are large and, as they are uncrowded, they do not aestivate but reproduce at a low rate (Dixon 1970a). However, if the aphids were to continue reproducing in summers when abundant it is likely that few of the offspring would reach maturity and those that did would be extremely small (340 $\mu$g) (Chambers unpublished). Such small aphids can be produced in the laboratory and are capable of reproduction but are unable to fly upwards except at high temperatures (Mercer unpublished) and therefore would be at risk if dislodged from a tree. That these very small aphids have such a poor chance of survival possibly accounts for aestivation in the sycamore aphid in most years.

In summary, the duration of aestivation in the field is prolonged when aphids are numerous. This is a consequence, in part, of the small size of the aphid and of the crowding they experience during their nymphal and adult life, and of that experienced by the previous generation. However, under laboratory conditions, even when additionally subjected to high temperatures and disturbed by wind, it has not proved possible to induce aphids to aestivate for as long as observed in the field. Therefore, it is possible that the plant has a role in determining the duration of aestivation. This is supported by the field observation that aphids come out of aestivation earlier if feeding on prematurely senescent leaves. Whatever the causal mechanisms, aphids spend longer in aestivation when abundant than when rare and this is a major factor in determining the variability in reproductive performance in early autumn.

### (b) *Autumnal reproductive rate*

An abundance of aphids in spring causes the stunting of sycamore leaves and at leaf-fall the leaves are greener and richer in nitrogen (Dixon 1971a). This led to the hypothesis that variation in autumnal reproductive rate from year to year results from aphid-induced changes in the host plant (Dixon 1970a). This was tested by measuring the reproductive rate of aphids in autumn on previously infested and uninfested saplings. In addition, reciprocal transfers of adults were made in the field between trees where aphids were reproducing at high and low rates. Both these experiments indicated that reproductive rate is related to the size or quality of the aphids and that the previous history of the plant was not important (Dixon 1975b). This is further supported by the observation that the yearly differences in reproductive rate in early autumn are related to the weight of the aphid at that time (Fig. 5.9). Even as late as the first half of September aphids are smaller in those years when aphids were abundant in spring and summer ($r = 0{\cdot}85, p < 0{\cdot}01$).

Although the amount of nitrogen in the leaves at leaf-fall is greater when aphids are abundant in spring (Fig. 5.10), the amount is also greater in the middle of the year. Therefore, high levels of nitrogen in the leaves at leaf-fall does not necessarily mean poor recovery of nitrogen from the leaves in autumn. However, there is no consistent significant relationship between percentage nitrogen recovery from the leaves and the aphids' reproductive rate. In fact, in some years certain trees show a negative recovery of nitrogen. This raises questions about the significance of these total nitrogen determinations.

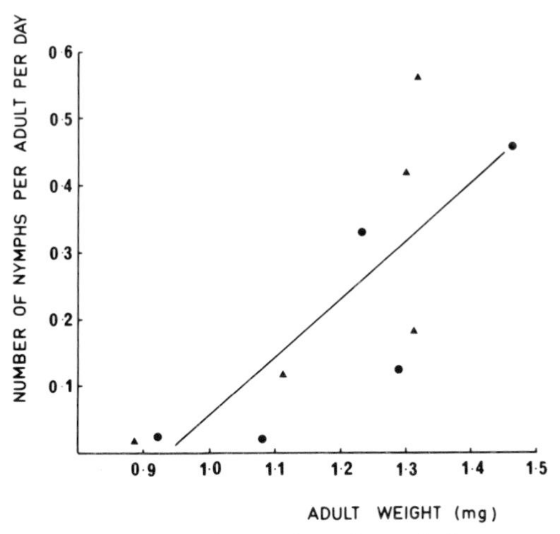


FIG. 5.9. Relationship between the number of nymphs born per adult per day in early autumn and the weight of the adults for two trees (●▲, $y = 0{\cdot}86x - 0{\cdot}81$, $r = 0{\cdot}79, p < 0{\cdot}01$).

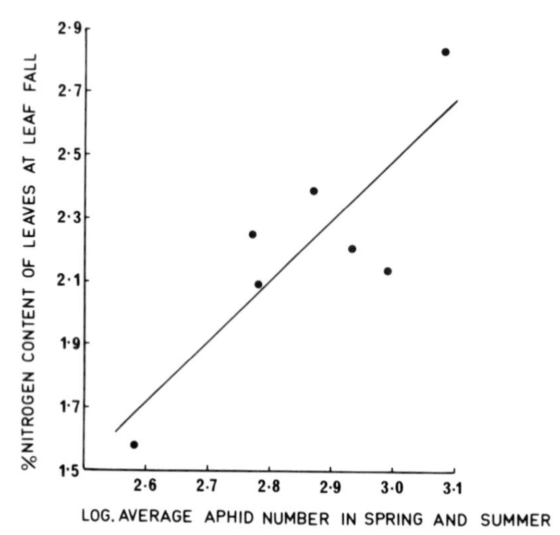


FIG. 5.10. Relationship between the percentage nitrogen content of the leaves at leaf fall and the logarithm of the average peak number of aphids present in spring and summer ($y = 1{\cdot}92x - 3{\cdot}23$, $r = 0{\cdot}84$, $p < 0{\cdot}02$).

Therefore, there appears to be no evidence in favour of an aphid-induced plant factor causing the differences in reproductive rate from year to year. These changes appear to result from changes in the quality of the aphid determined by its, and its mother's, experience of crowding.

However, experiments with other species of aphids have raised doubts about the interpretation of the above results. On reaching maturity aphids like the black bean aphid, *Aphis fabae* (Scop.), are committed to a certain fecundity dependent on their weight (Taylor 1975; Dharma & Dixon unpublished). Stresses imposed during adult life can result in slight changes in reproductive rate in time but more marked changes in birth weight of offspring. Unfortunately the latter was not measured for the sycamore aphid and any small differences in nutritive stress occurring over short periods of time may be obscured by the variation in reproductive rates between trees.

There is also some direct evidence for a plant factor. Aphids reared in autumn on previously infested saplings are 0·8 the weight of control aphids ($t_{(one\ tail)} = 1{\cdot}87$, d.f. 22, $p < 0{\cdot}05$). An interesting feature is that this effect in sycamore is marked only when heavy infestation occurs at the time of bud-burst. Heavy infestation after leaf growth is complete does not affect aphid development later in the year (Wellings unpublished) nor does it significantly increase the amount of nitrogen in the leaves at leaf-fall (Dixon 1971a). Therefore, while the reproductive rate may not be controlled by an aphid-induced plant factor, the quality of the aphid might be. Further experiments are now needed to determine whether there is a plant factor, and to resolve the relative importance of changes in the aphid and/or plant in regulating the numbers of the aphid.

In summary, although the aphid has a dramatic effect on the growth of sycamore, there is no convincing evidence that these changes are responsible for the differences observed in autumnal reproductive rate from year to year. Most of the evidence favours the effect being mediated through the aphid. There are indications, as with aestivation, that the effect on aphid numbers of the aphid-induced change in sycamore has been overlooked. However, whatever the causal mechanism, reproductive rates in early autumn are considerably lower in those years when the aphid is abundant in spring and summer.

## BETWEEN-YEAR EFFECTS

High numbers in spring are a consequence of high numbers the previous autumn (Dixon 1970b). However, the number present on the buds in spring are not correlated only with the number present the previous autumn but also with the time of leaf-fall. Early leaf-fall results in a heavy mortality of egg-laying females and a reduction in the number of eggs laid (Chambers unpublished), and leaf-fall is known to occur earlier in heavily infested saplings than in uninfested saplings (White 1970).

Bud-burst is delayed in springs following autumns when aphids are abundant. A ten-fold increase in the size of the autumn population is associated with the buds opening 5–6 days later the following year and a delay in egg hatch of 3–4 days (Chambers unpublished). Lack of synchronization between egg-hatch and bud-burst can result in very heavy mortality (Dixon 1976). By delaying egg-hatch in springs following autumns of high aphid abundance aphids can, in part, offset the tree's response. However, these statistical observations need to be confirmed experimentally.

Although leaf size in any year is determined by the number of aphids present in spring and does not appear to be influenced by the abundance of aphids the previous year, nevertheless leaves of saplings which were heavily infested the previous year tend to be paler (Dixon 1971a). However, in the presence of aphids they rapidly become a dark green.

Therefore, there is evidence to suggest that high aphid numbers in one year can affect the growth and development of sycamore the following year. If aphid-induced changes in sycamore are important in determining the subsequent well-being of the aphids, then the effect may become evident experimentally only when aphid infestations are maintained over two years. The very low autumnal numbers observed in 1961 and 1973 (Fig. 5.3) could be a response to the combined effect of the high numbers in the previous autumn and in the spring of those years. So far we have concentrated on the within-year effect as the carryover between years was not thought to be important. This followed from the observation that leaf size was determined by what went on within a year rather than between years and was thought of in terms of aphids competing with the tree for resources. However, it was appreciated that the effect could not be accounted for simply in terms of an energy or nutrient drain on the plant's resources (Dixon 1971a). There is now sufficient evidence to justify such long-term experiments.

## DISCUSSION

What has this study revealed? By starting in 1961 after the aphid had experienced a major perturbation in the form of the still autumn of 1960 it was possible to follow the recovery of the population over a long period of time. Fortunately, there was not another major disturbance until 1972. If disturbances occurred more frequently the unpredictable fluctuations in numbers would have proved confusing.

Although the study at Glasgow ceased in August 1974, the prediction was that as a consequence of the disturbance experienced in 1972 aphid numbers would achieve very high levels in the autumn of 1974. The reproductive rate in early autumn 1974 was high, as predicted, and extremely large numbers of aphids were caught in the suction trap at Dundee from August to October. Over the period 1967–77 the number caught in autumn 1974 is the second highest, with autumn 1972, as expected, registering the highest catch. These catches also reveal that the autumn of 1975 was a relatively poor year for aphid flight whereas 1976 was considerably better. Thus, after the disturbance of 1972, the suction trap catches indicate that the aphid continued to oscillate in numbers from year to year with decreasing amplitude in the same way as after the 1960 disturbance (Dixon 1970b).

A major factor in the overcompensated response leading to the collapse of the population in autumns of years when aphids are abundant in spring is the low fecundity of the aphids in those autumns. Changes in autumnal fecundity from year to year are associated with differences in the size of the aphid. This and the reproductive rates of aphids transferred between trees indicate that the quality of the aphid determines fecundity. The size of the aphid is determined to a large extent by the crowding that it and its parents experience. High numbers of aphids have a dramatic effect on the growth of sycamore which one would expect to affect the aphid. However, the precise role of sycamore in these changes in fecundity still needs to be resolved.

In this account I have not discussed the affect of the many parasites and predators (Dixon 1973) on the numbers of the aphid. Although they do not have a regulatory role, the action of certain of the predators and parasites could be important in determining the degree of overcompensation that occurs. Anthocorids are at their most voracious (Dixon & Russel 1972) and parasites actively parasitize and kill aphids (Hamilton 1973, 1974) when the aphid comes out of aestivation. As the numbers of these predators and parasites are determined by the number of aphids present early in the year, large numbers of young aphids are killed in autumns when aphid fecundity is low. This results in a more overcompensated response than would otherwise occur.

Parallel studies on the lime aphid, *Eucallipterus tiliae* (L.), have revealed the same inverse relationship between spring and autumn numbers. A simulation study of the population dynamics of this species has shown that density-dependent migratory activity is the most important factor contributing to the overcompensated response but that the action of predators makes the response more extreme (Dixon 1971b; Barlow 1977; Dixon & Barlow 1979). However, an aphid-induced plant factor is needed to account for changes in numbers of the lime aphid in some years (Barlow 1977). There is now evidence that such a plant factor exists (Dixon & Barlow 1979). Similarly, the walnut aphid, *Chromaphis juglandicola* (Kalt.), in California shows signs of an inverse relationship between spring and autumn numbers. This is possibly attributable to the combined effects of qualitative changes in the aphid and plant associated with high aphid numbers and the mortality inflicted by predators. In the absence of predators the aphid can possibly maintain its numbers (Sluss 1967; Dixon 1977). The introduction of the Iranian parasite, *Trioxys pallidus* (Hal.), into California in 1965 resulted in the control of the walnut aphid (Messenger 1975). However, there are no details of how this parasite responds to changes in aphid numbers. Studies on the green spruce aphid, *Elatobium abietinum* (Walker), also indicate that an aphid-induced deterioration in both the quantity and quality of food available in summer is important in causing the collapse of population outbreaks of this species (Kloft & Ehrhardt 1959; Parry 1974, 1976).

All the long-term studies of tree-dwelling aphids indicate that aphid-induced changes in the host plant are implicated in subsequent changes in aphid numbers. Therefore, more attention needs to be paid to defining the role of the plant in this interaction, despite the fact that some of the experiments may prove tedious to carry out.

## ACKNOWLEDGMENTS

I am grateful to Nigel Barlow and Richard Chambers for helpful comments and to Paul Wellings for criticism of the manuscript. Financial support for this study was provided by the Natural Environment Research Council.

## REFERENCES

**Barlow N.D. (1977)** A simulation study of lime aphid populations. *Unpublished Ph.D. thesis, University of East Anglia.*

**Dixon A.F.G. (1966)** The effect of population density and nutritive status of the host on the summer reproductive activity of the sycamore aphid, *Drepanosiphum platanoides* (Schr.). *Journal of Animal Ecology*, **35**, 105–112.

**Dixon A.F.G. (1969)** Population dynamics of the sycamore aphid, *Drepanosiphum platanoides* (Schr.). (Hemiptera: Aphididae): Migratory and trivial flight activity. *Journal of Animal Ecology*, **38**, 585–606.

**Dixon A.F.G. (1970a)** Quality and availability of food for a sycamore aphid population. *Animal Populations in Relation to their Food Resources* (Ed. by A. Watson) pp. 277–287. Blackwell Scientific Publications, Oxford.

**Dixon A.F.G. (1970b)** Stabilization of aphid populations by an aphid induced plant factor. *Nature*, **227**, 1368–1369.

**Dixon A.F.G. (1971a)** The role of aphids in wood formation. 1. The effect of the sycamore aphid, *Drepanosiphum platanoides* (Schr.) (Aphididae), on the growth of sycamore, *Acer pseudoplatonus* (L.). *Journal of Applied Ecology*, **8**, 165–179.

**Dixon A.F.G. (1971b)** The role of intra-specific mechanisms and predation in regulating the numbers of the lime aphid, *Eucallipterus tiliae* L. *Oecologia Berlin*, **8**, 179–193.

**Dixon A.F.G. (1973)** *Biology of Aphids.* Edward Arnold, London.

**Dixon A.F.G. (1974)** Changes in the length of the appendages and the number of rhinaria in young clones of the sycamore aphid, *Drepanosiphum platanoides. Entomologia experimentalis et applicata*, **17**, 1–8.

**Dixon A.F.G. (1975a)** Seasonal changes in fat content, form, state of gonads and the length of adult life in the sycamore aphid, *Drepanosiphum platanoides* (Schr.). *Transactions of the Royal Entomological Society of London*, **127**, 87–99.

**Dixon A.F.G. (1975b)** Effect of population density and food quality on autumnal reproductive activity in the sycamore aphid, *Drepanosiphum platanoides* (Schr.). *Journal of Animal Ecology*, **44**, 297–304.

**Dixon A.F.G. (1976)** Timing of egg hatch and viability of the sycamore aphid, *Drepanosiphum platanoides* (Schr.), at bud burst of sycamore, *Acer pseudoplatanus* L. *Journal of Animal Ecology*, **45**, 593–603.

**Dixon A.F.G. (1977)** Aphid ecology: life cycles, polymorphism, and population regulation. *Annual Review of Ecology and Systematics*, **8**, 329–353.

**Dixon A.F.G. & Barlow N.D. (1979)** Population regulation in the lime aphid. *Journal of the Linnean Society (Zoology)*, **67** (in press).

**Dixon A.F.G. & Logan M. (1972)** Population density and spacing in the sycamore aphid, *Drepanosiphum platanoides* (Schr.), and its relevance to the regulation of population growth. *Journal of Animal Ecology*, **41**, 751–759.

**Dixon A.F.G. & McKay S. (1970)** Aggregation in the sycamore aphid, *Drepanosiphum platanoides* (Schr.) (Hemiptera: Aphididae), and its relevance to the regulation of population growth. *Journal of Animal Ecology*, **39**, 439–454.

**Dixon A.F.G. & Russel R.J. (1972)** The effectiveness of *Anthocoris nemorum* and *A. confusus* (Hemiptera: Anthocoridae) as predators of the sycamore aphid, *Drepanosiphum platanoides*. 11. Searching behaviour and the incidence of predation in the field. *Entomologia experimentalis et applicata*, **15**, 35–50.

**Hamilton P.A. (1973)** The biology of *Aphelinus flavus* (Hym.: Aphelinidae), a parasite of the sycamore aphid, *Drepanosiphum platanoides* (Hemipt.: Aphididae). *Entomophaga*, **18**, 449–462.

**Hamilton P.A. (1974)** The biology of *Monoctonus pseudoplatani*, *Trioxys cirsii* and *Dyscritulus planiceps*, with notes on their effectiveness as parasites of the sycamore aphid, *Drepanosiphum platanoides*. *Annales de la Société entomologique de France* (NS), **10**, 821–840.

**Janzen D.H. (1977)** What are dandelions and aphids? *American Naturalist*, **111**, 586–589.

**Kennedy J.S. & Crawley L. (1967)** Spaced-out gregariousness in sycamore aphids, *Drepanosiphum platanoides* (Schrank) (Hemiptera, Callaphididae). *Journal of Animal Ecology*, **36**, 147–170.

**Kloft W. & Ehrhardt P. (1959)** Zur Sitkalauskalamität in Nordwestdeutschland. *Waldhygiene*, **2**, 47–49.

**Messenger P.S. (1975)** Parasites, predators and population dynamics. *Insects, Science and Society* (Ed. by D. Pimentel) pp. 201–223. Academic Press, New York.

**Mordwilko A. (1908)** Beiträge zur Biologie der Pflanzenläuse, *Aphididae* Passerini. *Biologisches Zentralblatt*, **28**, 631–639.

**Parry W.H. (1974)** The effect of nitrogen levels in Sitka spruce needles on *Elatobium abietinum* (Walker) populations in north-eastern Scotland. *Oecologia Berlin*, **15**, 305–320.

**Parry W.H. (1976)** The effect of needle age on the acceptability of Sitka spruce needles to the aphid, *Elatobium abietinum* (Walker). *Oecologia Berlin*, **23**, 297–313.

**Sluss R.R. (1967)** Population dynamics of the walnut aphid *Chromaphis juglandicola* (Kalt.) in northern California. *Ecology*, **48**, 41–58.

**Taylor L.R. (1974)** Monitoring change in the distribution and abundance of insects. *Report of the Rothamsted Experimental Station for 1973, Part* **2**, 202–239.

**Taylor L.R. (1975)** Longevity, fecundity and size; control of reproductive potential in a polymorphic migrant *Aphis fabae* Scop. *Journal of Animal Ecology*, **44**, 135–163.

**Taylor L.R. (1977)** Aphid forecasting and the Rothamsted Insect Survey. *Journal of the Royal Agricultural Society of England*, **138**, 75–97.

**White P.L. (1970)** The effect of aphids on tree growth. *Unpublished Ph.D. thesis, University of Glasgow*.

# 6. COMPETITION AND THE STRUGGLE FOR EXISTENCE

J. P. GRIME

*Unit of Comparative Plant Ecology (NERC), Department of Botany, University of Sheffield*

## INTRODUCTION

In a most perceptive paper, Milne (1961) has pointed out that in the vocabulary of many biologists the word competition is used as a synonym for the 'struggle for existence' (Darwin 1859). Milne proposes that competition should be regarded merely as one component of the struggle for existence and effectively he suggests that the term should be used to describe the activity whereby organisms capture resources in crowded environments.

With some reservations, I wish to endorse Milne's view and to suggest that lack of precision in the use of the word competition has had a particularly limiting effect upon the development of theories in plant ecology. The main objective in this paper will be to examine some of the opportunities to define plant strategies and vegetation mechanisms which follow from Milne's approach to competition. It is inevitable that, in the space available, many of the arguments which follow will be truncated. A more detailed treatment of the subject, including a review of experimental evidence and extrapolations to heterotrophic organisms, will be presented elsewhere (Grime 1979).

In order to place competition in perspective it is necessary to dissect the struggle for existence into its main components. The first step in such an analysis is to classify the threats to existence.

## THREATS TO EXISTENCE

Environments differ widely with respect to the identity of the factors controlling rates of plant growth, regeneration and mortality, and in consequence the nature of the struggle for existence varies from place to place and with the passage of time. It has been suggested (Grime 1974, 1977) that threats to existence fall into three categories, as follows.

## 1 *Stress*

In certain environments such as those in arctic or arid regions or beneath forest canopies, the most potent threats to survival arise from factors such as shortages of light, water and mineral nutrients or suboptimal temperatures. These constraints, which may be classified under the general heading of 'stress', may be an inherent characteristic of an environment or they may be induced or intensified by activities of the vegetation such as continuous shading or sequestration of minerals in the biomass. In contrast to the circumstances associated with competitive exclusion (see below), we are concerned here with habitats in which severe stresses are operating more or less continuously throughout the year and affect all species present in the environment. Under these conditions there is little scope for characteristics of morphology or phenology to provide mechanisms of stress avoidance.

In addition to the direct constraints on growth and regeneration arising from stress factors, a major cause of mortality which may be expected to occur in stressed environments is that due to infrequent or inconspicuous predation. In circumstances where growth and reproduction are held at constantly low levels, the capacity for recovery from predation (or other forms of damage) is likely to be critically reduced.

## 2 *Disturbance*

A second threat to existence which may be recognized is that associated with frequent and severe destruction of the vegetation by herbivores, pathogens or man (trampling, mowing and ploughing), or by phenomena such as wind damage, fire or climatic fluctuations (frost, drought and floods).

Where constant and severe disturbance coincides with extreme stress (e.g. footpaths in mature forest, areas subject to soil erosion in arctic and alpine habitats and overgrazed grasslands and scrub in arid regions), the result may be the total elimination of vegetation. In more productive conditions, the influence of severe and repeated disturbance is to select plants with life histories short enough to exploit the intervals between successive disturbances.

## 3 *Competitive exclusion*

Where resources are plentiful and there is a low incidence of disturbance, conditions encourage the development of a large rapidly-expanding biomass dominated by perennial plants with the potential for high rates of resource capture. Despite the presence under such conditions of a large reservoir of resources, the effect of the activity of the plants during the growing season is to produce expanding zones of depletion, the most conspicuous of which are for light (expanding upwards from the soil surface) and for water and mineral nutrients (expanding downwards from the soil surface). In this type of vegetation, high rates of mortality during the growing season and low rates of reproduction are characteristic of those plants which are outstripped by their neighbours and become 'trapped' in the depleted zones.

## EVOLUTIONARY STRATEGIES

The second step in an analysis of the struggle for existence is to describe the avenues of adaptive specialization which have evolved in response to the three major threats to existence. Here it is convenient to follow the example of many zoologists and to attempt to recognize strategies, i.e. groupings of similar or analogous genetic characteristics which recur widely among species or populations and cause them to exhibit similarities in ecology.

In an attempt to define strategies, a complication arises from the need to consider different phases in the life-cycle of the same organism. Even in the same habitat, juvenile and mature stages within the same population may be subject to different forms of natural selection or alternatively, because of differences in size and function, they may respond in different ways to the same selection force. It follows that in order to understand the basic features of a plant's ecology it is necessary to examine the strategies adopted during two different parts of the life history—the established (mature) phase and the regenerative (immature) phase.

### *Strategies in the established phase*

In the discussion which follows, it has been convenient to limit analysis to the three extreme situations in which stress, disturbance and competition acting in isolation pose major threats to survival. In the majority of habitats, less extreme conditions prevail and the various intermediate or secondary strategies which occur may be classified by reference to a range of equilibria which may be accommodated in a simple triangular model (Fig. 6.1).

On the basis of existing published data and from observations in field and laboratory, it seems reasonable to assert that each of the three major threats to survival (severe stress, frequent disturbance and competitive exclusion) has been associated with the evolution of a distinct type of strategy in the established phase. In Table 6.1 plants conforming to the three primary strategies have been described as competitors, stress-tolerators and ruderals, and an attempt has been made to list some of their most consistent characteristics. Elsewhere (Grime 1979) these attributes have been reviewed in detail. Here attention will be confined to two particularly important groups of characteristics. The first is concerned with response to stress, whilst the second relates to life history.

TABLE 6.1. Some characteristics of competitive, stress-tolerant and ruderal plants

| | Competitive | Stress-tolerant | Ruderal |
|---|---|---|---|
| (i) MORPHOLOGY | | | |
| 1. Life forms | Herbs, shrubs and trees | Lichens, herbs, shrubs and trees | Herbs |
| 2. Morphology of shoot | High dense canopy of leaves. Extensive lateral spread above and below ground | Extremely wide range of growth forms | Small stature, limited lateral spread |
| 3. Leaf form | Robust, often mesomorphic | Often small or leathery, or needle-like | Various, often mesomorphic |
| (ii) LIFE HISTORY | | | |
| 4. Longevity of established phase | Long or relatively short | Long to very long | Very short |
| 5. Longevity of leaves and roots | Relatively short | Long | Short |
| 6. Leaf phenology | Well-defined peaks of leaf production coinciding with period(s) of maximum potential productivity | Evergreens with weakly-defined patterns of leaf production | Short period of leaf production in period of high potential productivity |
| 7. Phenology of flowering | Flowers produced after (or, more rarely, before) periods of maximum potential productivity | No general relationship between time of flowering and season | Flowers produced early in the life history |
| 8. Frequency of flowering | Established plants usually flower each year | Intermittent flowering over a long life history | High frequency of flowering |
| 9. Proportion of annual production devoted to seeds | Small | Small | Large |
| 10. Perennation | Dormant buds and seeds | Stress-tolerant leaves and roots | Dormant seeds |
| 11. Regenerative* strategies | V, S, W, $B_s$ | V, $B_\varphi$, W | S, W, $B_s$ |

* Key to regenerative strategies: V, vegetative expansion; S, seasonal regeneration in vegetation gaps; W, numerous small widely-dispersed seeds or spores; $B_s$, persistent seed bank; $B_\varphi$, persistent seedling bank.

TABLE 6.1 (continued)

| | Competitive | Stress-tolerant | Ruderal |
|---|---|---|---|
| (iii) PHYSIOLOGY | | | |
| 12. Maximum potential relative growth rate | Rapid | Slow | Rapid |
| 13. Response to stress | Rapid morphogenetic responses (root–shoot ratio, leaf area, root surface area) maximizing vegetative growth | Morphogenetic responses slow and small in magnitude | Rapid curtailment of vegetative growth; diversion of resources into flowering |
| 14. Acclimation of photosynthesis, mineral nutrition and tissue hardiness to seasonal change in temperature, light and moisture supply | Weakly developed | Strongly developed | Weakly developed |
| 15. Photosynthesis and uptake of mineral nutrients | Strongly seasonal, coinciding with long, continuous periods of vegetative growth | Opportunistic, often uncoupled from vegetative growth | Opportunistic, coinciding with vegetative growth |
| 16. Storage of photosynthate and mineral nutrients | Most photosynthate and mineral nutrients are rapidly incorporated into vegetative structure but a proportion is stored and forms the capital for expansion of growth in the following growing season | Storage systems in leaves, stems and/or roots | Confined to seeds |
| (iv) MISCELLANEOUS | | | |
| 17. Litter | Copious, often persistent | Sparse, sometimes persistent | Sparse, not usually persistent |
| 18. Palatability to unspecialized herbivores | Various | Low | Various, often high |

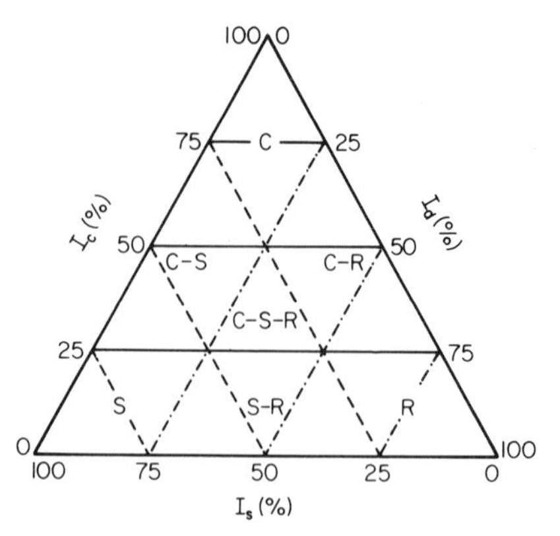


FIG. 6.1. Model describing the various equilibria between competition, stress and disturbance in vegetation and the location of primary and secondary strategies. $I_c$—relative importance of competition (——), $I_s$—relative importance of stress (---), $I_d$—relative importance of disturbance (-·-·-). Key to strategies: C—competitors, S—stress-tolerators, R—ruderals, C–R—competitive ruderals, S–R—stress-tolerant ruderals, C–S—stress-tolerant competitors, C–S–R—'C–S–R strategists'.

### *Three types of response to stress*

An important difference between competitive, stress-tolerant and ruderal plants is apparent in their response to major forms of stress. From a range of autecological or comparative studies (e.g. Salisbury 1942; Harper & Ogden 1970; Hickman 1975; Raynal & Bazzaz 1975; Kingsbury *et al.* 1976; van Andel & Vera 1977; Foulds 1978; Fenner 1979) there is evidence to suggest that the three strategies are distinguishable in both the form and the extent of their responses to shortages in light, water or mineral nutrients. Three types of response to stress (summarized in Table 6.1) have been recognized and these appear to be adapted respectively to conditions of stress, disturbance and competition. It would seem that the slow rates and relatively small extent of morphogenetic response and higher physiological adaptability which is associated with endurance of protracted and severe stress in stress-tolerant plants is of low survival value in environments where stress is a prelude to either competitive exclusion or to disturbance by phenomena such as drought. Similarly, the rapid and highly plastic growth responses to stress of competitive plants (tending to maximize vegetative growth) and of ruderals (tending to curtail vegetative growth and maximize seed production) are highly advantageous only in the specific circumstances associated, respectively, with competition and disturbance.

To summarize, therefore, we may conclude that the stress-response of the ruderal tends to ensure the production of seeds; those of the competitor maximize the capture of resources whilst those of the stress-tolerator allow the conservation of captured resources. This, of course, is an extremely simplified account of the immediate consequences of the three types of response. In the context of the entire life history, the three types of stress-response can be regarded as components of the rather different mechanisms whereby resource capture and fitness are maximized in different types of environment.

It is interesting to explore further the difference in method of resource capture exhibited by the competitors and the stress-tolerators. The effect of the stress-responses of the competitors, when coupled with the rapid turnover of leaves and roots characteristic of these plants, is to bring about a continuous spatial rearrangement of the absorptive surfaces which allows the plant to adjust to changes in the distribution of resources during the growing season. This type of growth response, although highly efficient in resource capture, involves high rates of reinvestment of captured resources and is clearly adapted to exploit productive but crowded environments where the effect of localized resource depletion by the rapidly-growing vegetation is to create within the habitat severe and continuously changing gradients in the distribution of light, water and mineral nutrients. In contrast with the 'foraging' growth responses of the roots and shoots of the competitor, resource capture in the stress-tolerator appears to be a more conservative activity primarily adapted to exploit temporal variation in the availability of resources in chronically unproductive habitats, i.e. the absorptive surfaces of the plant are associated with long-lived physiologically adaptive structures which, at least on an annual basis, tend to remain in the same location and to exploit temporary periods during which resources become available.

### *Three types of life history*

In the majority of ruderals, seed production is followed by the death of the parent plant. This phenomenon relates to an important principle which has interested several evolutionary theorists (e.g. Cole 1954; Williams 1966; Stearns 1976; Ricklefs 1977) and which concerns the partitioning of captured resources between parent and offspring and the optimizing of life histories by natural selection in various environments.

It is clear that when resources are expended upon reproduction at an early stage of the life history there is an increased risk of parental mortality. However, in environments as uncertain as those of the ruderal, high rates of mortality are inevitable and the cost of a marginally-increased rate of parental fatality is outweighed by the benefit of high fecundity. It is no surprise, therefore, that the result of natural selection in frequently and severely disturbed habitats has been the development of early and 'lethal' (Harper 1977) reproduction.

In comparison with the habitats of ruderals, the environments colonized by competitors and stress-tolerators are characterized by a low intensity of disturbance and this has two main effects. The first is to reduce the risks of mortality in long-lived plants whilst the second is to limit drastically the opportunities for seedling establishment. A high risk of mortality to juvenile members of the plant population is characteristic of the environments of both the competitor and the stress-tolerator but, despite this point of similarity, the evolution of life histories has taken a rather different course in the two types of habitat. Whereas in the competitors reproduction occurs at a relatively early stage in the life history and usually involves the expenditure each year of a considerable quantity of the captured resources, the stress-tolerators commence reproduction later and tend to show intermittent reproductive activity over a long life history.

In the crowded but productive environments colonized by competitive plants, seedling establishment is restricted because the habitat is occupied by a vigorously expanding mass of established vegetation. Hence, although successful competitors accumulate resources at rates sufficient to sustain abundant seed production, there is little opportunity for the population to expand its immediate frontiers by seedling establishment. This dilemma has evoked in competitive herbs, shrubs and trees several types of evolutionary response. One is evident in the high incidence of vegetative expansion (see later), a form of asexual reproduction which is viable in dense vegetation and allows the maintenance of high competitive ability in the regenerative phase. A second is the production annually of numerous (usually wind-dispersed) seeds which facilitate the colonization of new habitats. This form of reproduction is particularly common among the perennial herbs, shrubs and trees which appear in the early and intermediate stages of vegetation succession in fertile, undisturbed habitats. Seed production in these plants commences relatively early in the life history and appears to be related to the fact that as vegetation development proceeds and resource depletion occurs the environment becomes progressively less hospitable to the competitor. Early and continuous expenditure on wind-dispersed seeds in many competitors may be interpreted, therefore, as an indication that, in common with many ruderals, competitors tend to lead a 'fugitive' (Hutchinson 1951) existence.

Stress-tolerators occur in environments in which resources are severely restricted by absolute shortage or by the activity of the vegetation itself. In contrast to the populations of many competitors, it seems likely that those of the stress-tolerators often remain in continuous occupation of the same habitat for many years. The biomass remains fairly constant and opportunities for regeneration may be exceedingly rare and dependent upon senescence or occasional damage to established members of the populations. In these circumstances, low parental mortality and a conservative but sustained reproductive effort can make a major contribution to the maintenance and expansion of population size. It is not surprising, therefore, that we should find that in stress-tolerant trees, such as *Pinus aristata*, *Pinus sembra* and *Pseudotsuga menziesii* (Currey 1965; Smith 1970; Janzen 1971), the onset of reproduction is delayed and tends to occur only intermittently during a long life history.

## *Regenerative strategies*

No stage in a life history is exempt from the struggle for existence. This does not mean, however, that the offspring of plants experience exactly the same threats to survival as their parents. The small size of many offspring (particularly those originating from seeds) makes them relatively ineffective competitors for resources and also causes them to be vulnerable to the variety of stresses which can arise from close proximity to large established plants of the same or of different species. For this reason we may expect to observe some major differences in strategy between parents and offspring.

Before attempting a classification of the main regenerative strategies in plants, it is necessary to comment upon a broad distinction which has already been observed by many ecologists—that between vegetative reproduction and regeneration by seed. In terms of their capacity to influence the size, dispersion, survival and genetic adaptability of plant populations, there is no doubt that some broad distinctions can be drawn between the processes of vegetative reproduction and regeneration from seed. However, it is not possible to base a functional classification of regenerative strategies upon a simple dichotomy between vegetative reproduction and regeneration by seed. When the full range of plants is examined, it is clear that a great deal of convergent evolution in structure and in physiology has occurred between the two forms of regeneration, with the result that it is often appropriate to include examples from both categories within the same type of regenerative strategy.

Although information concerning mechanisms of regeneration by native plants is extremely fragmentary, certain basic forms of regeneration are now recognizable and these may be provisionally classified into five types. Under the headings which follow the essential features of each regenerative strategy are described and an attempt has been made to comment briefly upon some of the combinations which occur between established and regenerative strategies. The most common associations with the three primary established strategies have been noted in Table 6.1. In Figure 6.2 relationships involving the full range of established strategies have been summarized.

### *1 Vegetative expansion*

Under this heading it is convenient to assemble many of the regenerative mechanisms which involve the expansion and subsequent fragmentation of the vegetative plant through the formation of persistent rhizomes, stolons or suckers. The most consistent feature of vegetative expansion is the low risk of mortality to the offspring. This is achieved through prolonged attachment to the parent plant and mobilization of resources from parent to offspring.

The outstanding feature of the distribution of vegetative expansion in the

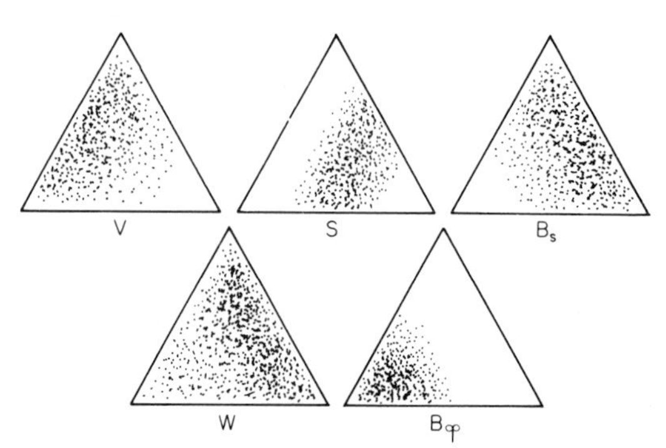


FIG. 6.2. Model describing the range of combinations occurring in flowering plants between regenerative strategies and strategies of the established phase of the life-cycle. V—combinations involving vegetative expansion, S—combinations involving seasonal regeneration, $B_s$—combinations involving a persistent seed-bank, W—combinations involving numerous widely-dispersed seeds or spores, $B_\varphi$—combinations involving persistent seedlings. The distribution of established strategies within the triangular model is the same as that described in Figure 6.1.

triangular model (Fig. 6.2) is its association with established strategies characteristic of relatively undisturbed habitats (competitors, stress-tolerant competitors and stress-tolerators). This pattern may be related to the fact that vegetative expansion involves a period of attachment between parent and offspring and is not viable, therefore, in circumstances in which the vegetation is affected by frequent and severe disturbance.

It seems likely that the role of vegetative expansion may be rather different in competitors and stress-tolerators. In productive, relatively undisturbed habitats, vegetative expansion is often an integral part of the mechanism whereby competitive herbs, shrubs or trees rapidly monopolize the environment and suppress the growth and regeneration of neighbouring plants. Under the very different conditions in severely stressed environments, however, the main advantage of vegetative expansion derives from the capacity to sustain the offspring under conditions in which establishment from an independent propagule is a lengthy and hazardous process.

## 2 *Seasonal regeneration in vegetation gaps*

In a wide range of habitats, herbaceous vegetation is subjected to seasonally predictable damage by phenomena such as temporary drought, flooding, trampling and grazing. Under these conditions the most common regenerative strategy is that in which areas of bare ground or sparse vegetation cover are created every year and are recolonized annually during a particularly favourable season. This form of regeneration reaches its highest frequency in the temperate zone where two main types can be distinguished with respect to the season in which recolonization occurs. In the first establishment takes place in the autumn whereas in the second it is delayed until spring.

Seasonal regeneration is particularly associated with competitive ruderals and stress-tolerant ruderals (see Figs 6.1 and 6.2), both of which are restricted to sites where the intensity of disturbance is sufficient to cause vegetation gaps to occur with high frequency and constancy. It would appear, therefore, that these two combinations arise from the fact that, in each case, the strategy in the regenerative phase and in the established phase is adapted to habitats subject to moderately severe, seasonally predictable disturbance.

Seasonal regeneration involves propagules which lack long-term dormancy, and in habitats subjected to more severe and/or less predictable forms of disturbance we may expect that this attribute will limit the regenerative capacity. It is probably for this reason that the incidence of exclusively seasonal forms of regeneration is low among ruderals *sensu stricto*.

In the British Isles there are a large number of herbaceous plants which fall into the category of competitive ruderals. In many of these plants seasonal regeneration is effected by large seeds which are dispersed from tall inflorescences which are either compound and spreading (e.g. the majority of Umbelliferae) or tall and flexuous (e.g. *Bromus sterilis*). The most likely adaptive value of such inflorescences is that they ensure that, despite their large size, the seeds of these plants become distributed within the habitat to an extent which ensures that a proportion fall onto bare soil or into areas where the density of vegetation and litter is insufficient to prevent establishment.

An essentially similar mechanism of gap exploitation can be recognized in many of the species (e.g. *Tussilago farfara*, *Trifolium repens* and *Ranunculus repens*) in which seasonal regeneration is primarily a vegetative process. In these plants, however, the dispersal of propagules is achieved by the proliferation and fragmentation of rhizomes and stolons. Although superficially similar, this type of seasonal regeneration differs from vegetative expansion in that establishment of the offspring coincides with (or is preceded by) the death or senescence of the parent and there is usually a higher rate of mortality among the offspring.

It seems reasonable to conclude that the strategy of seasonal regeneration represents a rather unsophisticated mechanism of gap exploitation in which, after propagules have been dispersed locally within the habitat and have germinated (or recommenced growth, in the case of fragments of rhizomes or stolons) simultaneously, survival is limited to those individuals which occur in gaps. It is evident that this method of regeneration depends upon the creation in the same habitat and in each year of a high density of suitable gaps. As we shall see later, where gaps occur more rarely or less predictably, seasonal regeneration tends to give way to other types of regenerative strategy.

*3 Regeneration involving a persistent seed bank*

When flowering plants are compared with respect to the fate of their seeds, two contrasting groups may be recognized. In one, most, if not all, of the seeds germinate soon after release, whilst in the other group many become incorporated into a bank of dormant seeds which is detectable in the habitat at all times during the year and may represent an accumulation of many years. These two groups are, of course, extremes and between them there are species and populations in which the seed bank, although present throughout the year, shows pronounced seasonal variation in size. Nevertheless, it is convenient to draw an arbitrary distinction between 'transient' and 'persistent' seed banks. A transient seed bank may be defined as one in which none of the seed output remains in the habitat in a viable condition for more than one year. In the persistent seed bank, some of the component seeds are at least one year old.

Regeneration involving a persistent seed bank commonly occurs in association with each of the established strategies with the exception of the stress-tolerators, where the seed bank is usually replaced by a bank of persistent seedlings (see Fig. 6.2).

Although some of the plants which develop persistent seed banks have mechanisms which facilitate seed dispersal by animals, most seed banks are located in close proximity to the parent plants and may even be situated directly beneath them or, in the extreme example of fire-adapted trees (Munz 1959; Zobel 1969; Vogl 1973), they may actually remain attached to the parent. It seems likely, therefore, that the main functional significance of a seed bank is related to the extent to which it allows regeneration *in situ*. This interpretation is consistent with the fact that seed banks are particularly common in proclimax vegetation such as grasslands, heathland and disturbed marshes, in which the process of vegetation change is cyclical rather than successional. Many forms of proclimax vegetation, especially those in farmland, are subject to alternating patterns of land use, with the result that in each area conditions suitable for particular established plants are available intermittently. In this situation the presence in the soil of a persistent seed bank allows the survival of populations during periods in which the management regime is unfavourable to the established plant.

Persistent seed banks occur in association with a wide range of primary and secondary strategies, and it is clear that in certain respects the role of a seed bank changes according to the established strategy with which it is associated. In ruderals such as the annuals of arable fields and marshland, the seed bank allows rapid recovery from catastrophic mortalities inflicted by cultivation, herbicide treatments or natural phenomena such as flooding, and also permits survival of periods of temporary stability during which the ruderals are excluded by perennial species. In competitive ruderals, stress-tolerant ruderals and C–S–R strategists, the role of the seed bank is similar to that which it exhibits in the ruderal except that here it appears to be mainly concerned with regeneration within gaps in perennial vegetation. In habitats dominated by competitors and stress-tolerant competitors, the intervals between major disturbances may be very long (> 15 years). Competitors and stress-tolerant competitors with persistent seed banks include many of the most familiar herbs and shrubs of vegetation types subject to occasional disturbance by factors such as burning or flooding. After destruction of heathland vegetation by fire, for example, the large seed banks of *Calluna vulgaris* enable this species to recover its dominant status relatively quickly (Whittaker & Gimingham 1962; Hansen 1976) and a similar phenomenon is involved in the rehabilitation of herbaceous species such as *Juncus effusus* and *Urtica dioica* in disturbed marshes and on river banks.

### *4 Regeneration involving numerous widely-dispersed seeds*

Combinations of this type cover a range similar to that involving buried seed banks but they occur also in association with stress-tolerators, where they are strongly represented in lichens, bryophytes, ferns and angiospermous epiphytes. There are major differences in ecology between species and populations which produce buried seed banks and the many grassland plants and trees which regenerate by means of numerous small wind-dispersed seeds. The latter appear to be fugitives adapted to exploit landscapes subject to spatially unpredictable disturbance. Where disturbance occurs as an exceptional event in an environment of moderate to high productivity, the site will remain open to colonization by small propagules for only a brief period. The persistence of the primary colonists during the subsequent process of vegetation succession will depend mainly upon the strategies adopted in the established phase. Although competitive trees, shrubs and herbs will often persist for a considerable period of time, ruderals tend to be rapidly displaced and depend for their survival upon the presence of freshly-disturbed sites within the dispersal range.

### *5 Regeneration involving a bank of persistent seedlings*

Regeneration involving a bank of persistent seedlings, alternatively described as 'advance reproduction' (Marks 1974), appears to be characteristic of plants adapted to circumstances in which the opportunities for seedling establishment occur infrequently and depend upon occasional mortalities in the population of established plants. As noted earlier (p. 130), seed production by stress-tolerators tends to be intermittent and the presence of the 'seedling bank' ensures that, despite this reduced reproductive effort, the capacity for regeneration from seed is maintained between seed crops.

### *Multiple regeneration*

In contrast to the strategies associated with the established phases of plant life histories, the five regenerative strategies are not primarily determined by inflexible 'design constraints' (Stearns 1976). They are not, therefore, mutually exclusive. For this reason it is not uncommon for the same genotype to exhibit two or more regenerative strategies (Table 6.2). Recognition of this has led to the concept of multiple regeneration (Grime 1979) and to the hypothesis that the ecological amplitude of a species may be determined, to some degree, by the number of its regenerative strategies. A preliminary comparison (Fig. 6.3) indicates that in the herbaceous flora of the Sheffield region there is, in fact, a positive association between number of regenerative strategies and the abundance of the species.

TABLE 6.2. The regenerative strategies exhibited by various flowering plants of common occurrence in the British Isles

| | Regenerative strategies | | | | |
|---|---|---|---|---|---|
| Species | Vegetative expansion | Seasonal regeneration | Persistent seeds | Numerous wind-dispersed seeds | Persistent seedlings |
| *Impatiens glandulifera* | | * | | | |
| *Dactylorrhiza fuchsii* | | | | * | |
| *Quercus petraea* | | | | | * |
| *Salix cinerea* | * | | | * | |
| *Ilex aquifolium* | * | | | | * |
| *Tussilago farfara* | | * | | * | |
| *Digitalis purpurea* | | * | * | | |
| *Bromus erectus* | * | * | | | |
| *Poa annua* (ruderal population) | | * | * | | |
| *Poa annua* (pasture population) | * | * | * | | |
| *Agropyron repens* | * | * | * | | |
| *Epilobium hirsutum* | * | * | * | * | |

## CONCLUSIONS

When an attempt is made to resolve the struggle for existence into its main components, it is convenient to use the word competition to describe the activity whereby certain types of plants achieve high rates of resource capture in productive, relatively undisturbed environments. When competition is defined in this way, it is possible to recognize contributions to fitness which are not so

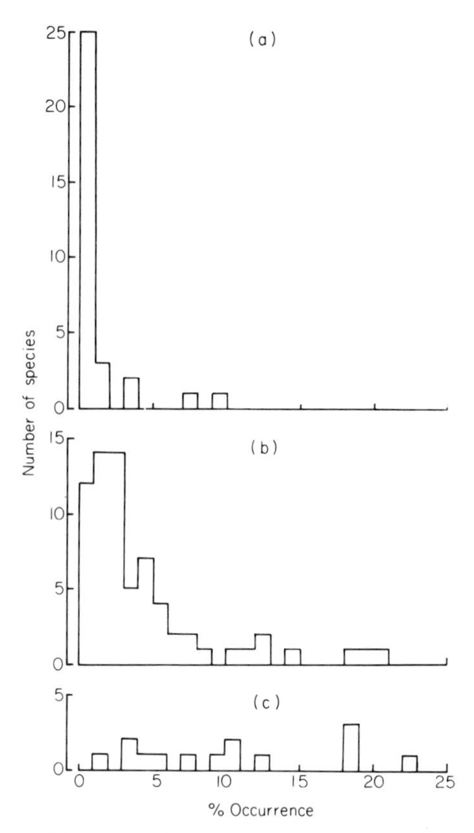


FIG. 6.3. Histograms illustrating the frequency of occurrence of grassland plants in the Sheffield area: (a) species with a single regenerative strategy, (b) species with two regenerative strategies, (c) species with three or more regenerative strategies. Frequency is based upon the percentage occurrence of the species in 2 748 $m^2$ quadrats distributed within the 32 major habitats occurring in an area of 2 400 $km^2$.

immediately and exclusively concerned with competition for resources but are of paramount importance in certain types of habitat. These alternative strategies are apparent in ruderal plants, where survival of frequent and severe disturbance is achieved by rapid completion of the life history and constant re-establishment, and in stress-tolerant plants, which are adapted to withstand conditions in which only low rates of growth and reproductive effort are possible.

A restricted use of the term competition is also helpful in attempts to characterize those parts of the struggle for existence which occur during the regenerative phases of plant life-cycles. Because they are a familiar feature of crop systems and are easily simulated in experiments, competitive interactions between seedlings have been studied extensively. In reality, however, competitions between offspring appear to be secondary in importance to the various mechanisms whereby offspring evade the potentially dominating effects of established plants.

## REFERENCES

**Cole L.C. (1954)** The population consequences of life history phenomena. *Quarterly Review of Biology*, **29**, 103–137.

**Currey D.R. (1965)** An ancient bristlecone pine stand in eastern Nevada. *Ecology*, **46**, 564–566.

**Darwin C. (1859)** *The Origin of Species by means of Natural Selection or the Preservation of Favoured Races in the Struggle for Life*. Murray, London.

**Fenner M.J. (1979)** A comparison of the abilities of colonizers and closed-turf species to establish from seed in artificial swards. *Journal of Ecology*, **66**, 953–964.

**Foulds W. (1978)** Response to soil moisture supply in three leguminous species. 1. Growth, reproduction and mortality. *New Phytologist*, **80**, 535–545.

**Grime J.P. (1974)** Vegetation classification by reference to strategies. *Nature*, **250**, 26–31.

**Grime J.P. (1977)** Evidence for the existence of three primary strategies in plants and its relevance to ecological and evolutionary theory. *American Naturalist*, **111**, 1169–1194.

**Grime J.P. (1979)** *Plant Strategies and Vegetation Processes*. John Wiley, London.

**Hansen K. (1976)** Ecological studies in Danish heath vegetation. *Dansk botanisk Arkiv*, **31**, 1–118.

**Harper J.L. (1977)** *Population Biology of Plants*. Academic Press, London.

**Harper J.L. & Ogden J. (1970)** The reproductive strategy of higher plants. I. The concept of strategy with special reference to *Senecio vulgaris* L. *Journal of Ecology*, **58**, 681–698.

**Hickman J.C. (1975)** Environmental unpredictability and plastic energy allocation strategies in the annual *Polygonum cascadense* (Polygonacene). *Journal of Ecology*, **63**, 689–701.

**Hutchinson G.E. (1951)** Copepodology for the ornithologist. *Ecology*, **32**, 571–577.

**Janzen D.H. (1971)** Seed predation by animals. *Annual Review of Ecology and Systematics*, **2**, 265–292.

**Kingsbury R.W., Radlow A., Mudie P.J., Rutherford J. & Radlow R. (1976)** Salt stress responses in *Lasthenia glabrata*, a winter annual composite endemic to saline soils. *Canadian Journal of Botany*, **54**, 1377–1385.

**Marks P.L. (1974)** The role of pin cherry (*Prunus pensylvanica* L.) in the maintenance of stability in northern hardwood ecosystems. *Ecological Monographs*, **44**, 73–88.

**Milne A. (1961)** Definition of competition among animals. *Mechanisms in Biological Competition* (Ed. by F.L. Milnthorpe) pp. 40–61. Cambridge University Press, London.

**Munz P.A. (1959)** *A California Flora*. University of California Press, Berkeley, California.

**Raynal D.J. & Bazzaz F.A. (1975a)** The contrasting life-cycle strategies of three summer annuals found in abandoned fields in Illinois. *Journal of Ecology*, **63**, 587–596.

**Ricklefs R.W. (1977)** On the evolution of reproductive strategies in birds: reproductive effort. *American Naturalist*, **111**, 453–478.

**Salisbury E.J. (1942)** *The Reproductive Capacity of Plants*. Bell, London.

**Smith C.C. (1970)** The coevolution of pine squirrels (*Tamiasciurus*) and conifers. *Ecological Monographs*, **40**, 349–371.

**Stearns S.C. (1976)** Life-history tactics: a review of the ideas. *Quarterly Review of Biology*, **51**, 3–47.

**van Andel J. & Vera F. (1977)** Reproductive allocation in *Senecio sylvaticus* and *Chamaenerion angustifolium* in relation to mineral nutrition. *Journal of Ecology*, **65**, 747–758.

**Vogl R.J. (1973)** Ecology of Knobcone pine in the Santa Ana Mountains, California. *Ecological Monographs*, **43**, 125–143.

**Whittaker E. & Gimingham C.H. (1962)** The effects of fire on regeneration of *Calluna vulgaris* (L.) Hull from seed. *Journal of Ecology*, **50**, 815–822.

**Williams G.C. (1966)** Natural selection, the costs of reproduction, and a refinement of Lack's principle. *American Naturalist*, **100**, 687–692.

**Zobel D.B. (1969)** Factors affecting the distribution of *Pinus pungens*, an Appalachian endemic. *Ecological Monographs*, **39**, 303–333.

# 7. TROPICAL RAIN FORESTS AND CORAL REEFS AS OPEN NON-EQUILIBRIUM SYSTEMS

JOSEPH H. CONNELL

*Department of Biological Sciences, University of California, Santa Barbara, California*

## INTRODUCTION

The great richness of species in tropical communities is well known. But what is often not appreciated is the considerable variation in species richness from site to site within tropical communities. In this paper I will discuss the reason for this variation within particular community types, directing my attention to tropical rain forests and coral reefs, which have usually been regarded as the epitome of complexity, diversity and stability. I suggest that the highly diverse sites in these communities are usually in a non-equilibrium state, in the process of moving towards an equilibrium of lower diversity. In ecological terms, they are in an intermediate stage of a succession towards a low diversity climax, which, in theory, is an equilibrium state of unchanging species composition.

During this process it is important to distinguish between open and closed systems. Caswell's (1978, p. 128) description applied to ecological systems is a useful summary: 'A closed system is one in which the population exists in a single, closed, roughly homogeneous volume of habitat. There is no migration into or out of this single habitat cell. An open system consists, in its simplest form, of a set of habitat cells coupled by migration. The flux of population between cells must be small enough that the cells retain some measure of independence but large enough that the cells are not totally isolated. Notice that an open population system, considered as a whole, may itself be closed. In this sense the distinction between open and closed systems might be described equally well as between subdivided and non-subdivided systems. . . . The most crucial distinction between closed and open systems is that in open systems local extinction is not an absorbing state. Recolonization of a local population that becomes extinct is now a possibility.'

In actual communities, cells are best defined according to the processes that influence the patterns being explained. Thus, in the population models of Andrewartha & Birch (1954) and later workers such as Caswell (1978), the cells are local populations that go extinct and are started again by immigration from other cells having extant populations. Extended to the community level, local populations of all species may be destroyed on a particular site, with the community being re-established by invasion from the surrounding populations. Here the cell is the empty gap, surrounded by occupied cells. Another definition of a cell uses the viewpoint of a predator that spends most of the time in cells defined by dense clumps of prey, ignoring or passing quickly through cells where prey are sparse. Or if the predator feeds on two species of prey, the predator may feed only in cells of one species, ignoring cells containing the other. In this paper I will use the term cell in all these different ways.

Similarly, the definition of 'equilibrium' depends upon the scale or pattern in which one is interested. In Caswell's (1978) usage, equilibrium applies to events within a cell. For example, if mechanisms exist that keep two competitors co-existing within a cell, the system is in equilibrium; I will use it in this sense in the present paper. However, another acceptable definition is that if two competitors persist indefinitely in an open system of many cells, but in any one cell one species always eliminates the other, the whole system can be regarded as being in 'regional' equilibrium, if the overall species composition does not change. Caswell (1978) has simulated a local open non-equilibrium system of two competitors and one predator and shown that all can persist with reasonable numbers of cells and degrees of migration between them.

I will begin my discussion with cells defined as gaps created by disturbances in forest or coral assemblages. The differences in the adaptations of species that invade earlier vs. later in these gaps were noted in the first studies of succession (Cowles 1899; Cooper 1913). What needs to be emphasized is the fact that once a gap is fully colonized, it is no longer open to further immigration by the 'early' species. This is because some of the adaptations which enable them to invade open sites (e.g. ability to germinate or attach in exposed conditions, high metabolic rates, etc.) make them ill-adapted to invade and survive in the very different conditions existing in occupied sites. Thus, for that particular set of species the system shifts from open to closed to further immigration so that these species then go extinct in that site.

As an example of such changes, consider a forest in which a major disturbance kills all organisms over a large area. The gap represents an empty cell surrounded by occupied cells from which colonists come to invade the empty one. (We assume that the whole area is physically homogeneous and that it remains so throughout the time span under consideration; the role of environmental heterogeneity will be addressed later.) If the area continues to be subjected frequently to major disturbances that create moderate or large-sized gaps (regime I, Table 7.1), the 'early succession' species will continue to invade these openings. The system will be an open non-equilibrium one in which early succession species persist by migrating from cell to cell. In contrast, if there are no more major disturbances for a while (regime II), the smaller gaps that are produced by minor disturbances can no longer be invaded by the early colonists

TABLE 7.1. The sequence during colonization of disturbed sites in a region that is physically homogeneous and without gradual climatic change

| Regimes of different scales of disturbance | General classes of species based upon their ability to invade open vs. occupied sites | | |
|---|---|---|---|
| | A<br>*Early colonizers*<br>unable to invade occupied sites | B<br>*Intermediates*<br>between early and late colonists | C<br>*Late colonizers*<br>unable to invade open, exposed sites |
| I. Disturbances moderate to large and frequent | Abundant, dominate cover | (1) *Open non-equilibrium*<br>Less common, subordinate | Absent or rare and near edges of gap |
| II. Disturbances small and frequent or, if infrequent, at an intermediate time after a large disturbance | (2) *Closed non-equilibrium if disturbances infrequent*<br>Early colonists not replaced die out | (1) *Open non-equilibrium if disturbances frequent*<br>Common, invade in gaps | Common, invade in gaps or shade |
| III. Disturbances very small and infrequent, or a long time after a large disturbance | (absent) | (2) *Closed non-equilibrium*<br>Intermediate species not replaced die out<br>(absent) | (3) *Open non-equilibrium* or *open equilibrium*<br>Co-existence due to compensatory mechanism<br>(4) *Closed equilibrium*<br>Without mechanisms listed in (3) above, single-species dominance |

that require the more open conditions of large gaps. For them the site has become a closed non-equilibrium one and they die out locally (step 2, Table 7.1).

The system will be an open non-equilibrium one for the intermediate and late colonists since small disturbances occur frequently, permitting the intermediate-succession species that require small gaps to persist by invading them. (Note that since cells are defined as gaps their size changes with the size of the disturbance.) If, however, gaps are infrequent (regime III), the shorter-lived species among the intermediates will not survive the longer intervals between gaps. For them the system has also become a closed non-equilibrium one, and they die out locally (step 2, Table 7.1).

With very infrequent small disturbances (regime III) only the most shade-tolerant late-succession species will persist, either in an open non-equilibrium or in an open or closed equilibrium. The open systems (step 3, Table 7.1) depend upon the existence of mechanisms that prevent elimination of species. Without these 'compensatory' mechanisms the species best adapted to the prevailing environmental conditions will replace all others in all the cells (step 4), in a closed equilibrium.

What are these compensatory mechanisms that could maintain co-existence in a physically uniform environment? Frequency-dependent mortality, for example switching by generalized predators (Murdoch 1969; Murdoch & Oaten 1975), will do so. It is interesting that where such switching has been found the different prey species are in separate patches, equivalent to different cells in an open system, rather than together in a single cell (op. cit.). The scheme proposed by Janzen (1970) and Connell (1970) for the maintenance of co-existence of tropical rain forest trees is an example of such frequency-dependent predation, in which attack on seeds or seedlings is proportionately heavier in the vicinity of the parent or in dense clumps of seedlings than elsewhere. Here the cells are quite small, being represented by the area immediately surrounding an adult tree, or by the dense patch of seedlings. As I will discuss later, other forms of mortality besides predation may compensate for the advantage of superior competitive ability.

Another mechanism, termed 'circular networks', was proposed by Jackson & Buss (1975) and Gilpin (1975). Instead of a linear and transitive hierarchy of competing species (species 1 eliminates species 2, 2 eliminates 3, implying 1 eliminates 3), the hierarchy is circular ($1 > 2 > 3$, but 3 eliminates 1 directly). The mechanism by which 3 eliminates 1 must, of course, differ from that used by 1 and 2 for such a system to work. If all three occur together in a cell, one will win, unless their competitive abilities are exactly balanced, which is unlikely. However, if they occur in pairs in different cells of an open locally non-equilibrium system one member of each pair will go extinct in any one cell, but the whole system will remain in regional equilibrium provided that the balance between rates of dispersal and extinction is the same for all species. Although the chances of this are small, the rate of competitive elimination may be much slower with circular networks than with linear hierarchies of competitors.

It now seems clear that all of these compensatory mechanisms require a set of cells connected by migration; in other words, an open system. Neither seems to operate effectively within a single cell, a closed system. Without such mechanisms the species best adapted to colonize and survive in cells occupied by its own or other species will increase at the expense of all the other species in the physically homogeneous habitat that we have assumed exists there. Once it replaces all others, the cells will all be occupied by the same species, and migration of identical propagules between cells becomes irrelevant. At that point the system is equivalent to a closed equilibrium of one large cell, since no other species can invade.

Thus, after a single large disturbance, the system progresses as shown in Table 7.1 from steps 1 to 4, from a single large cell through stages with smaller cells (represented by gaps where individuals have died) with some species having become locally extinct because the system has become a closed non-equilibrium for them, to end either in co-existence of several species in an open non-equilibrium or an open equilibrium, or in single-species dominance in a closed equilibrium. During the complete process, therefore, the scale of pattern changes from a large cell to small cells. If there are no compensatory mechanisms, it may eventually return to a large closed cell.

## TROPICAL RAIN FOREST TREES

How relevant is this theory to real ecosystems? In a recent paper (Connell 1978) I have addressed the problem of co-existence of many species in two tropical communities. For rain forest trees the evidence suggests strongly that if disturbance seldom occurs, succession will proceed until single-species dominance is achieved. The best evidence comes from the work of Eggeling (1947), who classified different parts of the Budongo forest of Uganda into three stages: colonizing, mixed and climax stands. Using observations made many years apart, he showed that the colonizing forest was spreading into neighbouring grassland. In these colonizing stands the canopy was dominated by a few species (class A in Fig. 7.1), but the juveniles (class B in Fig. 7.1) were of entirely different species. Adults of the class B species occurred elsewhere as canopy trees in mixed stands where a few species dominated (mainly ironwood, *Cynometra alexandrei*, which comprised 75 to 90% of the canopy trees). However, in these stands the understorey was composed mainly of juveniles of the canopy species. Thus, an assemblage of self-replacing species (that is, a climax community of low diversity) had been achieved. This is not a special case; the Budongo forest is the largest rain forest in Uganda and one-quarter of it is dominated by ironwood. Later and more extensive surveys showed that the

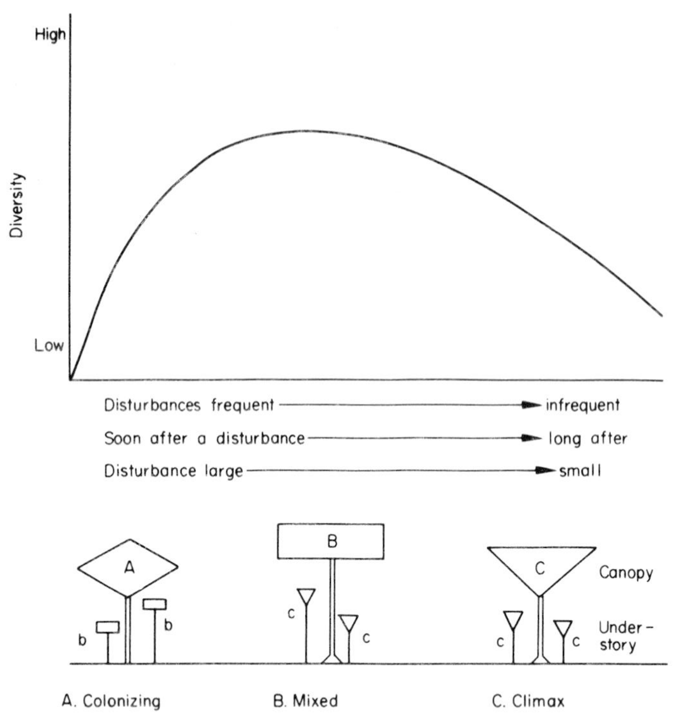


FIG. 7.1. The 'intermediate disturbance' hypothesis. The patterns in species composition of adults and young proposed by Eggeling (1947) for the different successional stages of the Budongo forest are shown diagrammatically at the bottom. (From Connell 1978.)

proportion so dominated in other forests in Uganda is even higher and have confirmed that, where *Cynometra* dominates the canopy, its juveniles also dominate the understorey (Langdale-Brown, Osmaston & Wilson 1964). Extensive stands of rain forests dominated by *Cynometra* or other single species occur in other areas of Africa, South and Central America and South-East Asia (Richards 1952; Whitmore 1975).

Another excellent example is the work of Jones (1956) in Nigeria. In this diverse tropical forest many of the larger trees, aged about 200 years, were dying. They probably became established in the first half of the eighteenth century in fields abandoned when the countryside was depopulated by the collapse of the Benin civilization. These trees had few offspring; most regeneration was by other species, shade-tolerant and of moderate stature. This mixed forest was, in fact, an 'old secondary' forest that had invaded after agriculture had stopped, being in about the same state as Eggeling's mixed forest in Uganda. In both Nigeria and Uganda high diversity was found in a non-equilibrium intermediate stage in the forest succession.

In most studies of highly diverse stands of tropical rain forests, there exists a set of species having many large trees and few offspring (Richards 1952; Whitmore 1975). Assuming that mortality of juveniles is not directly correlated with abundance, which it was not in two Queensland forests over a period of 9 years as shown in Table 7.2, this means that these species will not replace themselves and will be succeeded by other species that have abundant offspring. Thus, these high diversity forests are in a non-equilibrium intermediate stage in succession after a disturbance.

In a small experimental clearing made in a Queensland rain forest, the pattern of colonization changed in 12 years from an initially rather uniform cover of a few species of early colonists to a patchy mosaic of many different species (Webb, Tracey & Williams 1972). This later subdivision was the result of the properties of the local species, particularly the re-sprouting of survivors of the initial clearing and the proximity of adults that served as sources of seed. Thus, even in the earliest stages of succession, the site was being subdivided into smaller 'cells' by biological activity.

In some instances tropical forests appear to have reached a closed equilibrium state of low diversity. In the climax stands in Budongo with 75 to 90% ironwood in the canopy, there was a great abundance of offspring of ironwood in the understorey. This situation is characteristic of almost all 'single-dominant' forests studied (Richards 1952). However, it is also possible that some mixed high diversity forests might exist as open non-equilibrium systems (step 1, Table 7.1) because small disturbances occur frequently, or as open non-equilibrium or equilibrium systems (step 3) due to the operation of the compensatory mechanisms described above.

I will first discuss the compensatory mechanisms. For example, if the mortality of seeds or seedlings were greater in the vicinity of adults of the same

TABLE 7.2. Mortality of young trees ($\geq 0{\cdot}2$ m $< 6{\cdot}1$ m tall) in relation to their abundance for two rain forests in Queensland. Not all species had enough young trees to analyse; only those whose adults are capable of reaching the canopy and that had at least six young trees are included. The mortality rate between 1965 and 1974 was plotted against the original numbers mapped in 1965; the least-squares regression slope and correlation coefficient are shown. Neither correlation coefficient is significantly different from zero at $p < 0.05$. (From Connell 1978.)

| | Number of species | Regression of % mortality on abundance | |
|---|---|---|---|
| | | Slope | $r$ |
| Tropical, North Queensland, 17°S lat. | 49 | 0·039 | 0·217 |
| Subtropical, South Queensland, 28°S lat. | 46 | 0·002 | 0 |

TABLE 7.3. Mortality of seeds or seedlings near vs. far from adult trees of the same species; all known published field experiments or observations in tropical forests are listed

| Location, vegetation type | Plant species | Experimental treatment | % mortality (unless otherwise designated) | | Authority |
|---|---|---|---|---|---|
| | | | Near adult | Far from adult | |
| Queensland, 17°S lat. Evergreen rain forest | *Cryptocarya corrugata* | For each treatment 450 seeds laid on ground in nine half square m plots, 50 seeds/plot. | | | Connell 1970 |
| | | % mortality after 9·5 months: | 99·8 | 99·5 | |
| Same, 28°S lat. | *Eugenia brachyandra* | For each treatment 600 seeds laid on ground in twelve half square m plots, 50 seeds/plot. | | | This paper |
| | | % germinations in first/second year: | 14·8/2·5 | 14·0/2·3 | |
| | | Average length of life in months in three cohorts of seedlings during the first year after germination (number at start): | 4·7 (37)<br>5·4 (25)<br>5·7 (27) | 4·9 (38)<br>4·0 (22)<br>3·9 (24) | |
| Same, 17°S lat. | *Planchonella* sp. nov. | In each treatment (near vs. far) four plots of 2 square m, each having 98 seeds. % mortality of seedlings over 3 years | | | Connell 1970 |
| | | Trenched plots: | 66, 70 | 14, 27 | |
| | | Untrenched plots: | 51, 55 | 26, 43 | |

| | | | | | |
|---|---|---|---|---|---|
| Puerto Rico Evergreen rain forest | *Euterpe globosa* | % non-viable seeds found beneath crown vs. 1·5 to 2·5 m farther away. N for 70/71: near 1 765/1 804; far 337/111 — 1970: / 1971: | 95 to 100<br>0 to 11 | 83 to 100<br>0 to 20 | Janzen 1972a |
| Costa Rica, Guanacaste Deciduous forest | *Scheelea rostrata* | % mortality of cleaned palm nuts in piles of 50 in cages. N = near 690, far 681 | 33·8 | 35·7 | Wilson & Janzen 1972 |
| | | | Near (≤ 12 m)<br>Line 3 — Line 1 | Far (13 to 24 m)<br>Line 3 — Line 1 | |
| Same | *Sterculia apetala* | Numbers of herbivorous bugs appearing on seeds placed one per m on lines radiating from parent tree. Number/seed after the following intervals: | | | Janzen 1972b |
| | | 6 (line 3) and 13 (line 1) min: | 2·6 — 6·7 | 0·3 — 0·9 | |
| | | 43 (line 3) and 63 (line 1) min: | 5·2 — 17·2 | 1·8 — 3·7 | |
| | | 213 (line 3) and 228 (line 1) min: | 5·9 — 10·7 | 10·6 — 9·0 | |
| Same | *Spondias mombin* | % mortality of seeds found on ground at base of tree vs. under edge of crown of 12 trees (N = 1 200 per treatment) | 50 | 45 | Janzen 1975b |
| Same | *Dioclea megacarpa* (vine) | Observations of % of 22 shoot tips ≤10 m eaten vs. 24 tips > 10 m from adults | 86·4 | 16·7 | Janzen 1971 |

species than elsewhere, or in dense clamps than in isolation, this frequency-dependent mortality would constitute an open non-equilibrium system promoting co-existence of many species (Janzen 1970; Connell 1970). However, the evidence concerning this mechanism is equivocal. As shown in Table 7.3, a series of experiments and observations have indicated that mortality of tropical forest seeds and seedlings can be either greater, equal or lower near adults of their own species than elsewhere. Table 7.4 indicates also that increased density may or may not increase mortality. The seed experiments in Tables 7.3 and 7.4 were necessarily all done on species with large seeds, often the shade-tolerant 'climax' species. Comparisons of all species in two forests (Table 7.2) showed no correlation of mortality with abundance. However, an analysis I had made (Connell 1970, Table 7.3) showed that tree seedlings and saplings occurring in single-species clumps (i.e. those having nearest neighbours of the same species) suffered greater mortality than those occurring intermingled with other species.

Since in the same forest some species show frequency-dependent compensatory mortality and some do not, this cannot produce an equilibrium state. Species in which the mortality of seedlings is less or equal nearer than farther from the parent will have an advantage over others with the same dispersal

TABLE 7.4. Effect of density on mortality of seeds; the *Cryptocarya* and *Scheelea* seeds were placed far from adults of the same species. All published observations and experiments known from tropical forests are included

| Location, vegetation type | Plant species | Experimental treatment | % mortality | | Authority |
|---|---|---|---|---|---|
| | | | Dense | Sparse | |
| Queensland, Australia Evergreen rain forest 17°S lat. | *Cryptocarya corrugata* | 50 seeds in each of nine ½-m plots vs. a seed every meter along a line 104 m long. Mortality after 9·5 months | 100 | 99 | Connell 1970 |
| Costa Rica, Guanacaste Deciduous forest | *Acacia farnesiana* | % mortality of seeds on ground in dense stand of trees vs. in stand with trees sparse, intermixed with other species. N = 36 trees from dense stand, 11 from sparse stand, 30 to 100 pods per tree examined, 8 to 10 seeds per pod | 79·7 | 79·6 | Janzen 1975b |
| Same | *Scheelea rostrata* | Piles of 50 (in cages) vs. isolated pairs of cleaned nuts placed above litter. N: dense, 681; sparse, 82 | 35·7 | 6·1 | Wilson & Janzen 1972 |

capacity but in which the mortality of seedlings is greater near the parent. In summary, the evidence indicates that this form of compensatory mechanism is apparently not effective in maintaining high diversity in the tropical forests studied.

The original hypothesis (Janzen 1970; Connell 1970) invoked herbivores as the agents most likely to be responsible for higher mortality of seedlings in denser clumps or nearer adults of the same species. Since herbivorous bugs did discover seeds more quickly if they were placed nearer the adults than farther away (Table 7.3, *Sterculia* during the first hour only), and since shoot tips of seedlings were much more heavily eaten closer to adults (Table 7.3, *Dioclea*), this seems a reasonable mechanism. Yet, except for leaf-cutter ants, herbivores are seldom commonly observed in tropical rain forests (Elton 1973). However, this does not signify that their effects are slight, since damage may be difficult to detect (especially that caused by small animals and micro-organisms, or by sucking insects above ground) and may occur sporadically in time and space. Nevertheless, the results of the experiments and observations shown in Tables 7.3 and 7.4 indicate that herbivores are not generally effective in the manner proposed in the original hypothesis. Only one species (*Planchonella*) of the six in which mortality was estimated (Table 7.3) had greater mortality nearer adults than farther away. Also, only one of the three species in Table 7.4 had greater mortality at higher densities.

There are several possible reasons why herbivores are not effective in this way. First, it requires a high degree of specialization. Herbivores must choose to attack seeds or seedlings of a certain species proportionately more strongly near adults of the same species, ignoring all other juveniles. Although it has been shown that herbivorous insects in the tropics are more specialized on particular families of plants (Scriber 1973), species-specificity of tropical herbivores has rarely been unequivocally demonstrated. To do so would require that large samples of all possible species of host plants in a local area be examined in detail. The less complete the survey, the more specialized the herbivores will appear to be. While exhaustive surveys have been done in the temperate zone, notably by the Canadian Forest Insect Survey (Watt 1964), I know of no published evidence from tropical forests that even approaches this degree of completeness. Thus, tropical data will tend to indicate greater specialization due to inadequacy of sampling. The only exception may be that of Janzen (1975a), although no data have yet been published to indicate whether the sampling of the plant species was adequate. He states (op. cit., p. 42) 'of 111 species of bruchid beetles collected to date breeding in seeds of Costa Rican deciduous forest tree seeds, each of 102 of them seems to have only one host plant and each is different from that of the others.' He goes on to point out that general surveys of herbivore diets may be misleading because, although a species may have different hosts in different regions, it still may be species-specific in any one locality. In summary, until some evidence is published, I must conclude that tropical herbivores have not been shown to be sufficiently specialized to produce the degree of compensatory mortality needed for this mechanism to apply generally.

Secondly, deleterious effects of herbivores on plants are undoubtedly reduced by the attacks of their predators, parasitoids, parasites and pathogens, and by the chemical and mechanical defences of the plants. As indicated in Figure 7.2, there are several links in the food web of animals in forests, such that the effects of natural enemies may either increase or decrease the deleterious effects of herbivores. As Lawton & McNeill (1979) have pointed out, natural enemies probably cannot control the abundance of herbivorous insects if the latter have very high rates of natural increase. They suggest, however, that if these rates are reduced by the defences of the plants themselves, the natural enemies will have a much greater chance of controlling the numbers of herbivores. It seems that the combined forces of plant defences plus natural enemies have been enough to reduce the effects of herbivores on many of the young trees in the tropical forests studied. Thus, herbivores have apparently not been effective in maintaining species diversity of tropical forest trees in the manner suggested by Janzen (1970) and Connell (1970).

Circular networks are less likely in organisms as similar as trees in a forest stand. As described above, a very different competitive mechanism must operate for at least one of the links in a circular network. Consequently, the

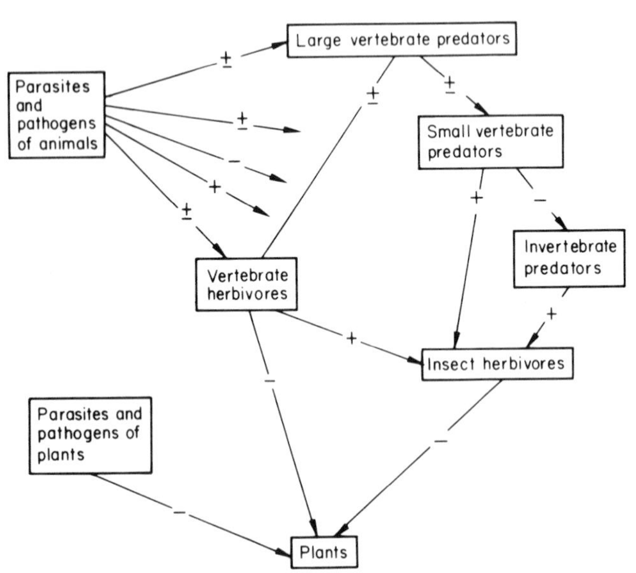


FIG. 7.2. Effects of other organisms on plants in a forest. The sign (+/−) on each arrow indicates the eventual effect of that interaction on plants. Thus, a predator that ate an insect herbivore would have a positive effect on the plant.

likelihood of a sufficiently different mechanism would be increased if the proponents were themselves very different (Jackson & Buss 1975). In their original examples, the organisms exhibiting circular networks were sessile marine invertebrates in different phyla. Since plants compete for light, soil nutrients and water and may interfere with each other using allelopathic chemicals, this mechanism remains a possibility.

Thus, compensatory mechanisms have not yet been found which could be effective in maintaining an equilibrium system in the absence of disturbance as shown by step 3, Table 7.1. Unless some such mechanisms do operate, diversity should decrease as shown in step 4, unless small disturbances occur frequently enough (step 1). Then the forest as a whole will be an open, locally non-equilibrium system of empty and filled cells which might achieve a regional equilibrium ensuring persistence of certain less shade-tolerant species among the more tolerant ones (Grubb 1977).

In summary, the highest diversity of tropical rain forest trees should occur either at an intermediate stage in succession after a large disturbance or with smaller disturbances that are neither very frequent nor very infrequent; either represents an open non-equilibrium. I have called this the 'intermediate disturbance' hypothesis (Connell 1978) and it is illustrated in Figure 7.1. With frequent large disturbances or soon after a single large disturbance, diversity will be low because only a few species have evolved the adaptations necessary for rapid colonization of such exposed, open sites. At an intermediate interval after a large disturbance, diversity should be high because many more species have had a chance to become established before the disappearance of the first invaders. If disturbances are frequent but small, diversity should also be high because species that are either moderately or very shade-tolerant (B and C types, Table 7.1) will co-exist. Eventually, if disturbances are both very small and infrequent, diversity will be lower, as the species which cannot become established in heavy shade or small light gaps will go extinct locally. The level of diversity will then depend upon the degree to which the compensatory mechanisms described above act (steps 3 vs. 4, Table 7.1).

Since the evidence so far indicates that such compensatory mechanisms are not generally effective in tropical forests, the high diversity often observed is more likely to be a consequence of the forest having an intermediate regime of disturbances, as indicated by II in Table 7.1. This represents an open, locally non-equilibrium system that may or may not be in regional equilibrium. If disturbances are frequent enough so that empty cells are formed at a sufficiently high rate to match the survival and seed dispersal rates of the less shade-tolerant intermediate species (class B, Table 7.1), then the forest is in regional equilibrium. Yet in most high diversity mixed rain forests there exists a set of species of large canopy trees which have few or no offspring within the forest, though they occur in permanently open sites along nearby roads (Richards 1952; Whitmore 1975). This indicates that they probably require very large openings to become established, so are members of class A in Table 7.1. These long-lived species apparently exist in an open, locally non-equilibrium system which requires occasional large disturbances to maintain their presence in regional equilibrium.

## REEF-BUILDING CORALS

Corals share many of the characteristics of terrestrial woody vegetation: most of their energy comes from photosynthesis of contained zooxanthellae, they are long-lived and have the same reiterated modular body pattern (Harper 1978). In form corals resemble bushes rather than trees. As in shrub stands, there is little understorey beneath corals. They spread laterally, interacting with neighbours at the borders and often becoming detached into separate portions as do shrub clones.

The relationship between disturbance and species richness is similar in corals to that in tropical forests. In my studies at Heron Island, Queensland, I found the greatest species densities in places either with fairly frequent disturbance by storms or near the surface on reef crests where desiccation at low tide occasionally caused local mortality. In the 15 years I have been studying this reef, three hurricanes have struck it, each causing different degrees of damage to the same permanently marked quadrats. One small region, subtidal and protected from storm waves by an adjacent reef, has escaped all damage and, though almost completely covered with living coral, has low species diversity.

In following the population dynamics of corals on particular sites (by photographing them at intervals), I have found (Connell 1973, 1976, 1978) that where storm damage is frequent they remain as open non-equilibrium systems (Fig. 7.3A), corresponding to regime I in Table 7.1. However, in the region protected from storm disturbance, the coral cover was mainly 'staghorn thickets', consisting of a few huge colonies of branching *Acropora* species that have overgrown all smaller neighbours. Judging from their size (up to 6 m in diameter), they are probably at least a century old (after the initial years of life most corals have rather constant growth rates, so age can be estimated with fair confidence; Connell 1973, Table III). Thus, they had not been seriously disturbed for at least a century, and possibly much longer. Since this area has also served for many years as a protected anchorage for small boats, there is a certain amount of small-scale damage from anchors, with the result that colonies of other species occur, having recently colonized these gaps. This region is thus an open non-equilibrium system, due to the continual provision of small gaps which may be invaded by corals of lesser competitive ability than staghorns (regime II, Table 7.1).

A few species of corals seems to fit into the class A of Table 7.1, 'early colonizers'. These are *Pocillopora damicornis*, *Porites* spp., *Cyphastrea microphthalma*, *Leptastrea purpurea* and *Acropora palifera*. The first three were the

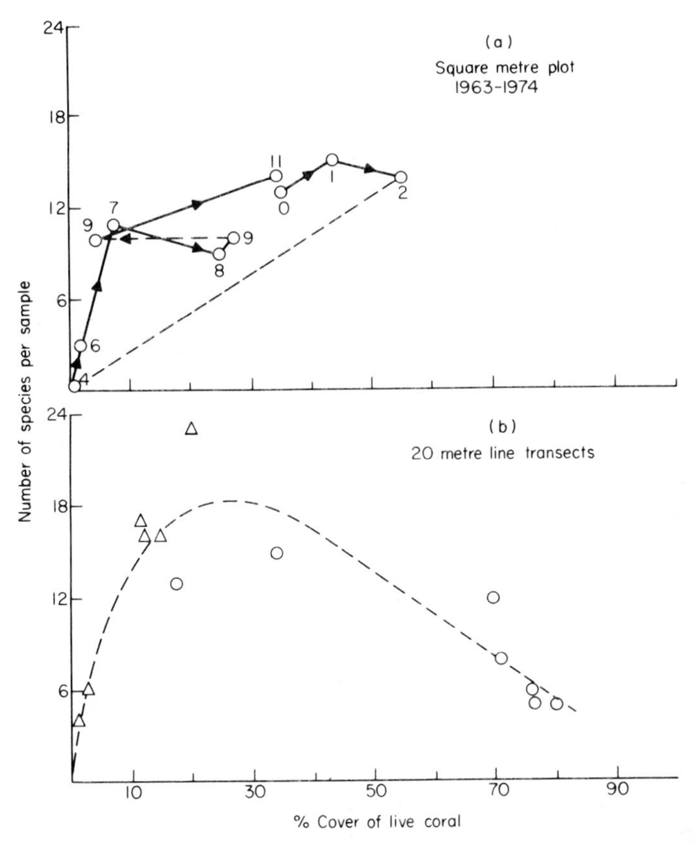


FIG. 7.3. Species diversity of corals in the subtidal outer reef slopes at Heron Island, Queensland. (a) Changes over 11 years on one of the permanently marked plots on the north slope. The number at each point gives the years since the first census at year 0 (no censuses were made in years 3, 5 and 10). The dashed lines indicate changes caused by hurricanes in 1967 and 1972. (b) Results from line transects done 3 to 4 months after the 1972 hurricane. △, data from the heavily damaged north slopes; ○, data from the undamaged south slope; the line was drawn by eye. Where disturbances had either great or little effect (very low or high per cent cover respectively) there were few species, with maximum numbers of species at intermediate levels of disturbance.
(From Connell 1978.)

only identifiable corals that colonized artificial surfaces over 11 months at Low Isles, Queensland (Stephenson & Stephenson 1933). On concrete surfaces on the reef flat at Heron Island, I found only the species of *Pocillopora* and *Acropora* listed, even after four years' exposure. In the undisturbed site of low diversity, only one of these species occurred on four line transects of 20 m each that were dominated by staghorn corals; it consisted of a single, very large, old colony of *Porites* sp. that towered above the staghorns around it. In the gaps other 'intermediate' species of lesser competitive ability than staghorns had colonized.

From this evidence I suggest that these undisturbed areas covered by a few species of competitively dominant staghorn corals are closed to invasion by these 'early colonizing' species of Table 7.1 but open to 'intermediate-stage' species. These sites have reached the stage of regime II of Table 7.1 and are now in an open non-equilibrium state, due to the occasional small-scale disturbances from small boat anchors. Their diversity is generally low (Fig. 7.3b) because the disturbances are few and small, and because staghorns are well adapted to colonizing small openings because fallen branches very quickly become attached to the substratum (observations by myself and R. Day).

Do compensatory mechanisms operate to maintain diversity of corals in the open non-equilibrium or equilibrium of step 3 in Table 7.1 ? In regard to circular networks, I found none in a study of interactions between corals (12 species, 55 colonies, 82 interactions observed over 9 years; Connell 1976). As in trees, corals may be too similar for such circular networks to occur commonly. Frequency-dependent mortality remains a possibility. Although predation by the starfish *Acanthaster planci* is often intense, evidence indicates that it acts in a reverse frequency-dependent way, attacking preferentially the rare species (Branham *et al.* 1971; Glynn 1974, 1976).

However, the physical environment can act in a way that compensates for the competitive advantage enjoyed by some species such as staghorn corals. I measured the mortality of corals over a 4-year period that included a hurricane at Heron Island in Queensland. As described above, I had ranked these species in competitive ability by observing dynamic interactions over a period of 9 years on permanent quadrats. On the part of the reef crest that was badly damaged by the hurricane, the mortality of those species of corals that ranked high in the competitive hierarchy was much greater than those ranked low (Table 7.5). In contrast, the high-ranked species on an undamaged part of the reef crest had a lower mortality than low-ranked species over the same period. The reason for the difference was that the long-branched form of the superior competitors, the staghorns, made them more susceptible than inferior competitors to being broken off in storms. I have also found (Connell 1973) that branching species are more susceptible to being attacked by boring molluscs and sponges, increasing their vulnerability to damage in storms. Thus, species of corals that ordinarily win in competition suffer proportionately more from storm damage, compensating for their advantage.

Another instance in which the physical environment acted in a compensatory way to promote co-existence was also observed on the reef crests at Heron Island. As the larger branching colonies grow, they spread horizontally, eliminating their neighbours. However, they also grow upwards, and since

TABLE 7.5. Changes in coverage of corals of different competitive abilities. In each area two square metres were censused. (From Connell 1976.)

| | Species whose competitive abilities are High (branching *Acropora* spp.) | | Low (encrusting spp., massive spp., *Pocillopora, Stylophora*) | |
|---|---|---|---|---|
| Degree of vulnerability to storm damage | Cover ($cm^2$) in 1965 | % change in cover, 1965–69 | Cover ($cm^2$) in 1965 | % change in cover, 1965–69 |
| High (north crest) | 10 655 | −39·1 | 3 071 | +18·6 |
| Low (south crest) | 7 027 | +47·3 | 1 900 | −12·5 |

these crests are very shallow, the highest parts of the colony, usually near the middle, are killed by desiccation in the air. These dead portions are then eroded down by boring organisms, so that they become suitable for colonization by new colonies. Thus, the fast-growing superior competitors, although they eliminate other species, cannot monopolize the space on these shallow crests. Consequently, diversity remains high there even though much of the space is occupied (see Plate 2).

In summary, on the reef at Heron Island I found sites with each of the disturbance regimes of Table 7.1. Hurricanes have been frequent and each of the three that occurred during the 15 years of my study caused different patterns of damage. In areas unprotected from hurricane damage, the system resembles an open non-equilibrium system maintained by large disturbances (regime I). In contrast, in a site that had been protected from hurricanes for a long time by an adjacent reef, most of the area was covered by large colonies of 'staghorn' corals, with small openings presumably caused by boat anchors. This site was, therefore, in regime II, most early colonists having been eliminated (step 2) and the intermediates persisting by invasion of the small openings in an open non-equilibrium. Thirdly, on certain portions of the reef crest, the fast-growing superior competitors suffer more mortality than the inferior ones, due to the desiccation or breakage described above. Since the disturbances are minor (regime III) and the mortality compensates for the competitive advantage of some species even when the colonies are closely intermingled (see Plate 2), this is the open equilibrium described in step 3. Finally, one would predict that in the subtidal where there is no desiccation stress and without small disturbances, the system would not be in open equilibrium. Since there apparently were no circular networks or frequency-dependent mortality, the species of staghorn most effective in competition should then eliminate all other species and the system would go to step 4, a closed equilibrium.

## THE ROLE OF ENVIRONMENTAL HETEROGENEITY

Up to this point we have assumed that the physical environment was homogeneous in space and time. If so, diversity could be high only under a regime of intermediate-scale disturbances, unless some compensatory mechanisms were operating as described. Even without these factors, if the environment were heterogeneous within a local site or if the climate were gradually changing, species adapted to these differences would co-exist (Connell 1978).

The problem is to estimate the relative contributions towards the maintenance of diversity of environmental heterogeneity vs. disturbance and compensatory mortality. Environmental heterogeneity obviously exists and, with enough imagination, can be used to account for all diversity. No matter how diverse the assemblage, co-existence at equilibrium of many species requires only that each be sufficiently specialized to slight differences in the environment. For example, adaptations of tropical trees to slight differences in soil topography were invoked by Cain (1969) to explain the co-existence of many species of trees in tropical rain forests.

However, in my opinion, it is impossible to test such a hypothesis. First, it would be extremely difficult to decide with any certainty whether the species were, in fact, specialized to the required degree. Secondly, it would be impossible to be sure that one had chosen the relevant 'niche axes' to study. For example, even if it were demonstrated that the heterogeneity of the soil was no greater in a high diversity site than in one dominated by a single species (as in Eggeling's (1947) studies in Uganda), the objection could still be made that perhaps the soil characteristics that were measured were not the ones to which the species were adapting.

Specificity to particular habitats in tropical trees and corals exists but is extremely unlikely to be narrow enough to permit 100 species of trees to co-

---

PLATE 2 (facing). Interactions between corals in the same area over 8 years on the north crest, Heron Island, Queensland. For each photograph the square metre frame, divided into 20 × 20 cm squares, was positioned on the permanent stakes visible on the right-hand side; it was removed between times.

Notice the expansion of the platform coral that was centred in C–2 in 1963. There was a 'stand-off' between it and its neighbour of another species in C–1 for the entire 8-year period. In contrast, over the first 6 years it expanded rapidly into squares A–2, B–2 and B–3, covering colonies of both soft and hard corals. There was also a stand-off for 2 years with its neighbour in C–3, then after some damage from a hurricane in January 1967 a rapid expansion over that neighbour into C–4. In A–3 it was stopped by the massive coral in 1969 and was dying back in 1971, owing to the fact that the massive coral had digested off the tissues of the growing tips, so preventing overgrowth. Lastly, notice the death of the central oldest portion in C–2 in 1969 and partial regrowth in 1971. In 1972 another hurricane killed all the corals on and surrounding this site, except for a small portion of the massive one in A–3.

(The photographs were taken in October 1963 and in August of the other three years.)

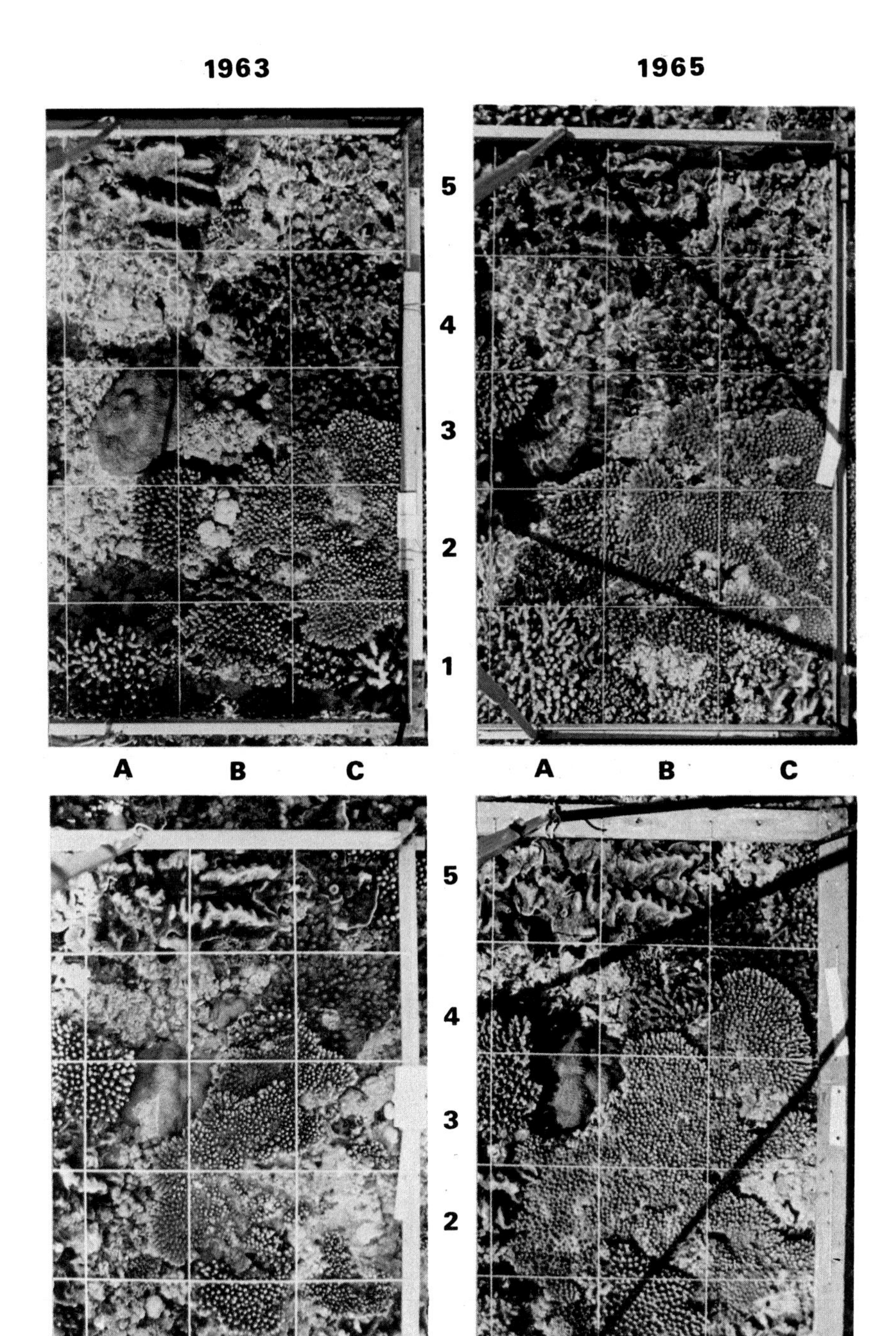
1963
1965
5
4
3
2
1
A
B
C
A
B
C
5
4
3
2
1
1969
1971

exist at equilibrium on a single hectare, or 10 species of corals on a single square metre. Whereas animals may specialize in feeding on certain prey (e.g. herbivorous insects on host plants, etc.), plants and sessile marine invertebrates have less opportunity for such fine specializations. Besides the well-known adaptations to different light intensities, associations between tropical trees and broad soil variations (parent material, drainage, etc.) or topography (tip-up mounds at the roots of fallen trees, ridges vs. slopes and stream courses, etc.) have been found (Richards 1952; Whitmore 1975; Williams *et al.* 1969; Austin, Ashton & Grieg-Smith 1972). Likewise, coral species show associations with broad zones such as reef flats, crests and with depth on outer slopes. However, none of these specializations seems sufficiently narrow to account for the high diversity shown. In addition, the individuals of different species are often intermingled in a complex pattern; if each species were highly specialized to habitat variation, this would require the environment to vary in a complex mosaic on a very small scale.

## DISCUSSION: DIVERSITY AND EQUILIBRIUM

Environmental heterogeneity within a site will permit co-existence of different species, but probably not to the degree seen in tropical rain forests and coral reefs. With no disturbances beyond the death of individual trees and of coral colonies, further diversity requires that the system be open, in the sense of being subdivided into cells connected by migration of propagules, predators, etc. Then mechanisms such as circular networks or frequency-dependent mortality caused by physical damage or by natural enemies, either specialists or generalists that 'switch', may maintain co-existence of more species in either an equilibrium or non-equilibrium state (Table 7.1, step 3).

In the corals, circular networks were looked for but not found; competitive interactions were hierarchical, not circular (Connell 1976). Frequency-dependent mortality caused by natural enemies apparently does not occur; the starfish *Acanthaster planci*, one of the most effective predators on corals, attacked the rare species preferentially in the two studies where the relevant data were collected (Branham *et al.* 1971; Glynn 1974). In certain habitats (such as the reef crest at Heron Island), physical extremes caused proportionately more mortality in species that had been demonstrated to be superior competitors. Such mortality, though not necessarily frequency-dependent, will certainly promote co-existence in an open equilibrium.

However, in the tropical trees where frequency-dependent mortality has been looked for, it has been demonstrated for only a few species. All the experiments done and observations made in tropical forests pertinent to this subject are given in Tables 7.3 and 7.4. They involved species with large seeds, a trait of shade-tolerant, late-colonizing 'climax' species. The data show that within the set of possible climax species most do not suffer frequency-dependent mortality and, as in corals, circular networks seems unlikely.

Thus, 'compensatory' mortality (Connell 1978) does not seem to be important in the tropical rain forests studied, and only in particular circumstances (reef crests) on coral reefs. One would then predict that, within a particular habitat, the one most effective competitor should eliminate all others as in step 4 of Table 7.1. Therefore, the principal way that high local diversity will be maintained is by a regime of sporadic disturbances that cause gaps into which different species invade, as shown by the regimes I and II of Table 7.1. This process can occur in several ways. The first, which I have called the 'Equal Chance' hypothesis (Connell 1978), would apply if all the species were equal in their abilities to disperse to a gap, become established and hold the space against further invasion. Aubreville's (1938) 'Mosaic Theory of Regeneration' was essentially this hypothesis, applied to a rain forest in Ivory Coast. The problem with it is that species need to be identical in their rates of increase and abilities to compete for sites. It seems unlikely that different species will be so alike or so precisely balanced for this hypothesis to apply.

The second way is that species may be adapted to occupy gaps of different sizes or at different times after they are formed. Grubb (1977) has called this sort of adaptation the 'regeneration niche'. Classes A, B and C in Table 7.1 represent three points in a continuum of sizes and/or times. If moderate to large-size gaps are formed regularly, this could remain an open non-equilibrium system of early- and intermediate-colonizing species (regime I). If, on the other hand, there is a long period without further disturbances, the succession of species represents an open non-equilibrium system, gradually becoming closed to earlier colonists as succession proceeds (step 2, Table 7.1).

Third, co-existence of competitors may be enhanced if disturbances happen frequently, even if the species are not equal so that some are always eliminated in competition within a cell, and even if there is no heterogeneity in the cells. In Caswell's (1978) model, co-existence of competitors in an open non-equilibrium system was prolonged if both the number of cells and the degree of 'connectedness' were increased. In forests or reefs, the number of empty cells increases if the disturbances cause only local damage (small gaps) and occur frequently. Connectedness increases with the degree of migration of propagules (seeds or larvae) between gaps. Co-existence in Caswell's (1978) model was prolonged when the migration rate of the inferior competitor was increased and when the rate of formation of new empty cells was raised by faster dispersal of a predator between cells. However, in forests or reefs, the dominant life forms (trees or corals) are immune to very small disturbances, in contrast to the more vulnerable annual herbs or algae. This puts a limit on the number of new empty cells that can be produced in these communities.

In summary, the species richness will be greatest if the disturbance regime is intermediate in both frequency and size of gaps formed, or if the site is near the midpoint in a succession. Such gaps are certainly common in the tropics, produced by lightning strikes, wind storms, landslips, attacks by boring organisms in corals (Connell 1973), predators, etc. Shifting cultivation by low-density human populations is also an intermediate disturbance of significance in tropical rain forests. Since tropical rain forest trees and reef-building corals often occur at high diversity, and since compensatory mechanisms do not seem to play a significant role in the stands studied, it seems clear that regimes of 'intermediate disturbances' probably account for much of their diversity.

## ACKNOWLEDGMENTS

This paper stems from work and discussions with many people over the past 15 years. Without the assistance of Geoff Tracey and Len Webb, the rain forest work could never have been done; many other people have helped in the field and laboratory. This paper benefited from criticisms by Margaret Connell, Rob Day, Sally Holbrook, Chris Onuf, Pete Peterson and Peggy Reith.

## REFERENCES

**Andrewartha H.G. & Birch L.C. (1954)** *The Distribution and Abundance of Animals.* University of Chicago Press, Chicago.

**Aubreville A. (1938)** La foret coloniale: les forets de l'Afrique occidentale francaise. *Annales Académie des Sciences Coloniales, Paris*, **9**, 1–245.

**Austin M.P., Ashton P.S. & Greig-Smith P. (1972)** The application of quantitative methods to vegetation survey. III. A reexamination of rain forest data from Brunei. *Journal of Ecology*, **60**, 305–324.

**Branham J.M., Reed S.A., Bailey J.H. & Caperon J. (1971)** Coral eating sea stars *Acanthaster planci* in Hawaii. *Science*, **172**, 1155–1157.

**Cain A.J. (1969)** Speciation in tropical environments: summing up. *Biological Journal of the Linnean Society*, **1**, 233–236.

**Caswell H. (1978)** Predator-mediated coexistence: a nonequilibrium model. *American Naturalist*, **112**, 127–154.

**Connell J.H. (1970)** On the role of natural enemies in preventing competitive exclusion in some marine animals and in rain forest trees. *Dynamics of Populations* (Ed. by P.J. den Boer & G.R. Gradwell) pp. 298–312. PUDOC, Wageningen.

**Connell J.H. (1973)** Population ecology of reef-building corals. *Biology and Geology of Coral Reefs*, Vol. 2, Biol. 1 (Ed. by R.E. Endean & O.A. Jones). Academic Press, New York.

**Connell J.H. (1976)** Competitive interactions and the species diversity of corals. *Coelenterate Ecology and Behavior* (Ed. by G.O. Mackie) pp. 51–58. Plenum, New York.

**Connell J.H. (1978)** Diversity in tropical rain forests and coral reefs. *Science*, **199**, 1302–1310.

**Cooper W.S. (1913)** The climax forest of Isle Royale, Lake Superior, and its development. *Botanical Gazette*, **55**, 1–44, 115–140, 189–235.

**Cowles H.C. (1899)** The ecological relations of the vegetation on the sand dunes of Lake Michigan. *Botanical Gazette*, **27**, 97–117, 167–202, 281–308, 361–391.

**Eggeling W.J. (1947)** Observations on the ecology of the Budongo Rain Forest, Uganda. *Journal of Ecology*, **34**, 20–87.

**Elton C.S. (1973)** The structure of invertebrate populations inside neotropical rain forest. *Journal of Animal Ecology*, **42**, 55–105.

**Gilpin M.E. (1975)** Limit cycles in competition communities. *American Naturalist*, **109**, 51–60.

**Glynn P.W. (1974)** The impact of *Acanthaster* on corals and coral reefs in the eastern Pacific. *Environmental Conservation*, **1**, 237–246.

**Glynn P.W. (1976)** Some physical and biological determinates of coral community structure in the eastern Pacific. *Ecological Monographs*, **46**, 431–456.

**Grubb P.J. (1977)** The maintenance of species-richness in plant communities: the importance of the regeneration niche. *Biological Reviews*, **52**, 107–145.

**Harper J.L. (1978)** *Population Biology of Plants*. Academic Press, London.

**Jackson J.B.C. & Buss L. (1975)** Allelopathy and spatial competition among coral reef invertebrates. *Proceedings of the National Academy of Sciences of the United States of America*, **72**, 5160–5163.

**Janzen D.H. (1970)** Herbivores and the number of tree species in tropical forests. *American Naturalist*, **104**, 501–528.

**Janzen D.H. (1971)** Escape of juvenile *Dioclea megacarpa* (Leguminosae) vines from predators in a deciduous tropical forest. *American Naturalist*, **105**, 97–112.

**Janzen D.H. (1972a)** Association of a rainforest palm and seed-eating beetles in Puerto Rico. *Ecology*, **53**, 258–261.

**Janzen D.H. (1972b)** Escape in space by *Sterculia apetala* seeds from the bug *Dysdercus fasciatus* in a Costa Rican deciduous forest. *Ecology*, **53**, 350–361.

**Janzen D.H. (1975a)** *Ecology of Plants in the Tropics*. Institute of Biology Studies in Biology, No. 58. E. Arnold, London.

**Janzen D.H. (1975b)** Interactions of seeds and their insect predators/parasitoids in a tropical deciduous forest. *Evolutionary Strategies of Parasitic Insects and Mites* (Ed. by P.W. Price) pp. 154–186. Plenum, New York.

**Jones E.W. (1956)** Ecological studies on the rain forests of southern Nigeria, IV. *Journal of Ecology*, **43**, 564–594; **44**, 83–117.

**Langdale-Brown I., Osmaston H.A. & Wilson J.G. (1964)** *The Vegetation of Uganda and its Bearing on Land-use*. Government of Uganda, Kampala.

**Lawton J.H. & McNeill S. (1979)** Between the devil and the deep blue sea: on the problem of being a herbivore. *Population Dynamics* (Ed. by R.M. Anderson, B.D. Turner & L.R. Taylor) pp. 223–244. Blackwell Scientific Publications, Oxford.

**Murdoch W.W. (1969)** Switching in general predators: experiments on predator specificity and stability of prey populations. *Ecological Monographs*, **39**, 335–354.

**Murdoch W.W. & Oaten A. (1975)** Predation and population stability. *Advances in Ecological Research*, **9**, 1–131.

**Richards P.W. (1952)** *The Tropical Rain Forest*. Cambridge University Press.

**Scriber J.M. (1973)** Latitudinal gradients in larval feeding specialization of the world *Papilionidae* (*Lepidoptera*). *Psyche*, **80**, 355–373.

**Stephenson T.A. & Stephenson A. (1933)** Growth and asexual reproduction in corals. *Scientific Report*, *Great Barrier Reef Expedition*, **3**, 167–217.

**Watt K.E.F. (1964)** Comments on fluctuations of animal populations and measures of community stability. *Canadian Entomologist*, **96**, 1434–1442.

**Webb L.J., Tracey J.G. & Williams W.T. (1972)** Regeneration and pattern in the subtropical rain forest. *Journal of Ecology*, **60**, 675–695.

**Whitmore T.C. (1975)** *Tropical Rain Forests of the Far East*. Clarendon, Oxford.

**Williams W.T., Lance G.N., Webb L.J., Tracey J.G. & Connell J.H. (1969)** Studies in the numerical analysis of complex rain forest communities. IV. A method for the elucidation of small-scale forest pattern. *Journal of Ecology*, **57**, 635–654.

**Wilson D.E. & Janzen D.H. (1972)** Predation on *Scheelea* palm seeds by bruchid beetles: seed density and distance from the parent palm. *Ecology*, **53**, 954–959.

# 8. COMMUNITY STRUCTURE: IS IT RANDOM, OR IS IT SHAPED BY SPECIES DIFFERENCES AND COMPETITION?

JARED M. DIAMOND
*Physiology Department, University of California Medical Center, Los Angeles, California 90024*

Biological communities consist of many species. Any assemblage of many non-identical objects has some properties that give it an appearance of structure, even if the objects differ only randomly and are non-interacting. For example, the frequency distribution of people's heights, of the numbers of pages in books and of gross national products of the world's nations is lognormal, and this fact is nothing more than a statistical consequence of effects of random variables on large heterogeneous collections of objects. To what extent does the structure of biological communities reflect merely the statistics of large numbers as opposed to non-random differences among species and to species interactions such as competition? For example, the species-abundance relation of most large biological communities, like the frequency distribution of gross national products, is lognormal and yields no 'biological' conclusions. However, certain successional communities and single-resource communities have different species-abundance relations that do yield 'biological' information (e.g. about how the species of these communities are dividing niches).

For numerous archipelagoes throughout the world, thorough surveys of resident bird distributions are available for dozens of islands in the archipelago. In this paper I shall compare these distributional lists with predictions based on the twin assumptions that bird species differ only randomly in their turnover rates and do not compete with each other. Few biologists today would expect these assumptions to prove valid. Our interest is less in the fact that these predictions fail than in what we can learn from the details of how they fail.

Table 8.1 summarizes the distributional data analysed in this paper.

## ISLANDS INHABITED PER SPECIES

In each archipelago some species inhabit many or all surveyed islands, other species few or only one. For instance, of the 56 land and freshwater bird species of the New Hebrides, the kingfisher *Halcyon chloris* is on all 28 surveyed islands,

TABLE 8.1. Data base: bird distributions on islands. For each named archipelago the table gives the total number of land and fresh-water bird species breeding in the archipelago (column 3) and the number of islands for which virtually complete lists of bird species are available (column 2)

| Archipelago | Number of islands surveyed | Number of bird species | Source |
|---|---|---|---|
| New Hebrides | 28 | 56 | Diamond & Marshall (1976) |
| Solomon Islands | 138 | 132 | Mayr & Diamond (1980) |
| Bismarck Archipelago | 39 | 152 | Mayr & Diamond (1980) |
| New Guinea | 41 | 528 | Diamond (unpublished) |
| Great Britain | 63 | 170 | Reed (1977), Diamond & Jones (1980) |

the warbler *Cichlornis whitneyi* on only one. In Great Britain the rock pipit is on almost all islands, the Dartford warbler only on Britain itself.

Suppose an archipelago has $N$ surveyed islands and $S_T$ bird species, that the $i$th island supports $S_i$ species and that the $j$th species is on $Z_j$ islands. From the distributional lists one can extract the vectors $S_i$ of 'species-per-island' and $Z_j$ of 'islands-per-species'. For each value of $Z$—e.g. $k$—there are $Y_k$ species with that value of $Z$. The total number of 'island populations' in the archipelago is $T = \sum_{i=1}^{N} S_i = \sum_{j=1}^{S_T} Z_j = \sum_k k Y_k$. The average value of islands-per-species is $T/S_T$.

Suppose that the species list for each island $i$ were actually a random draw of $S_i$ species from the archipelago species pool. Such random draws of $S_1, S_2, \ldots, S_N$ species yield $N$ 'random lists', from which one can calculate the vector $Y_k$ expected at random for comparison with the actual $Y_k$ vector. As illustrated in Figure 8.1, the variance of the actual $Y_k$ distribution is much greater than that of the randomly derived distribution for all five archipelagoes of Table 8.1. That is, if the island communities of these archipelagoes were randomly assembled, most species would be on close to the mean number of islands, and there would be no species that was confined to a few islands or else distributed over nearly all islands. In fact, each archipelago has numerous species confined to a single island (e.g. the grebe *Tachybaptus ruficollis*, the honeyeater *Meliarchus sclateri* and the white-eye *Zosterops murphyi* of the Solomon Islands) or occurring on virtually all islands (e.g. there are three species that occur on all 28 surveyed islands of the New Hebrides).

Why is the actual distribution of islands-per-species so non-random? Two factors play an obvious role. First, certain habitats, and hence species confined to them, may occur on only one or a few islands of an archipelago. For example, the montane warbler *Cichlornis whitneyi* and montane starling *Aplonis santovestris* are confined in the New Hebrides to the sole island with substantial areas above 3 000 ft; the duck *Aythya australia* is confined in the New Hebrides

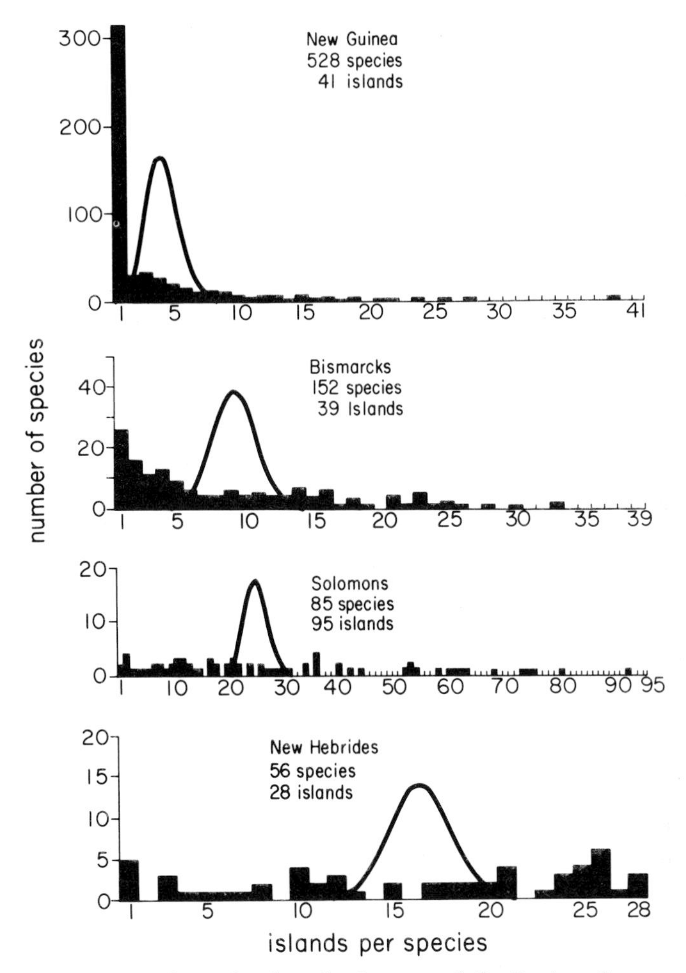


FIG. 8.1. Comparison of actual and randomly expected distribution of 'islands-per-species' in four tropical archipelagoes. For each species in each archipelago the number of surveyed islands inhabited by the species was counted up; this count is the value of 'islands-per-species'. For each archipelago the number of species with a given value of 'islands-per-species' was then counted, and the distribution of these numbers is given as a histogram for each archipelago. The Gaussian curve gives the distribution expected if each island had the same number of species as in reality but the species were drawn at random from the species pool. On each graph are written the number of surveyed islands and the total number of species in the archipelago. Of the four archipelagoes depicted, the total species number, and the fraction of the archipelago's area represented by the largest island, decrease from top to bottom; note correlated shifts in the histograms. 'Solomons' in this case means only the New Georgia group, not the whole Solomon Archipelago.

to seven islands with lakes. Second, as will be discussed in the next section, the value of 'islands-per-species' is set by immigration-extinction equilibria: species with high immigration rates and low extinction rates reach and persist on almost all islands, while species with the opposite characteristics occur on few islands.

## INCIDENCE FUNCTIONS

Given that the *i*th island of an archipelago supports $S_i$ species and that the *j*th species occurs on $Z_j$ of the archipelago's $N$ islands, are these $Z_j$ islands randomly distributed with respect to their $S_i$ values? This question can be tested as follows. Group the islands into sets whose members share similar values of species number $S$. For a given species, plot the fraction of each set's islands on which the species occurs (termed the species' incidence, $J$) against the average $S$ value for the set. If island communities were randomly assembled with respect to $S$, these incidence functions or graphs of $J$ against $S$ would approximate to straight lines through the origin. Such is rarely the case in practice (Diamond 1975). For most species $J$ is 0 for low $S$, rises steeply over some middle range of $S$ and is 1 for high $S$. The $S$ range over which $J$ rises from 0 to 1 varies greatly among species (Figs 8.2a and 8.2b). Since island species number $S$ generally increases monotonically with island area $A$, $J$ may also be plotted against $A$ and the resulting graph interpreted in terms of a species' 'area requirement' (Fig. 8.3; see Diamond 1978a for further discussion). For a few species (so-called supertramps) in species-rich tropical archipelagoes, $J$-vs.-$S$ or $J$-vs.-$A$ assumes a different form: $J$ decreases to 0 at high $S$ or $A$ (Fig. 8.4).

The explanation for the forms of Figures 8.2 and 8.3 may again be sought partly in terms of habitat requirements. Some habitats (e.g. coastal, aerial) occur on any island regardless of size, while other habitats (e.g. mountains, lakes, rivers) occur only on large islands. Species restricted to the latter habitats must conform to Figure 8.2b ($J = 0$ except at high $S$ or $A$); species of the former habitats *may* conform to Figure 8.2a ($J > 0$ on small, species-poor islands as well as large, species-rich ones). Whilst this explanation offers a starting point for understanding, it fails to explain the innumerable instances of species absent from islands with suitable habitat. Nor does it explain why species differ enormously in the value of the ratio $A/A_i$ (where $A$ is island area or area of suitable habitat and $A_i$ is the territory size of one pair) required to maintain a population—that is, why species differ enormously in minimum population size. For example, in the Solomon Islands the territorial cuckoo-shrike *Coracina papuensis* occurs on most islets large enough to hold a single pair, while the equally territorial cuckoo-shrike *Coracina holopolia* is confined to islands large enough to hold at least several hundred pairs. Some of these differences in $A/A_i$ arise because some species are socially solitary and others

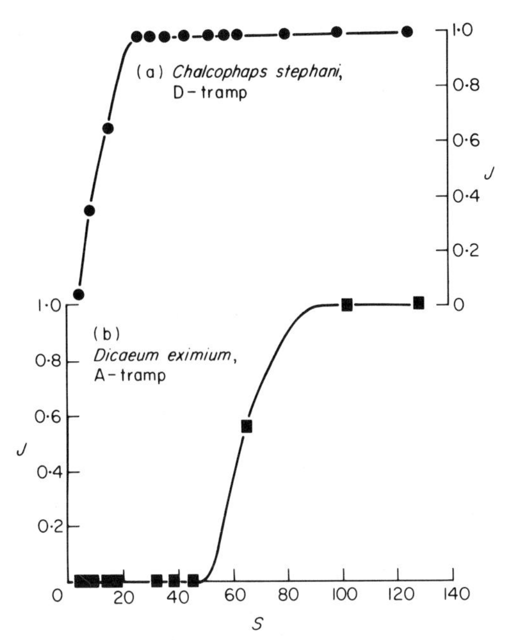


FIG. 8.2. Incidence functions for the pigeon *Chalcophaps stephani* and the berrypecker *Dicaeum eximium* in the Bismarck Archipelago. To construct this figure, islands of the Bismarck Archipelago were divided into groups such that the value of the total species number $S$ of all islands in a given group fell within a narrow range: e.g. the groups 2–5, 6–9, 10–15, 16–22, .... The ordinate $J$ is the incidence of the given species (i.e. the fraction of the islands in the group on which the species occurs), and the abscissa $S$ is the average species number for the islands of the group. Thus, $J = 1{\cdot}0$ or $J = 0$ means that a species occurs on all islands or on no island respectively that has approximately the indicated species number. Each point is based on up to thirteen islands. (From Diamond 1975.)

live in flocks, but such social factors make only a minor contribution to the observed range of $A/A_i$.

Considerations of immigration and extinction rates help us understand these problems. Populations do not survive forever but are subject to turnover: they risk extinction due to population fluctuations and may be refounded by immigrants (Mayr 1965; MacArthur & Wilson 1967; Diamond 1969; Jones & Diamond 1976; Diamond & May 1977). The larger the island, the larger the saturating population size and the lower the risk of extinction. Thus, in the steady state at any instant a species will inhabit some fraction, $J$, of islands of a certain area; this fraction will increase with immigration rates, $I$; and it will decrease with extinction rates, $E$, and hence increase with area. In general, $J = I/(I + E)$. Assume a hyperbolic relation between $E$ and $A$ ($E = a/A$, where $a$ is a constant, or $E = a/(A - A_i)$, where $A_i$ is the territory size of one

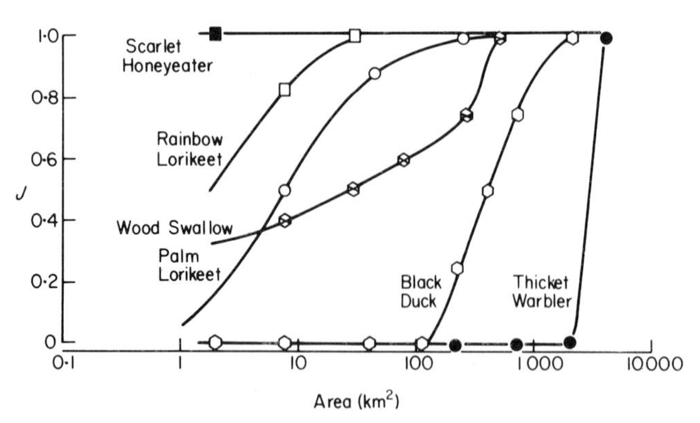


FIG. 8.3. Incidence functions for six bird species in the New Hebrides Archipelago. This figure was constructed in the same way as Figure 8.2, except that the abscissa is island area. That is, islands of the New Hebrides were divided into groups such that the area of all islands in a given group fell within a narrow range, and the ordinate $J$ is the fraction of the islands in the group in which the species occurs. Note that different species have very different 'area requirements'. For example, the scarlet honeyeater occurs on all islands down to an area of at least 2 km$^2$, while the black duck is confined to islands greater than several hundred square kilometres and the thicket warbler is confined to the largest island in the archipelago (3 937 km$^2$).

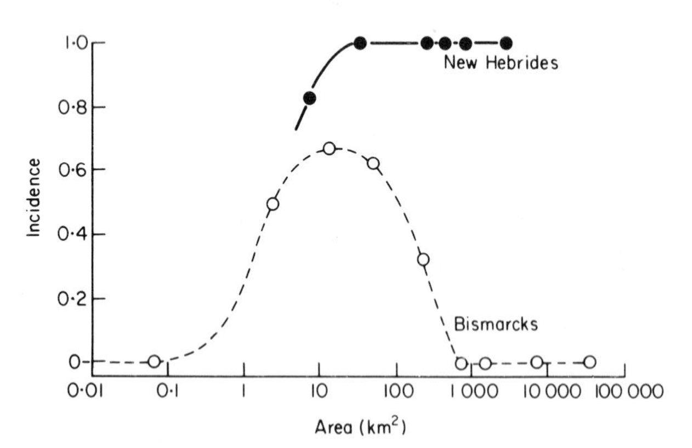


FIG. 8.4. Incidence functions of the pigeon *Macropygia mackinlayi* in the New Hebrides Archipelago and Bismarck Archipelago, constructed in the same manner as Figure 8.3. There are three times as many competing pigeon species on a Bismarck island as on a New Hebridean island of the same area. In the Bismarcks *Macropygia mackinlayi* has a 'supertramp' incidence function: i.e. it is competitively eliminated from large islands with many pigeon species and restricted to smaller islands. In the New Hebrides, where there are fewer competitors, *Macropygia mackinlayi* can maintain itself on large as well as on smaller islands.

pair); assume that $I$ increases weakly with $A$ ($I = b\sqrt{A}$, where $b$ is another constant); and recognize that the species–area relation often approximates the form $S = S_o A^z$. One thus obtains (Diamond & Marshall 1977) for the predicted $J$–$S$ or $J$–$A$ relation:

$$J = (b\sqrt{A})/[b\sqrt{A} + a/(A - A_i)] \tag{8.1}$$

or

$$J = b(S/S_o)^{1/2z}/\{b(S/S_o)^{1/2z} + a/[(S/S_o)^{1/z} - A_i]\} \tag{8.2}$$

Figure 8.5 illustrates predicted $J$–$A$ relations for different values of the ratio $b/a$, expressing the ratio of immigration rates to extinction rates. The forms of the predicted relations are similar to the observed forms of Figure 8.3. The higher the value of $b/a$, the greater is the likelihood of finding the species on a small island.

These considerations of immigration-extinction equilibria illuminate many features of island species distributions not explained by habitat requirements alone. For similar population densities and extinction probabilities, species that disperse readily over water are more likely to occur on small islands than

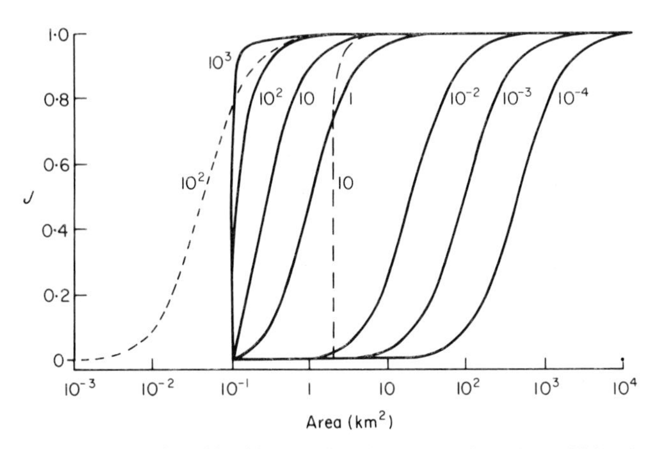


FIG. 8.5. Theoretical graphs of incidence $J$ (ordinate) as a function of island area (km²; abscissa) calculated from eqn (8.1). The underlying assumption is that incidence is set by an equilibrium between immigration and extinction rates: extinction rates decrease strongly and immigration rates increase weakly with increasing area, so that with increasing island area a species will be found to be present on the island an increasing fraction of the time. The number beside each curve is the ratio of the immigration coefficient $b$ to the extinction coefficient $a$. The territory size of a single pair, $A_i$, is 2 km² (— — —), 0·1 km² (———) or infinitely small (------). Note that incidence increases with increasing value of $b/a$; and that, especially for low values of this ratio, incidence may be very low on an island far larger in area than the territory size. (From Diamond & Marshall 1977.)

are sedentary species. Among the two Solomon Island cuckoo-shrikes contrasted in the previous paragraph but one, *Coracina papuensis* can often be seen flying over water gaps, *C. holopolia* never. For similar dispersal ability, species living at low density or with widely fluctuating populations are restricted to larger islands than species living at high density or with stable populations.

Competition also affects the form of incidence functions, by reducing population size and increasing extinction probability on an island of given area. Figures 8.4 and 8.6 provide two clear examples. Figure 8.6 illustrates that a parrot which occurs in three Pacific archipelagoes has a lower incidence, for given island area, in the archipelago with fewer competing species. Competition is essential to interpreting the supertramp incidence pattern of Figure 8.4 ($J$ decreasing at high $S$ or $A$), because $E$ must decrease and $J$ increase at high $A$ in the absence of competition. Supertramps prove to be vagile species of poor competitive ability. They are outcompeted on species-rich islands, but their high immigration rates enable them to maintain populations on small islands where their more sedentary competitors quickly go extinct between rare arrivals. A species with a supertramp incidence pattern ($J \rightarrow 0$ at high $S$ or $A$) in an archipelago shared with many competitors has a normal incidence pattern ($J \rightarrow 1$ at high $S$ or $A$) in an archipelago free of competitors (compare patterns for the Bismarck Archipelago and New Hebrides in Fig. 8.4). Thus, Figures 8.4 and 8.6 both illustrate that a species' probability of occupying an island of a given area is a function of the number of competing species.

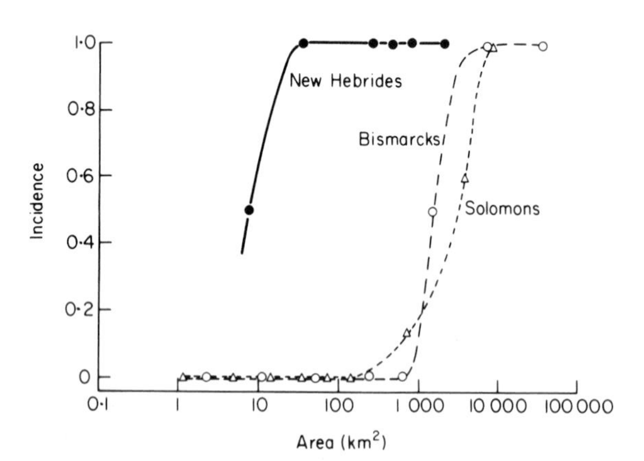


FIG. 8.6. Incidence of the parrot *Vini* [*palmarum*] in three archipelagoes of the tropical southwest Pacific. This figure was constructed in the same way as Figure 8.3. There is only one other parrot species in the New Hebrides but ten others in the Solomons and twelve others in the Bismarcks. Because of this increased competition, the incidence of this parrot is lower on a Bismarck or a Solomon island than on a New Hebridean island of the same area.

## TURNOVER FREQUENCIES

Direct evidence that bird species differ non-randomly in their immigration and extinction rates comes from turnover studies. Breeding censuses of an island in consecutive years show that populations turn over—i.e. occasionally go extinct and are occasionally re-established by immigration. Different species on a given island differ in turnover frequency. Some island populations have been breeding residents in every recorded year of ornithological observation and have exhibited no turnover. Other species immigrate and go extinct several times within a decade.

Are these species' differences in turnover frequency non-random? Turnover frequencies $\nu$ can be extracted from consecutive breeding surveys on numerous Californian (Jones 1975; Jones & Diamond 1976) and European (Reed 1977; Diamond & Jones 1980) islands. To depict the actual distribution of $\nu$'s over species, count each immigration or extinction as one turnover event; sum up for each species $j$ the number of turnover events $e_j$ on a given island during the available span of census years; rank the $e$ values in descending sequence; and plot the accumulated sum of $e$ values against the accumulated number of species $S$, beginning with the species that turns over most rapidly. Figure 8.7 exemplifies such a plot.

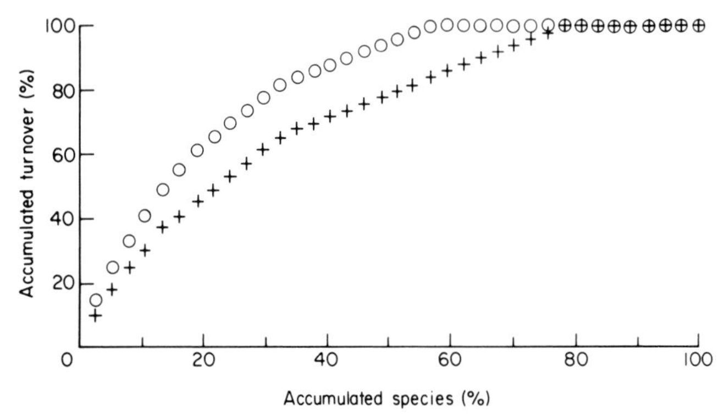


FIG. 8.7. Species' differences in turnover frequency on the British island of Bardsey, based on annual breeding censuses from 1954 to 1969. For each species $j$ the number of turnover events $e_j$ (immigrations or extinctions) during the sixteen census years was calculated, and species were ranked in descending sequence of $e_j$. The open circles give the actual accumulated number of turnover events as a percentage of total turnover events for all species (ordinate) plotted against the accumulated number of species as a percentage of the total species number. The crossed symbols give the curve expected if the observed total number of turnover events in each year were randomly assigned to species. The deviation between the actual and randomly predicted means that more species turn over rapidly, and more species exhibit no turnover, than would be the case if species differed only randomly.

Even if all species were identical, unequal $e$ values would be expected for purely statistical reasons. To generate the randomly expected $e$-vs.-$S$ plot, count up the total number of turnover events actually observed on an island in a given year, assign these randomly among species that breed on the island, do this for all census years, then add up the number of events $e_j$ for each species over the available span of years, and again plot accumulated $e$ against accumulated $S$ (Fig. 8.7).

For all islands studied, the observed and randomly expected curves differ in the same way as those depicted for the British island of Bardsey in Figure 8.7: the observed curve lies above the randomly expected curve and reaches the upper asymptote (100% of turnover events accumulated) at a lower $S$ value.

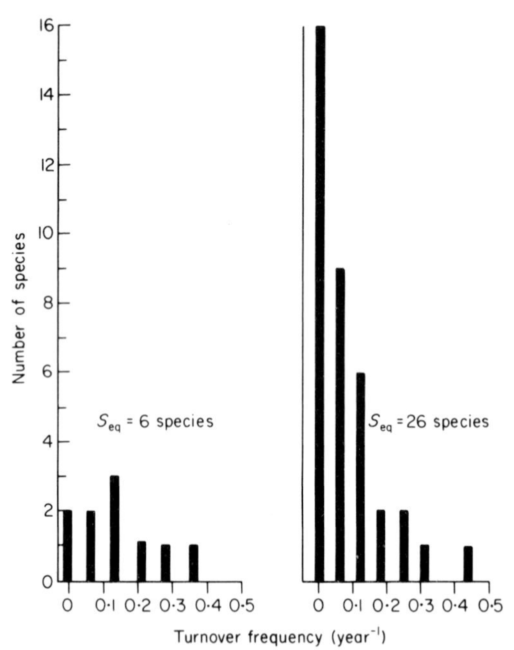


FIG. 8.8. Species' differences in turnover frequency on the British islands of Hilbre (left) and Bardsey (right). Annual breeding surveys on each island for 16 or 17 consecutive years were analysed. For each species that bred on the island during this period the turnover frequency was calculated as the number of turnover events (immigrations or extinctions) over this 16- or 17-year period, divided by 16 or 17 years. Populations were then grouped according to turnover frequency, and the bars indicate the number of species with a given turnover frequency. For example, a frequency of 0·2 year$^{-1}$ means that a species exhibited three cases of turnover (immigration–extinction–immigration or else extinction–immigration–extinction) on the island in 15 years. On average the number of breeding species is six on Hilbre and twenty-six on Bardsey. Note that the larger island of Bardsey has many more low-turnover and zero-turnover species than does Hilbre.

This means that the observed distribution of turnover frequencies is wider than expected at random: some species turn over more rapidly than expected, and more species exhibit no turnover. The reasons are that species differ greatly in their dispersal tendencies and immigration rates; and that species also differ in extinction rates because species differ in abundance and in magnitude of population fluctuations.

An alternative way of depicting species differences in turnover frequency is illustrated for two British islands in Figure 8.8, which gives histograms of species' turnover frequencies. On each island some species breed every year without interruption ($\nu = 0$), while others turn over nearly every other year ($\nu = 0{\cdot}5$/year). The larger, more species-rich island and the smaller, more species-poor island of Figure 8.8 have similar numbers of high-turnover species, but the larger island has more zero- and low-turnover populations. Comparison of numerous other islands shows this conclusion to be a general one: large islands differ from small islands mainly in having additional low-turnover populations. The reason is that on a small island the population of even the most abundant species may be so small that it occasionally suffers extinction. Only large islands can hold populations large enough to be secure against extinction for long times. The same conclusion applies to turnover in evolutionary time, as assessed from the degree of endemism of an island avifauna (Mayr 1965).

## TEMPORAL CORRELATIONS BETWEEN SPECIES

When I began turnover studies, I naïvely expected them to yield clear evidence for competition. I thought that if one noted species simply as present or absent and analysed the presence–absence records for a pair of ecologically similar species on an island, competition would result in breeding years of the two species coinciding less often than expected at random. I thought that if one analysed population fluctuations in the two species, competition would express itself in the abundances of the two species being negatively correlated. Both expectations proved invalid, for interesting reasons.

The first prediction was tested on ten islands that had been censused in 11–27 consecutive years. Presence–absence records for all pairwise combinations of species on an island were analysed to determine whether breeding years of the two species coincided more or less often than expected at random, using the $\chi^2$ test at the $p > 0{\cdot}05$ level as the test of significant deviations from randomness. All species pairs of which one or both member bred in every census year were eliminated from consideration. Of the remaining 1 519 pairs, 1 446 failed the $\chi^2$ test, while 43 coincided in their breeding years more often and 30 less often than expected at random. That 73 pairs, or 5% of the 1 519 pairs, pass the $\chi^2$ test at the 5% level is a statistical tautology for a random assemblage. That is, turnover events in these island communities do not correlate significantly between species.

This conclusion is reinforced by detailed consideration of the cases passing the $\chi^2$ test. If such cases involved ecologically similar species tending to replace each other repeatedly in a sequence of breeding seasons (e.g. *A*/–, –/*B*, *A*/–, *A*/–, *A*/*B*, –/*B*, –/*B*, *A*/–, where *A* and *B* are the species and each entry corresponds to one year), then competition would be suspected to play a role. In fact, most cases passing the $\chi^2$ test involve pairs of ecologically unrelated species (e.g. the pairs little owl–reed bunting and cuckoo–swallow on the British island of Skomer). The species do not replace each other repeatedly; usually one species breeds in the earlier census years while the other species breeds in the later census years (e.g. *A*/–, *A*/–, *A*/–, *A*/–, *A*/*B*, –/*B*, –/*B*, –/*B*). The most probable explanation of these cases has nothing to do with competition but involves environmental events that affect ecologically unrelated species and hence produce correlations between breeding records of the species. Examples of such events include: progressive increases in temperature favouring warm-adapted species and eliminating cold-adapted species; progressive changes in island habitats favouring species of one habitat over another; and the harsh winter of 1962–63 in Britain eliminating populations of numerous cold-sensitive species on many islands.

The second prediction was tested on nine islands that had been censused in 16–38 consecutive years, and for which the number of breeding pairs of each species as well as mere breeding presence or absence was known for each year. For all pairwise combinations of species on the island, the correlation coefficient between the numbers of pairs of the two species in the same year was calculated. Species whose population fluctuations correlated strongly often proved to be ecologically unrelated. For example, on the British island of Havergate the correlation coefficient between swallow and pheasant is $+0{\cdot}87$(!), between swallow and skylark $+0{\cdot}91$, between oystercatcher and reed bunting $-0{\cdot}85$, although no two species could be less likely to be competitors than the members of these pairs. Between competing species the correlation is more often positive than negative. Strong correlation coefficients (values greater than $+0{\cdot}5$ or less than $-0{\cdot}5$) usually reflect long-term population trends, as also true for presence–absence records, rather than irregular year-to-year fluctuations. For example, on the British island of Bardsey there is a strong positive correlation coefficient between meadow pipit and wren (0·78) and between meadow pipit and linnet (0·90), because populations of all three species took several years to recover from the harsh winter of 1962–63.

Why do these island studies fail to confirm the expectation that competing species will tend to fluctuate oppositely in abundance and even replace each other? Because this expectation was naive (Roughgarden 1975; May 1976). It tacitly assumed that population fluctuations arise initially from random causes (e.g. demographic accidents) in a constant environment, and that a random increase in one species depresses its competitor's abundance in that same year. In fact, a major cause of population fluctuations is environmental fluctuation which alters species' carrying capacities. Competitors are species that utilize similar resources and are therefore likely to respond similarly to environmental fluctuations. Hence the decrease in abundance of species *A* due to an increase in its competitor *B* is likely to be offset by an increase in *A*'s carrying capacity arising from the same environmental change that caused *B*'s increase in the first place. Whether the result of environmental fluctuations will be positive, negative or no correlation between *A*'s and *B*'s abundances depends on the competition coefficient between the species and on whether the species' carrying capacities are correlated positively, negatively or not at all. In addition, the strongest correlation between abundances of two competitors may not involve their abundances in the same year but their abundances a year or more out of phase—e.g. if one of the species has a low population growth rate or if an abundant population of *A* does not prevent *B*'s nesting attempts that year but causes high mortality in *B*'s nestlings (the breeders of next year).

Thus, competition does not lead to simple and unequivocal predictions about correlations between the temporal patterns of abundance of two species. In retrospect, it is not surprising that turnover studies and annual breeding censuses have failed to provide clear evidence of competition.

## SPATIAL CORRELATIONS BETWEEN SPECIES

Negative correlations between spatial distributions of species have contributed to the renewed appreciation of the importance of interspecific competition in recent decades. Familiar examples include checkerboard distribution patterns and spatial niche shifts. By checkerboard distribution is meant a pattern in which two competing species occupy islands or geographical areas to the mutual exclusion of each other, in an irregular geographical array. Figure 8.9 presents a typical example: two very similar flycatcher species of genus *Pachycephala* inhabit 29 islands of the Bismarck Archipelago in an irregular array, *P. pectoralis* on 11 islands, *P. melanura* on 18 islands, there being no island that supports both species. Spatial niche shifts include patterns in which species *A* and *B* occupy mutually exclusive habitats in areas of sympatry but one or both species occupy a wider range of habitats in areas of allopatry. Since Lack's (1944) early recognition of niche shifts in the two chaffinch species of the Canary Islands, numerous subsequent cases have been discovered through field observation or experimental manipulation (see Diamond 1978b for summary).

Naturally, spatial correlations between competitors' distributions are most easily recognized when only two species are involved. Yet most species have

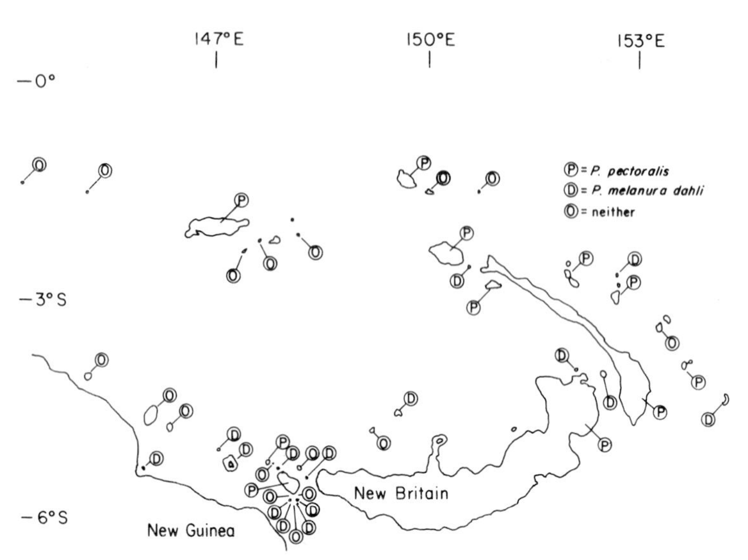


FIG. 8.9. Checkerboard distribution of *Pachycephala* flycatcher species in the Bismarck region. Islands whose flycatcher faunas are known are designated by P (*Pachycephala pectoralis* resident), D (*Pachycephala melanura dahli* resident) or O (neither of these two species resident). Note that most islands have one of these two species, no island has both and some islands have neither. (From Diamond 1975.)

several important competitors (so-called diffuse competition) rather than a single dominant competitor. Although checkerboard distributions and niche shifts involving two species provide neat examples to illustrate competition, the distribution or niche of a species must far more often be correlated with combined presences of several species. How can one begin to trace out these complex correlations produced by diffuse competition?

Figure 8.9 provides a starting point for recognizing checkerboard distributions involving species combinations. Although the two flycatchers used in constructing Figure 8.9 divide 29 Bismarck islands between themselves, 21 other Bismarck islands have neither of these species. These 21 'vacancies' are initially surprising, as both species are good colonizers and one (the supertramp *Pachycephala melanura*) is among the first species to colonize defaunated islands. However, all 21 of the vacant islands are occupied by various combinations of other flycatcher species. The two *Pachycephala* species belong to a group of seven, ecologically related, gleaning flycatchers: *Pachycephala pectoralis*, *Pachycephala melanura*, *Monarcha cinerascens*, *Monarcha verticalis*, *Monarcha chrysomela*, *Myiagra hebetior* and *Myiagra alecto*, abbreviated respectively P, D, C, V, R, H and A. Except for P–D and possibly D–R, each species co-exists with each other species on some island. However, Bismarck islands are found to support only certain combinations of these species. If an island has only one species, it is always C, D or rarely A, never P, V, R or H. Islands with two species have the combination AC, CD, CP or AD, never any of the remaining 18 two-species combinations that are theoretically possible. The only three-species combinations observed are APV, CPR, ACP or ACV (31 other combinations never observed); the observed four-species

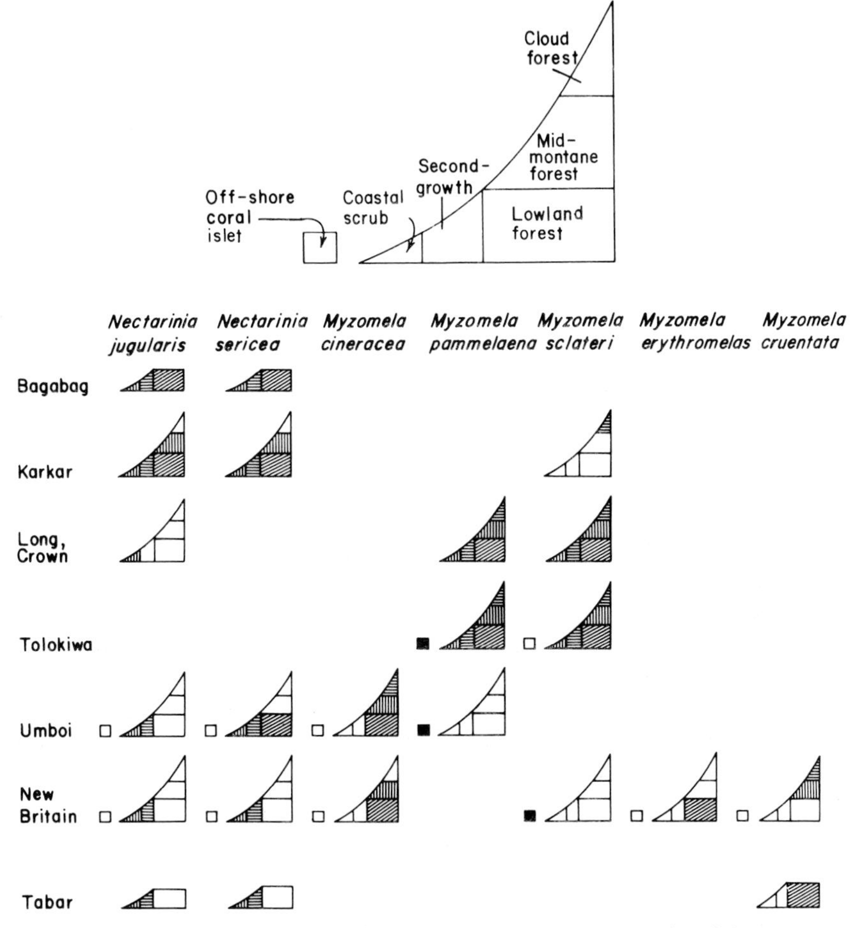


FIG. 8.10. Inter-island variation in habitat partitioning among species of the myzomelid and sunbird guild. The sketch at the top shows how the major habitats are pictured. For each island the habitats occupied by each guild species are shown as variously shaded. Note that the range of habitats occupied by each species varies greatly from island to island, depending on what other species are present. Bagabag and Tabar Islands are not high enough to support mid-montane forest or cloud forest. (From Diamond 1975.)

combinations, AHPV, CHPV, ACDV, ACDH, ACPR or ACRV (29 other combinations never observed); the sole five-species combination, AHPRV (20 other combinations never observed); and no island supports six or all seven species. Most of the 21 vacant islands of Figure 8.9 support the flycatcher combinations ACV, AC or C, which can co-exist with D or P on larger islands but resist invasion by D or P on smaller islands.

Why are these combinations 'permitted' and other combinations 'forbidden'? Although species' differences in dispersal play a role, a major reason involves resource utilization and competition: the permitted combinations may be those that leave the fewest resources utilized, that are therefore relatively immune to displacement of one component species by an invading competing species, and that still provide each component species with enough resources to maintain a stable population (Diamond 1975).

Figure 8.10 illustrates niche shifts related to combinations of species rather than to presence or absence of a single species. The seven competing species involved belong to the sunbird (Nectariniidae) and honeyeater (Meliphagidae) families and serve as the Bismarck Archipelago equivalent of the New World's hummingbirds: they are small, curve-billed birds that visit flowers for nectar and insects. The niche of each species varies greatly from island to island. For example, *Myzomela sclateri* may be present in all habitats (Long and Crown Islands), confined to cloud forest (Karkar) or confined to small offshore islets (New Britain); *Nectarinia sericea* may be in all habitats (Bagabag), only in lowland and midmontane forest (Karkar), only in lowland forest (Umboi), only in coastal vegetation and forest edge (New Britain) or only in forest edge (New Guinea); and *Myzomela cineracea* may be at all elevations in forest (Umboi), only in lowland forest (widespread New Guinea allospecies) or only in lowland savannah (south New Guinea allospecies). These niche shifts in each species are related to the particular combination of other species sharing each island, such that each habitat has at most two (rarely three) of the seven species. Examples of the 'assembly rules' underlying these permitted combinations are that two *Nectarinia* species can share a habitat but two similar-sized *Myzomela* species cannot; a *Myzomela* species can share a habitat with one but not both *Nectarinia* species; and one *Nectarinia* species can share a habitat with one small *Myzomela* but not with one small and one large *Myzomela*.

## ACKNOWLEDGMENTS

It is a pleasure to acknowledge my debt to Michael Gilpin, H. Lee Jones, Robert May and Timothy Reed for valuable information and discussions, and the National Geographic Society and Lievre Fund for support of field work.

# REFERENCES


**Diamond J.M. (1969)** Avifaunal equilibria and species turnover rates on the Channel Islands of California. *Proceedings of the National Academy of Sciences of the United States of America*, **64**, 57–73.

**Diamond J.M. (1975)** Assembly of species communities. *Ecology and Evolution of Communities* (Ed. by M. Cody & J. Diamond) pp. 342–444. Harvard University Press, Cambridge, Massachusetts.

**Diamond J.M. (1978a)** Critical areas for maintaining viable populations of species. *The Breakdown and Restoration of Ecosystems* (Ed. by M. Woodman) pp. 27–40. Plenum Press, New York.

**Diamond J.M. (1978b)** Niche shifts and the rediscovery of interspecific competition. *American Scientist* **66**, 322–331.

**Diamond J.M. & Jones H.L. (1980)** *Dynamics of Species Communities.* Princeton University Press, Princeton (in press).

**Diamond J.M. & Marshall A.G. (1976)** Origin of the New Hebridean avifauna. *Emu*, **76**, 187–200.

**Diamond J.M. & Marshall A.G. (1977)** Distributional ecology of New Hebridean birds: a species kaleidoscope. *Journal of Animal Ecology*, **46**, 703–727.

**Diamond J.M. & May R.M. (1977)** Species turnover rates on islands: dependence on census interval. *Science*, **197**, 266–270.

**Jones H.L. (1975)** Studies of avian turnover, dispersal, and colonization of the California Channel Islands. *Unpublished Ph.D. thesis, University of California, Los Angeles.*

**Jones H.L. & Diamond J.M. (1976)** Short-time-base studies of turnover in breeding birds of the California Channel Islands. *Condor*, **76**, 526–549.

**Lack D. (1944)** Ecological aspects of species-formation in passerine birds. *Ibis*, **1944**, 260–286.

**MacArthur R.H. & Wilson E.O. (1967)** *The Theory of Island Biogeography.* Princeton University Press, Princeton.

**May R.M. (1976)** *Theoretical Ecology: Principles and Applications.* Saunders, Philadelphia.

**Mayr E. (1965)** Avifaunal turnover on islands. *Science*, **150**, 1587–1588.

**Mayr E. & Diamond J.M. (1980)** Colonization and speciation in Northern Melanesian birds. *Bulletin of the Museum of Comparative Zoology at Harvard College* (in press).

**Reed T.M. (1977)** *Island biogeography theory and the breeding landbirds of Britain's offshore islands.* Unpublished Honours dissertation, Cambridge University.

**Roughgarden J. (1975)** A simple model for population dynamics in stochastic environments. *American Naturalist*, **109**, 713–736.

# 9. OPTIMAL FORAGING IN PATCHY ENVIRONMENTS

RICHARD J. COWIE[1] AND JOHN R. KREBS[2]

[1] *Animal Behaviour Research Group, Department of Zoology, Oxford*
[2] *Edward Grey Institute, Department of Zoology, Oxford*

## INTRODUCTION

Two of the major biological factors affecting population dynamics are competition and predation. Competition is a symmetrical process in that both competitors are adversely affected and this tends to result in the evolution of mechanisms which reduce the interaction. Predation, on the other hand, is directional in that only the predator benefits from the association. This leads to an evolutionary 'race' resulting in the development of efficient predators and elusive prey. In this paper we will be examining one of the ways in which predators are adapted to forage efficiently. In particular, how they cope with the exploitation of patchily distributed prey populations.

A common feature of many animal populations is that individuals are not distributed homogeneously throughout the environment but occur in groups. There are a number of reasons why this is so, which include: their resources may be clumped causing them, for example, to congregate in areas of high food availability or suitable habitat; living in groups may provide protection from predators, e.g. the selfish herd effect (Hamilton 1971), or reduce the risk of being found (e.g. Brock & Riffenburg 1960); they may need to come together in groups to reproduce (e.g. Parker 1978). However, for whatever reason animals occur in groups or clumps, this grouping poses a problem for potential predators. In order to exploit clumped prey efficiently a predator needs to decide (1) which areas or 'patches' to visit and (2) when to move from one patch to another, where we define the term 'patch' as the area containing the clump of prey items. This could be a natural unit such as a pine cone containing insect larvae or a bush laden with berries or a heterogeneity detected by quadrat sampling, for example earthworms in a lawn. In some cases these decisions can be made simply on the basis of visual cues or previous experience. For example, a heron might know from previous experience which ponds are worth fishing or it might be able to tell at a glance whether there are enough fish present to warrant staying. However, if, as is often the case, the prey are cryptic or hidden, the two questions are resolved into one, namely—how long is it best to remain in each individual patch?

A number of solutions have been proposed for this problem and they can be classified broadly into three types: (1) number expectation (e.g. Gibb 1962)—the predator leaves each patch after a certain number of prey has been found; (2) time expectation (e.g. Krebs 1973)—the predator should leave the patch after a certain amount of time has been spent there; (3) rate expectation (e.g. Charnov 1973)—the predator should leave the patch when the rate of food intake (usually estimated as prey capture rate) falls to a critical threshold level.

In each case the particular threshold which an animal uses may be fixed by natural selection or it may be flexible and capable of being adjusted by the animal. A fixed threshold model may be feasible if the quality of an environment is very predictable, but it would be very inefficient if the number of prey per patch varied widely in different environments. For similar reasons it seems fairly clear that a rule for leaving based on the capture rate would be more efficient and adaptive than a number or time expectation. These latter mechanisms also make no allowance for the time taken to travel between patches, which can be an important consideration. At this stage, therefore, we will consider one of the rate expectation models in more detail—one which is based on a relatively new body of literature called optimal foraging theory.

The essence of the optimal foraging approach is the idea that individual predators which are more efficient at capturing prey will have more resources at their disposal and are therefore more 'fit' than those which are less proficient. Evolution should, therefore, favour efficient predators, and we assume that it results in animals behaving in a way which maximizes their inclusive fitness. This allows us to postulate what short-term goal an animal may be using in order to maximize its fitness in the long term. Most authors (e.g. MacArthur & Pianka 1966; Charnov 1973) have agreed that maximizing the net rate of food (energy) intake is a reasonable short-term goal for animals while they are foraging. The rationale behind this is either that predator populations are directly limited by food (Gibb 1960) or that feeding competes with other essential activities such as watching for predators and defending a territory. We can, therefore, ask the question 'Do animals behave in a way which maximizes their net rate of energy (or food) intake while foraging?' It is this hypothesis which can be tested, not the assumption that an animal's foraging is optimal.

In order to test the idea that an animal is maximizing its rate of food or energy intake, we first have to determine the optimal solution for any particular set of circumstances. In the case of the clumped food resources, it is as follows.

The capture rate within a patch is likely to be a decreasing function of the time spent there, due to effects such as depletion of the available prey, random search or to the predator's locating the most easily found prey first. This means that if the predator stays in a patch for too long its capture rate will fall below that which could be achieved by moving to the next one. Conversely, if it does not stay for long enough, it will spend too large a proportion of its time travelling.

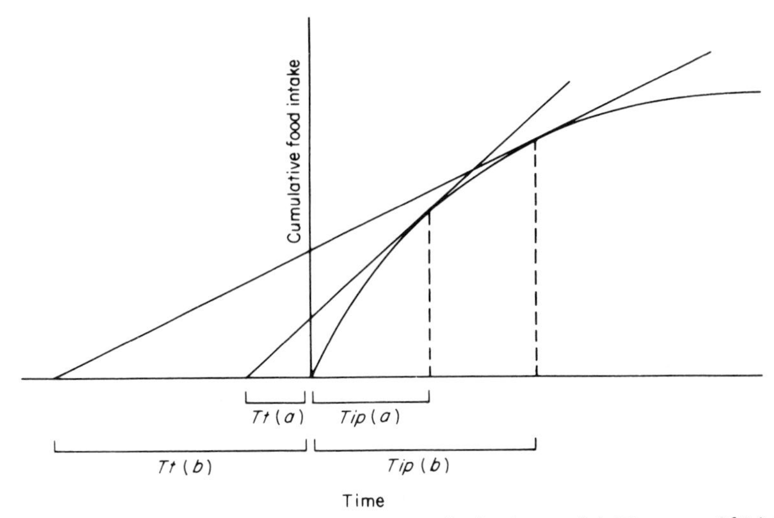


FIG. 9.1. A graphical representation of the marginal value model. The curve (*fTp*) is the average depletion curve for all the patches in an environment. For a given travel time (*Tta*) the predator should spend an average time (*Tipa*) in each patch.

The optimal solution to this problem has been formalized by Charnov (1973, 1976), who has termed his model the marginal value theorem. It is represented graphically in Figure 9.1. If the curve *f*(*Tp*) represents the *average* cumulative energy intake for all patches in the environment, then the predator whose behaviour is optimized should leave each patch when its capture rate in that patch drops to the average capture rate of the environment as a whole. Since the average capture rate includes the time taken to travel between patches, this occurs at the point where a tangent constructed from the origin touches the curve. Thus, a longer travel time will lower the average capture rate and result in the predator staying for longer in each individual patch. The critical rate at which the predator should leave is termed the marginal capture rate and this provides the basis of a rate expectation model.

## WAYS OF TESTING THE PATCH DEPLETION MODEL

Having proposed an optimal solution, we can now examine a predator's foraging strategy and see how well it is predicted by the model. A good fit is likely to indicate that our original assumptions were correct and that the model provides a good description of the animal's behaviour.

There are basically three ways of testing the patch depletion model, because only two variables, time and net energy intake, are involved. This means that it is possible to generate predictions about: (1) the time spent in each patch; (2) the energy intake per patch; (3) the rate of energy intake when the predator leaves a patch.

Since the marginal value model is a rate expectation model, we will begin by examining the last of these predictions (i.e. the marginal capture rate) as a means of testing the patch model.

Ideally, we would like to be able to measure the capture rate directly at the moment a predator leaves a patch, although in practice this is impossible because, in most cases, the energy intake comes in a series of discrete steps (prey items) and we do not know over what time period to calculate an animal's capture rate. To overcome this problem Krebs, Ryan & Charnov (1974) introduced the concept of a giving-up time (GUT), which they defined as the time between the last capture and the time at which the predator leaves a patch. They argued that if the prey were of a single type and had a relatively short handling time, then the intercatch interval could provide an estimate of the capture rate within a patch. Due to depletion, the longer a predator stays in a patch, the longer the intercatch interval becomes, and they suggested that when the predator exceeded a certain threshold without finding any prey (the giving-up time) it should leave. Since the marginal capture rate should equal the average capture rate according to the marginal value model, this provides the prediction that in an environment with a high average capture rate the GUT should be shorter than that in a poorer environment. Krebs, Ryan & Charnov (1974), in their study of black-capped chicadees (*Parus atricapillus*), found that this was the case and concluded that the birds were behaving as predicted by the optimal foraging model.

At this stage, however, we would like to sound a note of caution regarding the use of GUT's in testing foraging models. Since the GUT is based on the average intercatch interval, no matter what rule a bird is using in deciding when to leave a patch, it will still have a shorter GUT in the rich environment when the intercatch interval is, by definition, short. Cowie (1979) has termed an example of this the 'random door-slamming model'. In this model the predator leaves a patch when an independent random event occurs, in this case, when the aviary door slams shut in the wind. If the bird is foraging in a rich environment at the time, it is likely to have made a capture more recently than it would have done in a poor environment, i.e. it has a shorter giving-up time. In Table 9.1 we compare the predictions of this simple model with the giving-up times observed by Krebs *et al.* in four different environments. There is reasonable agreement, although the random departure predictions are too short, suggesting that this model provides an alternative explanation of their results. It does not, however, predict that the giving-up time should have been the same for all patch types within an environment as found by Krebs *et al.* We are not trying to imply that the birds were behaving at random, but simply that a number of different models can make similar predictions about giving-up

TABLE 9.1. A comparison of the giving-up times observed by Krebs, Ryan & Charnov (1974) in four different environments with those predicted from their data by the door-slamming model

| Environment | Observed GUT (seconds) | GUT predicted by model (seconds) |
|---|---|---|
| 1 | 15·68 | 12·94 |
| 2 | 11·07 | 10·42 |
| 3 | 11·38 | 9·94 |
| 4 | 7·56 | 7·05 |

times. In conclusion, it seems that giving-up times are so closely linked to the average capture rate that they have little predictive value when testing models of this type.

This leaves the two alternative methods of testing the marginal value model —the average time spent in each patch and the average amount of energy removed. Because the latter in our experiments is likely to be a discrete variable, we propose that to calculate the optimal time that should be spent in each patch yields the best prediction.

## EVIDENCE FOR OPTIMAL FORAGING IN PREDATORS AND PARASITES

This was the approach used by Cowie (1977), who tested the model by manipulating the time taken to travel between patches whilst measuring the relationship between the net energy intake and the time spent in each patch. The optimal time that should be spent in a patch for any given travel time can be calculated from this information.

Cowie performed experiments on six wild-caught great tits (*Parus major*) which were allowed to forage in an environment containing a number of artificial patches. These consisted of small pots made from 4-cm lengths of plastic drainpipe (6·5 cm diameter) sealed at one end. They were filled with sawdust and six quarter-sections of mealworm were hidden in each. The pots were arranged on five artificial trees in an aviary (4·6 × 3·7 m). Each bird was subjected to six ten-minute trials in two types of environment. These environments were identical in all respects except for the time and energy costs of travelling between patches. Each pot was covered with a cardboard lid which the bird had to remove before it began foraging. There were two types of lid; one rested on top of the drainpipe and could easily be flipped off, the other fitted just inside the rim and had to be prized out. Each experimental environment contained patches with lids that were either all 'hard' or all 'easy' to remove, these differences corresponding to 'long' and 'short' travel times respectively. Cowie found that the rate of energy intake fell off exponentially with the amount of time spent in each patch. This allowed him to derive an expression which describes the relationship between the travel time and the optimal time which should be spent in each patch:

$$T_t = 1/m(\exp(mT_p(\text{opt})) - 1) - T_p(\text{opt}) \qquad (9.1)$$

where $T_t$ = travel time, $m$ = a constant and $T_p(\text{opt})$ = optimal time in patch.

Figure 9.2 shows the average time spent per patch by each bird in relation to travel time. The dotted curve shows the optimal solution calculated using the above equation. Although it provides a reasonable fit to the data, Cowie also calculated the energetic costs of searching and travelling as the marginal value model is couched in terms of *net* energy intake. When these adjustments are made to the optimal solution, a much better fit is obtained (solid line) and the observed data are in good agreement with the model.

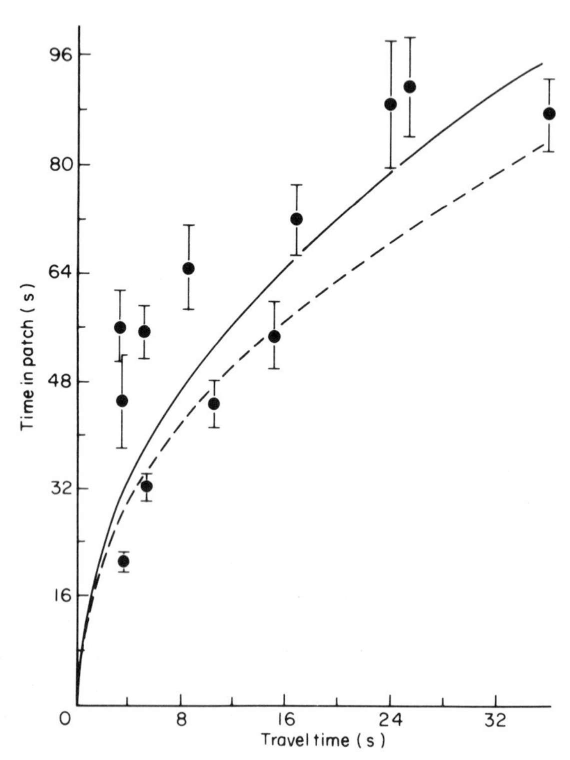


FIG. 9.2. The relationship between the mean travel time and the mean time spent in each patch for each bird in both experimental treatments (Cowie 1977). The unbroken curve shows the optimal solution after making adjustments for energy expenditure. The vertical bars represent the standard error of the mean for each bird.

There are few other data which allow more critical tests of the model, although there are many results describing gross predator behaviour relative to clumps of prey. Typically, in these experiments, the travel time is not manipulated, but the predator is presented with a number of patches of different quality. The graphical solution presented earlier referred only to one type of patch, but if there are several patches of different quality within an environment, the same departure rule should apply to all of them. That is, a patch should be abandoned when the capture rate falls to the average for that environment. This makes it possible to predict the amount of time a predator should spend in each patch.

A good example of this approach is that of Cook & Hubbard (1977), who looked at oviposition strategies in the ichneumonid parasitoid *Nemeritis canescens*. Although this is not strictly foraging, it is a very similar problem in that the wasp should try to maximize its encounter rate with suitable hosts. Cook and Hubbard assumed that the marginal encounter rate should be the same for all patches when the parasitoid leaves and that the number of prey located in each patch could be described by the 'random predator' equation of Rogers (1972). They compared their predictions on the optimal allocation of time with data obtained by watching individual wasps searching for 6 hours in an arena containing five patches of different host density. Their results are shown in Figure 9.3. At the highest density patches the wasps' behaviour was very close to that predicted by optical foraging theory, although they spent more time than predicted in the very lowest density patches.

Similar experiments have also been performed on birds. For example, Smith & Sweatman (1974) and Zach & Falls (1976a,b) performed experiments on captive great tits (*Parus major*) and ovenbirds (*Seiurus aurocapillus*) respectively. In both cases the birds were allowed to forage in an arena containing several areas of equal size (six in the former study, four in the latter) which differed in the number of prey hidden in them. Although the birds were exposed to a series of short trials, they were allowed to continue with these until they had reached a stable allocation of foraging time between the different areas. The results of both studies are presented in Figure 9.4. As with *Nemeritis*, both sets of data deviate from the optimal time allocation in that the birds spent too much time foraging in the low quality patches. Why should this be so?

One of the assumptions of Cook and Hubbard's model is that there is a fixed total hunting time and that the animal has some knowledge of the profitability of different areas. If so, the wasps might be able to determine the total number of encounters they would be likely to make and then reduce each patch in turn, starting with the most profitable (perhaps detected by scent, Waage 1977), to the final marginal level of prey density. However, this seems unlikely and a more plausible alternative would be to stay in the best patch until the profitability dropped to that of the second best and then to oscillate between the two. When the profitability of these fell to the third best, then oscillate

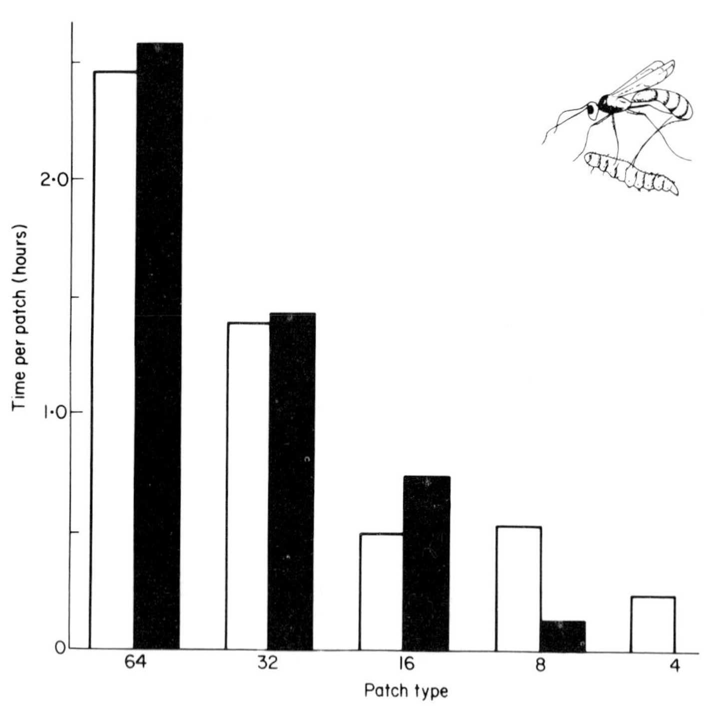


FIG. 9.3. The amount of time spent on each patch type (unshaded) by *Nemeritis* compared to the time predicted by Cook & Hubbard's (1977) model (shaded). (From Hubbard & Cook 1978.)

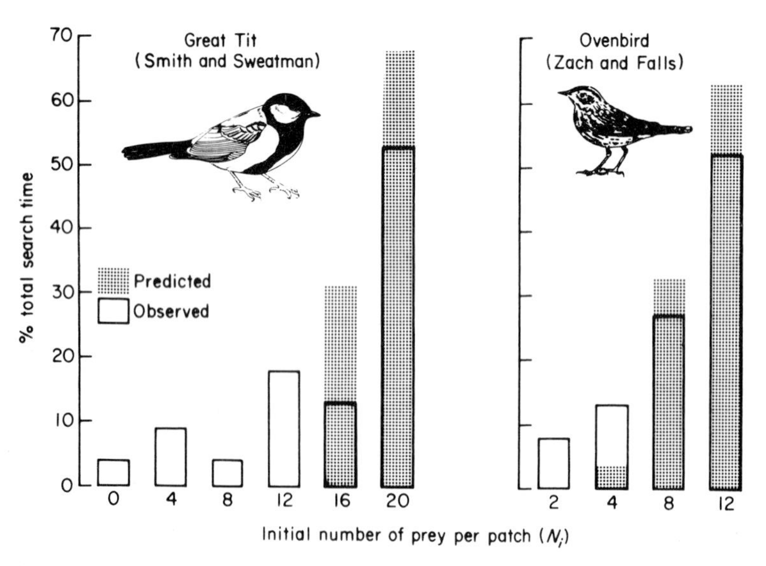


FIG. 9.4. The application of Cook & Hubbard's (1977) model of time allocation to great tits and ovenbirds (see text).

between all three and so on. Although this latter strategy would produce optimal behaviour if the total foraging time was not known by the animal, it still requires some knowledge of the relative profitability of patches. This may explain why *Nemeritis* (in Hubbard and Cook's data) spent longer than expected in the low density patches—because it was sampling them rather than exploiting them. Similarly, although in both avian studies the birds had previous experience of their environments, Oster & Heinrich (1976) have shown that in an unpredictable environment it is still a good idea to invest some time in sampling less profitable areas in case the distribution of prey changes. The question of sampling, and how much time to invest in it, now appears to be emerging as one of the key questions in optimal foraging theory and one which we will now consider in more detail.

## SAMPLING AND OPTIMAL FORAGING

The models and experiments we have discussed so far describe the behaviour of omniscient predators in stable, predictable environments. However, predators in the real world are often faced with unknown and fluctuating habitats. It is implicit in optimal foraging models that predators have the flexibility to cope with changing average rates of food intake, and that they can therefore acquire information about average habitat quality. Sampling an unknown environment could be done in two general ways. One strategy is to divide the total foraging time into an initial period of pure sampling, followed by an exploitation period. The alternative is to exploit continuously and use recent experience to make decisions about how long to stay in future patches. One specific mechanism of this type would be a sliding 'memory window' which allows the predator to average its recent experience in deciding when to leave the current patch. These two alternatives are not exclusive: predators could gradually switch during a foraging period from pure sampling to combined sampling and exploitation and finally to pure exploitation.

A direct demonstration that great tits use the 'sample-then-exploit' strategy of acquiring information is the experiment described by Krebs, Kacelnik & Taylor (1978). They presented captive great tits with an environment containing two patches which differed in profitability. The birds could determine which patch was the better of the two only by sampling, and the average results from nine birds showed that they visited the two patches for short bursts equally at the beginning of an experiment before switching to almost pure exploitation of the better patch. There was no depletion within patches, so that having located the good patch the optimal strategy was to search only in that patch.

These results are qualitatively as expected from the sample-then-exploit hypothesis, but more interesting is the question of how much sampling to do before making a decision as to which patch to exploit. This problem is exactly analogous to the classical 'two-armed bandit' problem, in which a player is faced with two fruit machines which offer rewards at two different unknown rates. The goal is to maximize the total rewards obtained in a given session, the length of which is determined by how many coins the player has at the start. It is easy to see that there is a trade-off. If the player spends too little on sampling, he might choose to exploit the wrong machine; if he spends too much on sampling, he misses the opportunity to get a higher pay-off by exploiting the good machine. One can also see intuitively that the optimal solution to this trade-off problem will depend not only on how different are the reward rates in the two machines (the bigger the difference, the less sampling is needed to detect it), but also on the total number of coins ($N$) available. The smaller the value of $N$, the less it pays to sample. Taking an extreme case, if the player has two coins, the best strategy is to guess which is the good machine and play it twice. No future benefit could be gained from the information acquired by sampling each machine once. The player with a large supply of coins pays more in sampling before committing the remainder to exploitation.

The value of $N$ can also be viewed in more naturalistic terms as corresponding to the periodicity of environmental fluctuations. An environment which stays stable over long periods corresponds to a large value of $N$, so one can make the qualitative prediction that predators should spend less time sampling in environments which fluctuate rapidly. The same argument can be used for the hypothetical 'memory window' method of acquiring information: in rapidly fluctuating environments the 'window' should be short; in more slowly changing environments it should be longer.

Figure 9.5 shows the results of Krebs *et al.*'s experiment. The two patches consisted of operant devices in which a great tit hopped on a perch to deliver a small piece of mealworm. The rewards were delivered by means of a large horizontal perspex disc drilled with holes around its upper circumference. Each hole contained a reward, and the disc was contained in a metal box with only one hole accessible at a time through a small window in the wooden top of the box. Hopping on the perch rotated the disc a small step so that the next hole was aligned with the window, the proportion of hops which rotated the disc being set in a variable ratio schedule. (On average $x\%$ of hops delivered a reward, with a pseudo-random distribution about the mean.) As we have mentioned, at the beginning of an experiment the birds worked equally in the two patches, visiting each for short bursts. Subsequently, they switched to exploit the better patch. Figure 9.5 shows the number of hops (corresponding to coins in the two-armed bandit problem) spent sampling before a decision to exploit, for various degrees of difference between the reward rates in the two patches. The data fit closely to the curve predicted by a simulation of the two-armed bandit problem (Krebs, Kacelnik & Taylor 1978) for $N = 150$. This value of $N$ happens to correspond to the modal value of $N$ for the experiments,

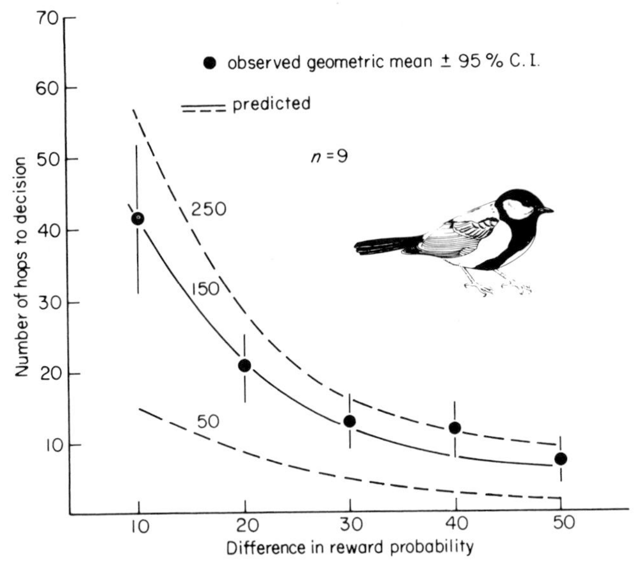


FIG. 9.5. The results of the experiment of Krebs, Kacelnik & Taylor (1978). The graph shows the average results for nine birds (solid dots). The number of sampling hops prior to switching to exploitation is plotted against the difference in reward probability between the two patches (e.g. 30% difference means that one patch gave rewards on 40% of hops while the other gave 10% rewards). As explained in the text, the two patches were operant devices. The predicted results were obtained by a dynamic programming simulation of the two-armed bandit problem. The optimal amount of sampling is that which maximizes the total rewards obtained in an experiment. The observed data points fit quite closely to the predicted optimum for $N = 150$.

so it appears that the birds are doing the right amount of sampling for the environment in which they are living.

This experiment provides direct evidence for the use of a 'sample-then-exploit' strategy of acquiring information. We have no conclusive evidence on the question of whether great tits also use a 'memory window', but we will refer to this again briefly in the next section.

## BEHAVIOURAL MECHANISMS UNDERLYING OPTIMAL FORAGING

The results we have described seem to indicate that both birds and insects are capable of behaving in a way which approaches the predictions of optimal foraging theory. However, these studies only indicate that the animals are capable of optimizing; they do not tell us what behavioural mechanisms are involved. In fact, Oaten (1977) has suggested that, since a stochastic version

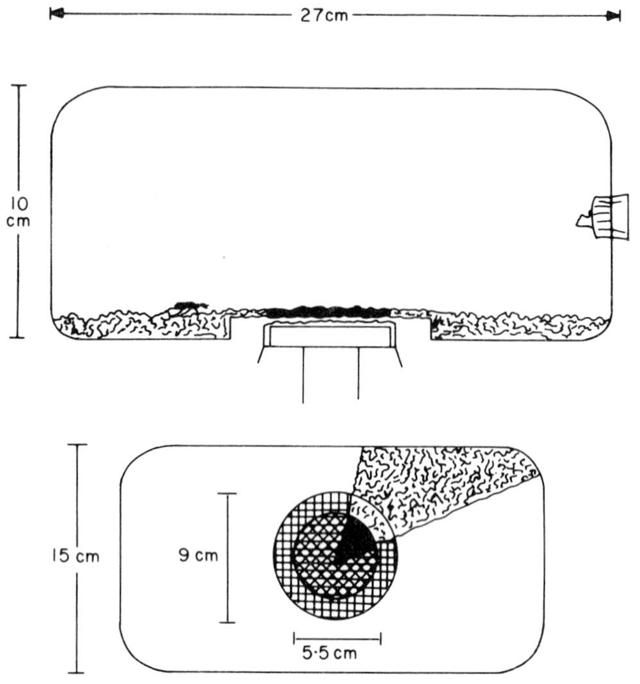


FIG. 9.6a. The apparatus used by Waage (1977, 1979) to control oviposition timings. The floor of the plastic container was covered with wheat middlings and in the centre of this surface was a circular patch (5·5 cm diameter) of host-contaminated middlings (without hosts) which overlayed a stretched terylene screen. Oviposition of wasps through this screen, while probing on the patch, was controlled by raising larvae-filled or empty dishes beneath the patch.

of the marginal value model would require very complex calculations by the animal, it may be that simple behavioural mechanisms have evolved to approximate the optimal solution. *Nemeritis* illustrates this well.

Waage (1977, 1979) investigated the locomotory responses which influence the allocation of patch time in the ovipositing wasps. He found that the time at which a *Nemeritis* leaves a patch depends on the waning of an 'edge response' and that this edge response is due largely to an increased rate of turning when the parasitoid comes to the edge of an area containing a contact chemical produced by the hosts. He also found that the wasps stay longer if they oviposit successfully, although this effect is only apparent if the host chemical is present. Furthermore, by using a cleverly constructed piece of apparatus (Fig. 9.6a) which allowed him to control the timing of ovipositions, Waage showed that the effect of an oviposition depends on its timing, so that five ovipositions spaced out in time result in a longer stay than the same number clumped at

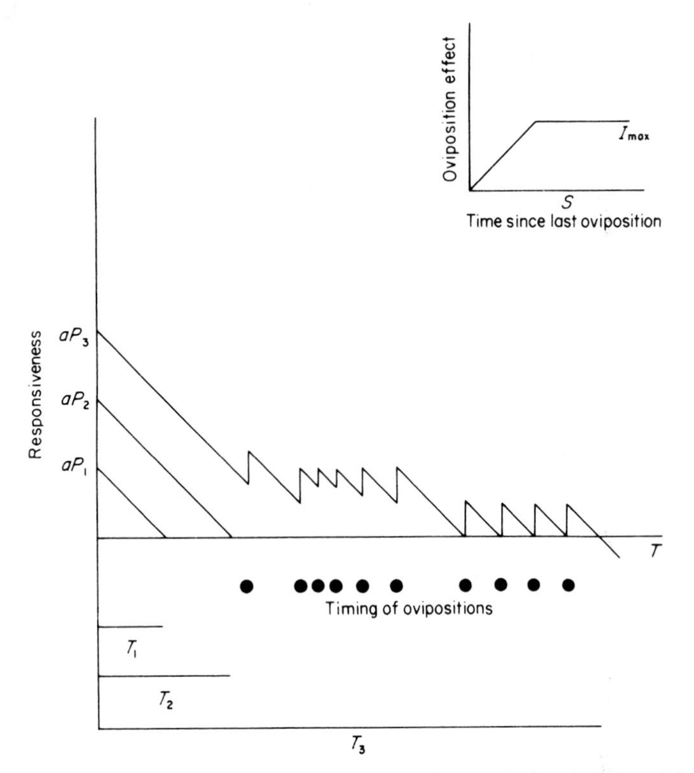


FIG. 9.6b. Graphical representation of the model for patch time in *Nemeritis* (Waage 1977, 1979), incorporating the effects of patch stimuli ($aP$) and oviposition stimuli ($I$). The patch times $T_1$, $T_2$ and $T_3$ are for the hypothetical patch densities $P_1$, $P_2$ and $P_3$ respectively. The inset indicates how the time since the last oviposition influences the value of the increment ($I$).

the start of a visit. He demonstrated that these results could be explained by a simple behavioural mechanism determining patch time which takes into account the decay of the edge response and the incremental effect of ovipositions (Fig. 9.6b). The intensity of the chemical stimulus, perceived as the wasp enters the patch, sets some hypothetical level of responsiveness ($aP$) which decays to a threshold ($r$), whereupon the turning response to the patch edge is no longer elicited and the patch is abandoned. An oviposition while on the patch affects this decay by adding an increment of responsiveness and the value of this increment is affected by the time since the last oviposition (see inset). This very simple mechanism will produce similar results to the model of Cook & Hubbard (1977) in that the marginal rate of oviposition will be similar for a range of patch densities. However, Waage also showed that this marginal threshold is little affected by previous experience in the short term.

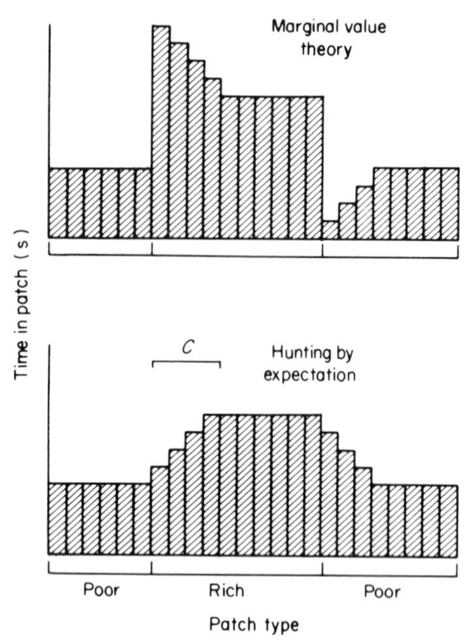


FIG. 9.7a. The predicted response to a change in the quality of the environment, made by the marginal value and time expectation models, in conjunction with a memory window (Cowie 1979). The distance $C$ represents the size of the hypothetical memory window.

This means that the behaviour may not always result in optimal foraging as it does not take into account variations in the overall prey density. However, it may be that the life of the adult wasps is too short to make that much flexibility worthwhile and that they need a quick 'rule' for patch exploitation which they can employ without first indulging in time-consuming sampling of the environment.

Although a simple, relatively inflexible mechanism may be suitable for insect parasitoids, this is clearly not so in the case of birds which may face a wide range of conditions within their lifetime. Moreover, the travel-time experiment (Cowie 1977) showed that great tits are capable of making accurate adjustments in their behaviour dependent on the characteristics of the environment in which they are foraging. How do they do it?

Earlier evidence seems to rule out a number expectation (Krebs, Ryan & Charnov 1974; Zach & Falls 1976b), although not a time expectation. In order to test whether the birds were basing their decision to leave on a time or rate expectation, Cowie (1979) performed another experiment on great tits which involved subjecting the birds to two environments with widely different prey densities. The experimental set-up was similar to that described earlier,

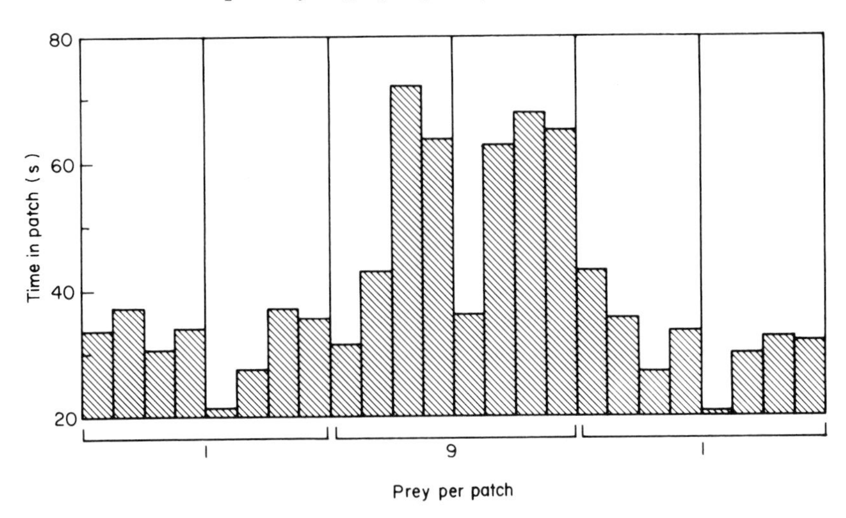


FIG. 9.7b. The mean response of six great tits which were subjected to two trials in a poor environment, followed by two in a rich and two back in the poor. The data for each trial are divided into quartiles. (Cowie 1979.)

although in this case the travelling time remained constant. The birds were allowed to forage for two trials in a poor environment (1 prey/patch), followed by two trials in a rich environment (9 prey/patch) and then two trials back in the poor environment. Figure 9.7a shows two possible ways in which a bird might respond if it is basing its decision to leave on a rate or time expectation combined with some kind of memory window. If using the marginal capture rate when it switches from the poor to the rich environment, it should respond by spending a very long time in the first few patches it encounters, as it will be using a low marginal-rate expectation acquired in the poor environment. This should subsequently be modified by experience in the rich environment until a new optimum level is achieved. When the bird moves from the rich environment to the poor, the reverse effect should be observed. If, on the other hand, the birds are hunting by time expectation, we might expect a different response. In this case a transition is followed by a gradual adjustment of the time expectation as the birds become acquainted with the quality of the new environment. Most important, there is no abrupt change in the time spent per patch following the transition. Figure 9.7b shows the average result which Cowie obtained from testing six birds. The data in each trial are divided into quartiles to make it possible to combine data for different birds. These histograms suggest that the overall response may be more similar to that predicted by the hunting-by-time expectation model because, following the transition, there is a gradual adjustment of the total patch time. This means that there is a tendency for the birds to stay longer after a good-to-poor transition and shorter after a poor-to-good one. Although these results eliminate the marginal capture rate used in conjunction with a memory window, they do not rule out completely the possibility that the birds are using a sample-then-exploit strategy, as discussed in the previous section. There is a tendency in the data for the first quartile to be shorter than the others; this could be due to sampling or to the fact that the birds were nervous at the start of each trial. It is significant, however, that in the rich-to-poor transition the first quartile is longer than the others and this leads us to suspect that a memory window is involved. In fact, Cowie (1979) has also performed some other experiments which indicate that the birds do base their decision to leave on a time expectation. However, it must be emphasized that this time expectation is flexible and can be adjusted depending on the overall food density in the environment. Thus, as in the case of *Nemeritis*, it may provide the animals with a simple behavioural mechanism which they can use to approximate the optimal solution.

Breck (1978) has suggested on theoretical grounds that a time expectation might provide a good proximate rule for a predator to use in order to achieve the ultimate goal of maximizing net gain rate. He compared three models in which a predator based its decision to leave a patch either on a giving-up time, or time expectation, or a random strategy. He found that the capture rate under the time expectation mechanism was less sensitive to stochastic variations in the inter-capture intervals than the rate produced by a giving-up time, and that if the prey were randomly distributed amongst the patches, a time expectation was the best mechanism for a predator to use. To illustrate this point Breck re-analysed the data of Cowie (1977) and showed that a flexible time expectation produced the highest mean capture rate in each environment. However, he did not analyse in detail the crucial question of how a predator might arrive at the optimal time expectation, although some averaging of gain rate must presumably be involved.

The alternative proximate rule, discussed by Breck (1978), which is based on a giving-up time does not seem to be applicable to great tits. Cowie (1979) found that great tits often leave a patch immediately after finding a prey item (a giving-up time of zero). Figure 9.8 shows that this result is consistent with the marginal value model. In Figure 9.1 we assumed that the average cumulative gain curve was continuous, but in the experiments with great tits the gain within a patch occurs in steps (i.e. prey items), and if the optimal patch time is calculated graphically by fitting a tangent, the solution is for the predator to leave immediately after making a capture (Fig. 9.8). Cowie (1979) took this idea one stage further and suggested that it would be more important for a predator to 'leave after a find' if the steps (prey) were large than if they were small. He tested this prediction by performing an experiment which involved the discs already described in the context of the two-armed bandit problem. In this case the discs were programmed with a progressive reward schedule which meant that the longer a bird stayed at a disc, the more hops it had to perform to receive the next reward. By manipulating these reward schedules, Cowie presented six great tits with two different environments in which the total energy intake with respect to the time spent at a disc was identical, but in one environment the prey were four times as big and were encountered at intervals four times as long. Cowie found that in both environments the birds frequently left after capturing a prey item, but that by the end of the environment the birds were doing so more frequently in the 'large prey environment'.

To summarize this section, there is growing evidence that whilst animals are capable of exploiting patches in an optimal fashion, they do so by using simple 'rules of thumb', which they can use to come close to the true optimal solution. The conclusion that predators use simple behavioural mechanisms in exploiting patches should not be considered as evidence against optimal foraging. Mechanisms such as those described for *Nemeritis* provide a causal account of behaviour which complements the functional interpretation provided by optimality theory.

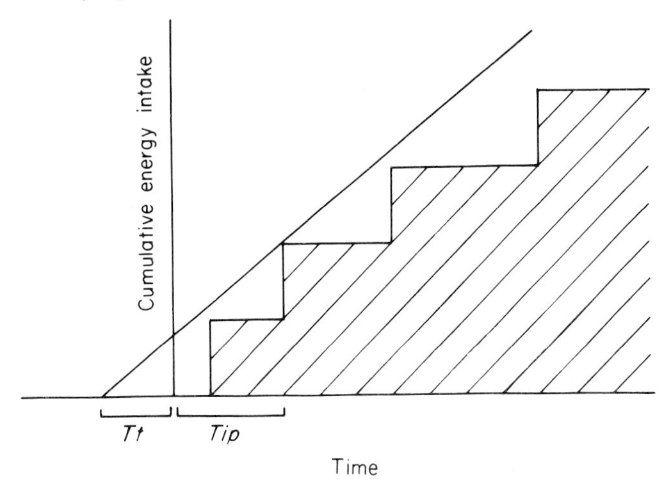


FIG. 9.8. In Cowie's (1979) experiments the intake within a patch occurred in steps, each step corresponding to a piece of mealworm. A graphical solution equivalent to Figure 9.1 shows that the optimal predator should leave a patch immediately after capturing a prey item. (*Tt* is the average travel time and *Tip* is the optimal patch time).

## THE MARGINAL VALUE MODEL IN OTHER CONTEXTS

Earlier (p. 187 *et seq.*) we described how the marginal value model (Fig. 9.1) has been tested in experiments using foraging birds or searching parasitoids. In this section we briefly describe three other examples in which the same model has been examined.

### *Copulation time in dungflies*

Male dungflies (*Scatophaga stercoraria* L.) assemble at fresh cowpats, where they compete for incoming females. The females mate and then lay their eggs in the dung. When a male captures a receptive female, which could be either through finding a new arrival or by taking over a female which is already *in copula*, the pair usually emigrate from the pat (to avoid interference) and complete sperm transfer in the surrounding grass before returning to lay the eggs. In spite of the fact that the male guards the female at this stage, takeovers are common and the second male also mates with the female. In a most ingenious series of experiments, Parker (1970) was able to find out how many eggs a male could expect to fertilize after a takeover as a consequence of copulating for different lengths of time. The experiments involved allowing a single female to mate with two successive males, one of which had been sterilized. The second male's sperm always fertilizes the eggs (as a result of sperm competition), so that by counting the proportion of eggs hatching Parker was able to calculate the amount of sperm transferred after different copulation times. If, for example, the second male was sterile, the longer he copulated, the smaller was the proportion of eggs hatching. These experiments allowed Parker to plot the cumulative gain function (corresponding to Fig. 9.1) which, together with an estimate of the time a male spent searching for a new female, enabled him to predict successfully the observed copulation time.

### *Handling time in* Notonecta glauca

Cook & Cockrell (1978) showed that when *Notonecta* sucks food from larvae of *Culex molestus*, the dry weight extracted as a function of handling time is a negatively accelerating function of the type shown in Figure 9.1. When the *Notonecta* were presented with prey at five different densities (corresponding to five 'travel times'), the handling time varied inversely with prey density—a result qualitatively in agreement with the marginal value model.

### *Nectar extraction by bumblebees*

When bumblebees (*Bombus sonorus*) extract nectar from a flower of the desert willow (*Chilopsis lincaris*), they first suck nectar from a pool at the base of the corolla and then remove nectar from five grooves which radiate out from the corolla base (Whitham 1977). Groove nectar is harder to extract than pool nectar, so that when most flowers contain pool nectar the marginal value theory would predict that bees should ignore the groove nectar. This is exactly what Whitham found. In the early morning bees ignore the groove nectar, but as the morning progresses, and the nectar is depleted, the bees switch to taking groove nectar as well.

## OPTIMAL FORAGING AND STABILITY OF PREDATOR–PREY INTERACTIONS

As we have emphasized, optimal foragers concentrate their searching effort in patches of high prey density. Hassell & May (1973) have demonstrated theoretically that any mechanism producing aggregation of predators or concentration of foraging effort in places where prey are abundant has a powerful stabilizing effect on predator–prey interactions, by providing a partial refuge for prey in low density patches.

The first theoretical analysis of the stabilizing effect of predator aggregation assumed that predators allocated their foraging effort in relation to prey density by a simple exponential function (Hassell & May 1973), or by using a simple rule such as a fixed giving-up time (Hassell & May 1974; Murdoch & Oaten 1975). More recently Comins & Hassell (1979) have examined the stabilizing effect of optimal allocation of search effort by foraging parasitoids. They analysed a simplified optimal foraging model similar to that used by Cook & Hubbard (see p. 189). Comins and Hassell also assumed that transit time between patches is negligible, and that the parasitoids can recognize patch quality from a distance (for example, by scent) so that they are always in the currently best patch. In exploiting healthy hosts, all the parasitoids initially aggregate on the best available patch; when this is reduced to a level of profitability equal to the second best patch, the parasitoids divide themselves equally between the two patches and so on. The exact division of parasitoids between patches in later parts of the interaction is influenced by handling time. Although Comins and Hassell's model is based on the premise that *individual* parasitoids maximize their encounter rate with healthy hosts, while Cook and Hubbard in effect assumed that this was the goal of the total parasitoid *population*, the overall description of the allocation of searching effort by the parasitoid population is the same in the two models.

Figure 9.9a,b summarizes two results of Comins and Hassell's model. Figure 9.9a shows the consequence of optimal foraging for parasitoid searching efficiency (see figure caption for exact definition). If parasitoids search at random with respect to patch quality, searching efficiency is unaffected by parasitoid population density (broken line). If, however, parasitoids aggregate in high density patches, searching efficiency is better than random at low parasitoid density. As parasitoid density increases, prey exploitation of the originally good patch(es) increases (pseudo-interference—Free, Beddington & Lawton 1977) and the searching efficiency over all patches therefore declines. The optimally foraging parasitoids move to other patches when competition increases and so they never do worse than random search; when all patches are reduced to equal profitability, they do no better than random. The fixed aggregation strategy produces non-adaptive persistence by the parasitoids in

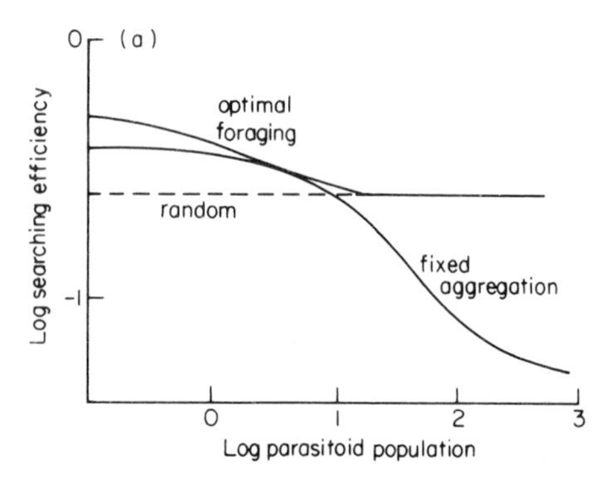


FIG. 9.9a. The relationship between overall parasitoid searching efficiency and population size. Searching efficiency ($a'$) is defined as follows: $a' = 1/PT(N/S(T))$, where $N$ is the total host population, $P$ the parasitoid population and $S(T)$ the number of surviving hosts at time $T$. If the parasitoids search at random, $a'$ is by definition a constant, but with fixed aggregation or optimal foraging searching efficiency declines with increased parasitoid numbers. The slope of this decline at any particular value of $P$ is the pseudo-interference constant ($m$) for that parasitoid population size.

the best patch(es) even when competition through increased parasitoid density makes these patches unprofitable. Optimal foraging is clearly better than fixed aggregation for the parasitoid.

Figure 9.9b shows the difference between optimal foraging and fixed aggregation for stability of predator–prey interactions. The graph is derived from Hassell & May (1973), who showed that interference (or pseudo-interference in this case) promotes stability in predator–prey models. They calculated which values of the interference constant ($m$) (defined in the figure caption) will result in stability with particular host rates of increase. The shaded region of Figure 9.9b covers all the stable combinations. The two solid lines show combinations of $m$ and host rate of increase which are produced by the optimal foraging model and by one example of a fixed aggregation model. The qualitative conclusion is that the two types of model have similar regions of stability. In summary, optimal foraging makes an important difference to parasitoid fitness but little qualitative difference to predator–prey stability.

## SUMMARY

We have emphasized three main points. Simple deterministic optimal foraging models describe with some success the behaviour of predators and parasitoids in laboratory experiments. Predators may, however, achieve the optimal end result by using simple behavioural mechanisms which give roughly the right

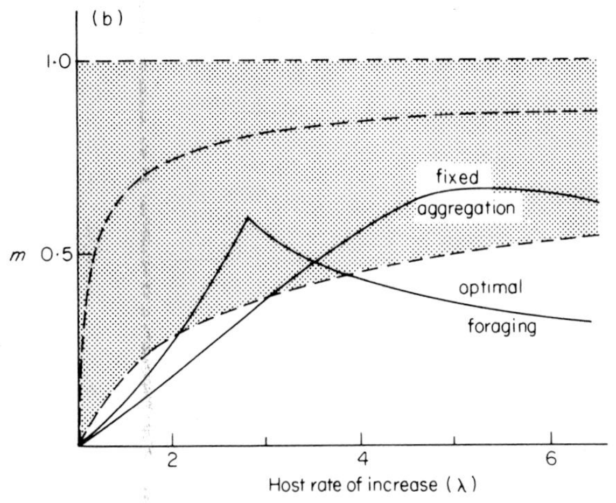


FIG. 9.9b. The effect of optimal foraging and fixed aggregation on stability of a parasitoid-host interaction. The plot shows combinations of the pseudo-interference constant ($m$) and host rate of increase ($\lambda$), resulting in local stability (Hassell & May 1973) (shaded region). The 'optimal foraging' line was derived by calculating from the optimal foraging model the equilibrium number of parasitoids for a particular host distribution and rate of increase, and reading off the corresponding value of $m$ from Figure 9.9a. The 'fixed aggregation' line was derived in a similar way using an exponential aggregation rule with an exponent value of 1. (From Comins & Hassell 1979.)

answer, two examples being the scent habituation mechanism of *Nemeritis* and the time expectation used by great tits. Learning is an important component of foraging in an unknown environment and we have shown that, in a simple experiment, great tits sampled an unknown environment before exploiting it, and that their division of effort into sampling and exploiting was as predicted by an optimal foraging model.

## ACKNOWLEDGMENTS

We thank the Science Research Council for support and Robin Cook, Steve Hubbard and Barbara Cockrell for permission to quote unpublished work. Nick Davies commented on the manuscript.

## REFERENCES

**Breck J.E. (1978)** Sub-optimal foraging strategies for a patchy environment. *Unpublished Ph.D. thesis, Michigan State University.*

**Brock V.E. & Riffenburg R.H. (1960)** Fish schooling, a possible factor reducing predation. *Journal du Conseil. Conseil permanent international pour l'exploration de la mer*, **25**, 307–317.

**Charnov E.L. (1973)** Optimal foraging: some theoretical explorations. *Unpublished Ph.D. thesis, University of Washington.*

**Charnov E.L. (1976)** Optimal foraging: the marginal value theorem. *Theoretical Population Biology*, **9**, 129–136.

**Comins H.N. & Hassell M.P. (1979)** The dynamics of optimally foraging predators and parasitoids. *Journal of Animal Ecology* (in press).

**Cook R.M. & Cockrell B.J. (1978)** Predator ingestion rate and its bearing on feeding time and the theory of optimal diets. *Journal of Animal Ecology*, **47**, 529–548.

**Cook R.M. & Hubbard S.F. (1977)** Adaptive searching strategies in insect parasites. *Journal of Animal Ecology*, **46**, 115–127.

**Cowie R.J. (1977)** Optimal foraging in great tits. *Nature, London*, **268**, 137–139.

**Cowie R.J. (1979)** Foraging behaviour of the Great Tit (*Parus major*) in a patchy environment. *Unpublished D.Phil. thesis, Oxford University.*

**Free C.A., Beddington J.R. & Lawton J.H. (1977)** On the inadequacy of simple models of mutual interference for parasitism and predation. *Journal of Animal Ecology*, **46**, 543–554.

**Gibb J.A. (1960)** Population of tits and their food supply in pine plantations. *Ibis*, **102**, 163–208.

**Gibb J.A. (1962)** L. Tinbergen's hypothesis on the role of specific search images. *Ibis*, **104**, 106–111.

**Hamilton W.D. (1971)** Geometry for the selfish herd. *Journal of Theoretical Biology*, **31**, 295–311.

**Hassell M.P. & May R.M. (1973)** Stability in insect host-parasite models. *Journal of Animal Ecology*, **42**, 693–726.

**Hassell M.P. & May R.M. (1974)** Aggregation in predators and insect parasites and its effect on stability. *Journal of Animal Ecology*, **43**, 567–594.

**Hubbard S.F. & Cook R.M. (1978)** Optimal foraging by parasitoid wasps. *Journal of Animal Ecology*, **47**, 593–604.

**Krebs J.R. (1973)** Behavioural aspects of predation. *Perspectives in Ethology* (Ed. by P.P.G. Bateson & P.H. Klopfer) pp. 73–111. Plenum Press, New York.

**Krebs J.R., Ryan J.C. & Charnov E.L. (1974)** Hunting by expectation or optimal foraging? A study of patch use by Chickadees. *Animal Behaviour*, **22**, 953–964.

**Krebs J.R., Kacelnik A. & Taylor P.J. (1978)** Test of optimal sampling by foraging great tits. *Nature, London*, **275**, 27–31.

**MacArthur R.H. & Pianka E.R. (1966)** On the optimal use of a patchy environment. *American Naturalist*, **100**, 603–609.

**Murdoch W.W. & Oaten A. (1975)** Predation and population stability. *Advances in Ecological Research*, **9**, 2–131.

**Oaten A. (1977)** Optimal foraging in patches: a case for stochasticity. *Theoretical Population Biology*, **12**, 263–285.

**Oster G. & Heinrich B. (1976)** Why do bumblebees major? A mathematical model. *Ecological Monographs*, **46**, 129–133.

**Parker G.A. (1970)** Sperm competition and its evolutionary effect on copula duration in the fly *Scatophaga stercoraria* L. *Journal of Insect Physiology*, **16**, 1301–1328.

**Parker G.A. (1978)** Searching for mates. *Behavioural Ecology: An Evolutionary Approach* (Ed. by J.R. Krebs & N.B. Davies) pp. 214–244. Blackwell Scientific Publications, Oxford.

**Rogers D.J. (1972)** Random search and insect population models. *Journal of Animal Ecology*, **41**, 369–383.

**Smith J.N.M. & Sweatman H.P.A. (1974)** Food searching behaviour of titmice in patchy environments. *Ecology*, **55**, 1216–1232.

**Waage J.K. (1977)** Behavioural aspects of foraging in the parasitoid *Nemeritis canescens* (Grav.). *Unpublished Ph.D. thesis, London University.*

**Waage J.K. (1979)** Foraging for patchily-distributed hosts by the parasitoid, *Nemeritis canescens* (Grav.). *Journal of Animal Ecology* (in press).

**Whitham T.G. (1977)** Coevolution of foraging in *Bombus* and nectar dispensing in *Chilopsis*: a last dreg theory. *Science*, **197**, 593–596.

**Zach R. & Falls J.B. (1976a)** Ovenbird (Aves: Parulidae) hunting behaviour in a patchy environment: an experimental study. *Canadian Journal of Zoology*, **54**, 1863–1879.

**Zach R. & Falls J.B. (1976b)** Do ovenbirds (Aves: Parulidae) hunt by expectation? *Canadian Journal of Zoology*, **54**, 1894–1903.

# 10. INVERTEBRATE GRAZING, COMPETITION AND PLANT DYNAMICS

J. B. WHITTAKER

*Department of Biological Sciences, University of Lancaster, Bailrigg, Lancaster*

## INTRODUCTION

Herbivory as an ecological phenomenon has not received as much attention as predation (of animals) or competition. Perhaps this is partly because the organisms concerned are occupying different trophic levels and require different techniques for their study. It is undoubtedly made more difficult by the plasticity of the plants and the fact that the plant may continue to live after being grazed (Harper 1977). For this reason I prefer not to use the term 'predator' for a grazing herbivore and will refer to herbivory or grazing. The fact that the latter has often been restricted to the activities of vertebrates seems to be no good reason for not extending it to invertebrates.

Apart from the special cases of crop pests, most studies of herbivory by invertebrates have made the herbivore itself the centre of attention. There are, therefore, plenty of studies of the dynamics of herbivores and a goodly amount of research into the responses of animals to changes in the food plant (Feeny 1968; McNeill 1973; van Emden & Way 1973; Cull & van Emden 1977; Horsfield 1977, etc.).

Relatively few studies, however, have been concerned with the dynamics of the plant when grazed by invertebrates even though it has been known for a long time that larger grazing animals may have a profound effect on individual plant species and on the structure of communities. However, large-scale defoliation of plants in natural communities by invertebrates seems to be a comparatively rare phenomenon (Hairston, Smith & Slobodkin 1960), although when it does occur it can be extremely dramatic, e.g. the virtually complete defoliation of mountain birch, *Betula tortuosa* (Ledeb.), in parts of north Finland by *Oporinia autumnata* Bkh. (Tenow 1972). This had led many authors (Williams 1954; Hairston, Smith & Slobodkin 1960) to conclude that invertebrate herbivory is not of very great significance in the dynamics of plants.

Enlightened observers of grazing by invertebrates have followed Wilson (1943) and Huffaker (1952, 1962) in their supposition that quite small amounts of damage to a plant may result in its demise by a mechanism which is more complex than direct destruction. Harper (1967) and Harris (1973b), in two excellent reviews of 'predation' of plants, emphasize that casual observations of damage to plants by invertebrates can give very little insight into the dynamic processes involved in plant–animal interactions. Relatively low densities of herbivores could be having a very significant effect on the dynamics of a plant if the dispersal rate and searching efficiency of the animal are high (Harper 1977). Janzen (1970, 1974), too, has shown how predation of seeds by invertebrates may have a significant influence on the abundance of species and diversity of plant communities.

It is no accident that much of the literature concerned with these dynamic processes is in the field of biological control of weeds. This is because only in well-documented successful control programmes is the history of attack by the invertebrate known and quantified. So, if invertebrates are known to be important in these cases, why do we so readily ignore them in considerations of natural plant communities? This question has been clearly posed by Harris (1973b). He summarizes his findings in the general statement that the main effect of insects is to produce rather stable plant communities of many species, each at low density. If this is only partly true, it highlights the need for much more study of the effects of invertebrates on the dynamics of plants.

I have not considered it my task in this paper to cover again the ground of recent reviews of plant–animal interactions (Harris 1973b; van Emden & Way 1973; Harper 1977), but rather to take one particular aspect of their arguments and to explore it in more detail; namely, the effects of grazing on plants which are themselves already suffering some additional stress. Grime (1973) draws attention to a mechanism by which grazing may alter the incidence of species of high 'competitive index' in certain kinds of vegetation. A most important influence on the vulnerability of a plant is its competitive ability (see Grime this symposium). This is often observed as an 'all-or-nothing' phenomenon in which the plant survives or not. However, the selective advantage resulting in the ousting of a competitor may be very slight and easily eroded by an additional selection pressure such as grazing. In these circumstances the level of grazing required may be quite low and result from low densities of herbivores which may not be readily visible.

In this paper I wish to consider grazing by invertebrate animals, not in isolation but in relation to food plants which may be competing with each other or with other species.

## THE SIGNIFICANCE OF INCREASED PLANT YIELDS UNDER DEFOLIATION

It must not be assumed that defoliation by invertebrates necessarily results in a reduction in the yield of a plant (Alcock 1964; Harris 1974; Conway 1976). There is no doubt that compensatory growth can occur (Harris 1973b, 1974) and that, at low herbivore pressure, the yield from a crop may be higher than in the absence of animals (Kincade, Laster & Brazzel 1970). From the point of view of the agriculturalist the yield of the marketable part of the crop over a short period of time may be all that matters. The relationship is more complex, however, when the long-term performance of a plant species is considered. Ellison (1960) points out that, although moderate levels of defoliation may be stimulatory if they occur at a time when carbohydrate reserves in the plant are high, a similar level of defoliation at other times may be damaging. Thus, Jameson (1963) argues that damage to young leaves may enhance yield, whilst damage to old leaves may reduce it. Johnson (1956) believes that the balance of opinion is that even light grazing is likely to be deleterious in the long run because the stimulation that occurs is presumably made at the expense of stored food reserves and root growth is retarded (p. 210).

From the point of view of my theme, it is irrelevant in general terms whether grazing increases or reduces yield since either may result in a shift in the competitive balance between the grazed plant and other species.

## ABOVE- AND BELOW-GROUND EFFECTS OF GRAZING

Casual assessment of the damage done to plants by grazing invertebrates is apt to be based on the quantities of material removed from the visible, above-ground parts of the plant. There is abundant evidence, however, that removing green material from plants by clipping or grazing often reduces root growth. Jameson (1963) reviews the relevant literature citing earlier commentaries by Weinmann (1948), Curtis & Clark (1950), Troughton (1957) and Ellison (1960). They conclude that whilst yield of above-ground parts of plants after clipping is somewhat unpredictable, root production is universally depressed. This observation is linked to studies of shrubs, trees (Kulman 1971) and herbaceous plants as well as grasses (Weinmann 1948).

Most studies of the effects of harvesting on root growth have relied on the artificial procedure of clipping the vegetation. Authors are well aware that there are dangers in this method, but insofar as it has usually been used to simulate vertebrate grazing, are usually agreed that it can give reasonable results (Culley, Campbell & Canfield 1933) if used with care. However, this assumption is not necessarily valid in the case of invertebrates and the effect of grazing on roots must be considered empirically. Although Skuhravy (1968) gives one case of defoliation increasing tuber yield, the majority of studies have shown that harvesting, clipping or grazing is likely to reduce the size of below-ground food reserves and roots in general (Bryant & Blaser 1961; Heady 1961; Jameson 1963). Dyer and Bokhari (1976) examined what happened when grass-hoppers, *Melanoplus sanguinipes* (Fabr.), were allowed to graze on blue grama grass (*Bouteloua gracilis* H.B.K. Lag) grown in hydroponic culture. They concluded that the most marked effect of above-ground grazing by these insects was to increase below-ground activity resulting from the translocation of material down into the roots. They also contend (although perhaps rather less convincingly) that roots of grazed plants may produce greater quantities of root exudates. If this is so, then the competitive influence of such plants may be greatly affected (Rovira 1969). Cussans (1974) emphasizes that weed populations may be influenced by toxic exudates from other plants, a subject which was reviewed by Holm (1971) and is now receiving more attention. If exudates are increased by grazing, then it may have even wider implications than previously thought. Weiss (1976), on the other hand, in his work on artificial defoliation of the spiny emex (*Emex australis* Steinh.) which simulated feeding by a weevil (*Apion antiquum* Gyll.), found that although total dry weight of plants was significantly reduced at higher levels of defoliation, stem weight was not. Root weight, however, was significantly depressed in all treatments (Table 10.1) except at the very end of the experiment when there was some recovery of roots. This effect is not confined to herbivores which defoliate, since Dixon (1971) found that aphids (*Eucallipterus tiliae* L.) feeding on lime (*Tilia x vulgaris* Hayne) prevented root growth.

Weiss (1976) also recorded a decrease in the root–shoot ratio as a result of defoliation and Brouwer (1962) showed the same effect for beans. This was also noted in the experiments by Bentley & Whittaker (p. 215), but the two plants concerned responded differently in that *Rumex obtusifolius* L. had an increased root–shoot ratio when grazed by a chrysomelid beetle, *Gastrophysa viridula* Degeer., whereas *R. crispus* L. had a decreased ratio (Table 10.2). Nevertheless, grazing always resulted in a net decrease in root material. In these experiments there was an interesting difference between the effects of clipping (simulated grazing) and the effects of true grazing. For example, clipping of *R. obtusifolius* resulted in a bigger percentage reduction of roots than of above-ground material whilst grazing produced a greater percentage loss above ground.

TABLE 10.1. Effect of defoliation (by removal of leaf discs or entire leaves) on total dry weight and root dry weight of *Emex australis* after 102 days (from Weiss 1976)

| % and method of defoliation | Total g plant$^{-1}$ | Stems g plant$^{-1}$ | Roots g plant$^{-1}$ | Total roots |
|---|---|---|---|---|
| 0 | 15·5 | 7·3 | 4·2 | 3·7 |
| 50 (discs) | 13·4 | 7·0 | 2·8* | 4·8 |
| 75 (discs) | 11·7* | 6·5 | 1·9* | 6·2* |
| 50 (leaves) | 6·5* | 6·9 | 2·0* | 3·3 |
| 75 (leaves) | 2·2* | 6·3 | 0·5* | 4·4 |

* Significantly different ($p = 0{\cdot}05$) from the control.

TABLE 10.2. Root–shoot ratios of *Rumex obtusifolius* and *R. crispus* when ungrazed or grazed by *Gastrophysa viridula*

| | *R. obtusifolius* | *R. crispus* |
|---|---|---|
| Ungrazed | 1·18 | 2·14 |
| Grazed | 3·57 | 1·69 |

These examples suggest that above-ground grazing by invertebrates, like that by vertebrates, usually results in significant loss of below-ground material. To this may be added the direct damage to roots by below-ground grazers such as nematodes, fly larvae, etc. In some instances this can be extreme, so that Coulson & Whittaker (1978), for example, found that Tipulidae feeding on roots of grasses in an upland limestone grassland ingested almost as much plant material as did the sheep and Homoptera grazing above.

## EXPERIMENTAL STUDIES OF GRAZING WITH COMPETITION OR OTHER STRESS

Despite numerous remarks relating grazing damage to the severity of competition (e.g. Goeden, Fleschner & Ricker 1967; Huffaker 1952; Harris 1973a; Mueggler 1967; Andres & Bennett 1975; Wilson 1943; Doutt 1960, etc.), there are very few quantitative studies of this phenomenon involving invertebrates. The few examples which follow will illustrate some of the mechanisms involved.

### *Gorse seedlings*

In 1931 Chater wrote an interesting account of the way in which animals (including invertebrates) may influence the survival of gorse seedlings. The data which he presents permit the construction of a partial cohort life table representing the mortality of observed seedlings (Table 10.3). Chater did not distinguish between the damage caused by rabbits and other grazing animals, but he did show separately that up to 14% of the mortality was caused by unidentified Lepidoptera and that *Anarsia spartiella* Schrank in particular was responsible for eating the growing points in June. If this occurred when the base of the shoot was unshaded by other plants, then the lower buds were stimulated to develop and the plant simply became more bushy. If, however, the seedlings were competing with grasses and *then* were grazed, the lower buds did not develop and death resulted. In these circumstances it is not the amount of damage which is vital, but the fact that it occurs at the same time as other stress.

TABLE 10.3. A cohort life table for seedlings of gorse (from Chater 1931)

| *lx* | *dx* | *dxF* |
|---|---|---|
| 851 | 71 | Eaten by rabbits and invertebrates |
| | 16 | Bird scratching or pulling |
| | 8 | Choked by grass |
| | 221 | Dried up |
| | 316 | |
| 535 | 54 | Rotted |
| | 12 | Improperly rooted |
| | 85 | Missing |
| | 151 | |
| 384 | | |

### *The interactive effects of nematode damage and competition between oats and barley*

A study by Sibma, Kort & deWit (1964) is perhaps the best example in the literature of how competition and grazing may interact.

Oat plants, which were susceptible to nematode damage, and barley plants (which were not) were sown in varying proportions, from 100% of oats to 100% of barley, in glasshouse cultures. Nematodes were added to half the treatments and this resulted in infestations of 500–600 animals per 200 ml of soil.

When the oats and barley were grown separately in the absence of nematodes, they produced 360 and 210 kernels per container respectively. On the basis of these findings, growing oats and barley in differing proportions at sowing ought to result in the frequencies illustrated by the dotted lines in Figure 10.1a. In fact, however, the results are as shown by the solid lines, indicating an enhanced yield of the oats and a depressed yield of barley resulting from interspecific competition.

When oats and barley were grown in conditions of interspecific competition in the presence of nematodes, the competitive advantage of the oats and the disadvantage of the barley were removed so that their respective performances were no longer distinguishable from those predicted if interspecific competition had been unimportant (Fig. 10.1b). This was attributed to the damage to roots caused by nematodes preventing the oats from occupying their usual share of the available soil and enabling the barley to occupy more than expected. The densities (500–600 animals/200 ml of soil) of nematodes involved in this experiment were not sufficient to affect the performance of plants competing intraspecifically, but the interactive effect of herbivory and interspecific competition

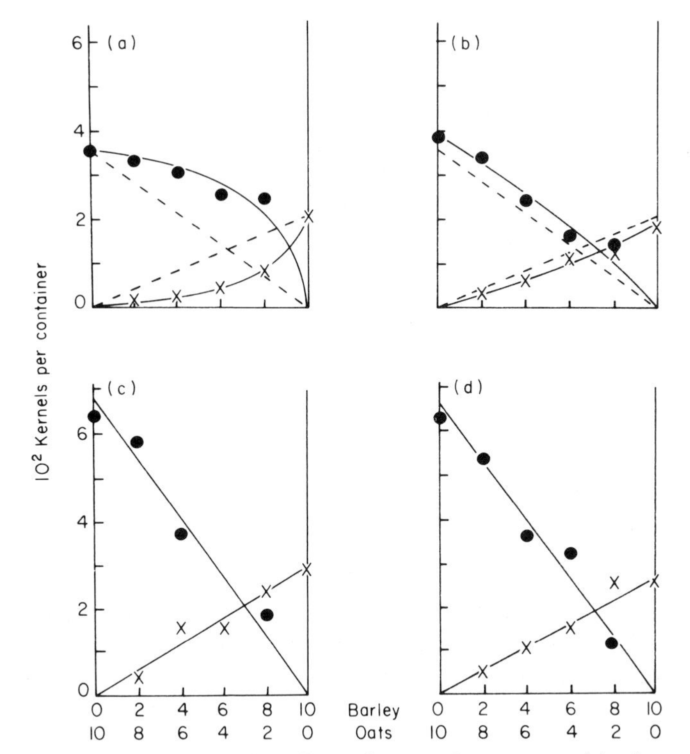


FIG. 10.1. Ratio diagrams showing effects of nematodes on competition between oats and barley in laboratory cultures. (From Sibma, Kort & de Wit 1964.) Data expressed as $10^2$ kernels per container.
(a) In the absence of nematodes and with sub-optimal supply of water.
(b) With 500–600 nematodes/200 ml of soil and with sub-optimal supply of water.
(c) In the absence of nematodes with an optimal supply of water and nutrients.
(d) In the presence of 500–600 nematodes/200 ml of soil and with an optimal supply of water and nutrients.
× × barley, ○ ○ oats, —— observed relationship, --- expected relationship in the absence of interspecific competition.

was sufficient to alter the outcome of the experiment completely. The effect of the herbivory was seen only if the plants were stressed as a result of the competition. In a similar experiment, where nutrients and water were not allowed to become limiting, the results in the nematode-infested and non-infested replicates could not be distinguished (Fig. 10.1c,d).

## *Soil fungi and Collembola*

Parkinson, Visser & Whittaker (1979) explored experimentally the hypothesis that grazing by a small collembolan, *Onychiurus subtenuis* Folsom, was

TABLE 10.4. Colonization of leaf discs by the test fungi (Basidiomycete 290 and Sterile Dark form 298) in the presence and absence of *O. subtenuis*

| Original fungal composition of leaf litter macerate | Number of Collembola added per vial | Number of plated leaf cores colonized after 5 days (mean ± S.E.) | | Number of plated leaf cores colonized after 10 days (mean ± S.E.) | |
|---|---|---|---|---|---|
| | | Basidiomycete 290 | Sterile Dark form 298 | Basidiomycete 290 | Sterile Dark form 298 |
| 100% sterile macerate | 0 | 0 | 0 | 0 | 0 |
| | 20 | 0 | 0 | 0 | 0 |
| 25% Basidiomycete 290 + 75% sterile macerate | 0 | 4·0 ± 0·6 | 0 | 11·2 ± 0·4 | 0 |
| | 20 | 5·2 ± 0·9 | 0 | 9·8 ± 0·6 | 0 |
| 25% Sterile Dark form 298 + 75% sterile macerate | 0 | 0 | 11·6 ± 0·4 | 0 | 11·4 ± 0·6 |
| | 20 | 0 | 9·8 ± 0·9 | 0 | 11·6 ± 0·4 |
| 25% Basidiomycete 290 + 25% Sterile Dark form 298 + 50% sterile macerate | 0 | 6·8 ± 0·2 | 5·4 ± 0·7 | 6·4 ± 1·1 | 5·2 ± 0·6 |
| | 20 | 1·4 ± 0·5 | 3·4 ± 1·3 | 6·6 ± 1·3 | 1·4 ± 0·6 |
| | 50 | 0·4 ± 0·4 | 2·6 ± 1·1 | 9·6 ± 0·7 | 1·6 ± 0·4 |

affecting the competitive saprophytic ability of fungi colonizing fallen aspen (*Populus tremuloides* Michx.) leaves in a forest at Kananaskis, Alberta.

After a preliminary demonstration that in the laboratory the collembolans differentially grazed Sterile Dark forms and avoided Basidiomycetes (Visser & Whittaker 1977), small simulated litter 'profiles' were set up containing macerated sterile aspen leaves above moist sterile sand. These were inoculated with Basidiomycete, Sterile Dark hyphae and Collembola according to the experimental design in Table 10.4 and a sterile aspen leaf disc was placed on top of each 'profile'. The colonization of these discs by fungi was examined at 5 and 10 days and the results are presented in Table 10.4 and Figure 10.2. When the fungi were not in competition, they were each able to colonize the aspen leaf disc whether or not they were subject to grazing by *O. subtenuis* (Fig. 10.2a,b). When in competition, however, the Sterile Dark fungus was initially the most successful colonizer in the absence of Collembola (Table 10.4), whereas the Basidiomycete was the most successful colonizer when *O. subtenuis* was present (Fig. 10.2c). Gut analyses of the Collembola showed that this switch in the competitive saprophytic ability of the two fungi was due to differential grazing by the collembolan. We concluded that the observed colonization of the aspen leaves in the field at Kananaskis may be explained by grazing of a small invertebrate affecting the outcome of a competitive relationship.

### Rumex *spp. and a chrysomelid beetle*

A further example concerns the distribution of two species of dock, *Rumex obtusifolius* and *Rumex crispus*, in several localities near to Lancaster. One of the study areas is a shingle bank in the River Lune. Although this seems to be an ideal habitat for *R. crispus*, it is in fact colonized by *R. obtusifolius*, which is heavily grazed by a chrysomelid beetle, *Gastrophysa viridula*.

Preliminary experiments (Bentley & Whittaker 1979) on the interactions between the two *Rumex* species and the beetle have consisted of growing the plants in conditions of intra- or inter-specific competition, with or without grazing by *G. viridula*. The beetle is capable of ingesting either *Rumex* species, but if given the choice will ingest twice as much *R. obtusifolius* as *R. crispus*. Levels of grazing by *G. viridula*, which had no significant effect on either species grown alone, resulted in extensive damage to *R. crispus* when the two species were competing interspecifically (Fig. 10.3). Damage to *R. crispus* was higher than expected because *R. crispus* responds by a reduction in the root–shoot ratio, thus making more material available for grazing, whereas *R. obtusifolius* responds by an increase in the root–shoot ratio, thus protecting material from grazing below ground (Table 10.2, Fig. 10.3). It is postulated that the grazing by *G. viridula* of *R. crispus* in the presence of *R. obtusifolius* may result in elimination of *R. crispus*.

These findings are being extended by field experiments at Lancaster,

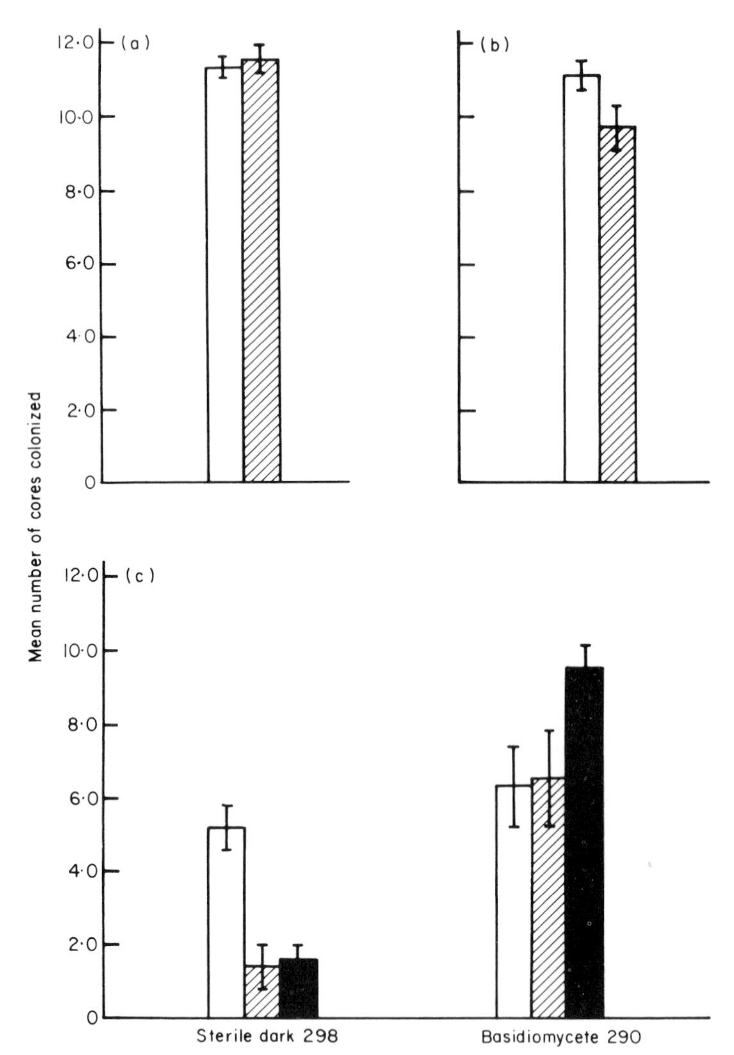


FIG. 10.2. Effects of grazing by a collembolan (*Onychiurus subtenuis*) on competition between two fungi. (From Parkinson, Visser & Whittaker 1979.) Data shown mean number of leaf cores ±S.E. colonized by Sterile Dark 298 and/or Basidiomycete 290 in the presence and absence of Collembola.
(a) Sterile Dark 298 alone.
(b) Basidiomycete 290 alone.
(c) Sterile Dark 298 and Basidiomycete 290 together.
Open column, 0 Collembola; diagonal shading, 20 Collembola; solid shading, 50 Collembola per vial.

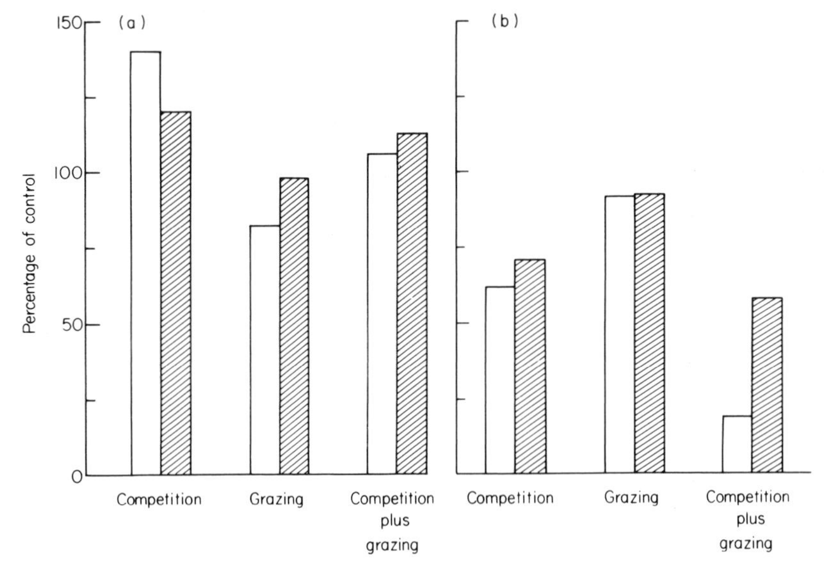


FIG. 10.3. Effects of light grazing by *Gastrophysa viridula* on competition between *Rumex obtusifolius* and *R. crispus*. (From Bentley & Whittaker 1979.) Mean leaf area (open columns) and mean root dry weight (shaded columns) of (a) *Rumex obtusifolius* and (b) *Rumex crispus*.

especially at the shingle bank location mentioned. This spring plants of *R. crispus* will be transplanted to the *R. obtusifolius* site on the shingle bank where grazing by *G. viridula* (with its headquarters on *R. obtusifolius*) will be permitted or prevented. It remains to be seen whether the natural absence of *R. crispus* on the shingle bank is a direct result of grazing by *G. viridula*.

## DISCUSSION

Time and again, in discussions about the biological control of weeds, success is attributed not just to the direct effects of grazing but to the combination of grazing and a change in the competitive status of the plant. Unfortunately, nearly all these remarks are qualitative but carry the weight of experienced observers. For example, it was deemed to be competition with smartweed (*Polygonum* sp.) and water primrose (*Jussiaea* sp.) in conjunction with insect damage (*Agasicles hygrophila* Selman & Vogt) which resulted in the successful control of alligatorweed, *Alternanthera philoxeroides* (Mart.) Griseb., in South Carolina (Weldon, Blackburn & Durdan 1973). Yet this is not likely to be nearly as important in aquatic situations as on grasslands where the growth of potential competitors of weeds is actively encouraged. Thus, the success of control of *Hypericum perforatum* L. is attributed to such an interaction in some areas (Wilson 1943), whilst Hollaway & Huffaker (1952), writing about the same plant, suggest that it is probably beetle grazing at a time when the *Hypericum perforatum* is subject to intense competition from range plants that prevents the weed re-establishing itself. Even the control of *Opuntia* by a cochineal insect (*Dactylopius* spp.) may be successful in some areas only if the grazing is accompanied by competition of the *Opuntia* with surrounding grass (Goeden, Fleschner & Ricker 1967).

In these, as in the experimental studies reviewed, quite moderate levels of grazing may be combined with otherwise ineffective competition to produce a severe effect on the target plant.

It is not obvious how grazing by invertebrates may affect competition in any particular case. Herbivory by invertebrates is often a subtle, highly selective thing which, as discussed above, may have the effect, at least in the short term, of increasing or decreasing plant yield. Reactions of the plant to grazing may include substantial modifications to root–shoot ratios and root activity and these may in turn affect the extent of the competition for light or between roots (see discussion of Grime 1973, and Newman 1973). This may in turn result in high or low vegetation diversity.

The problem of adapting models developed for animal populations to plants has been discussed at length by Harper (1977). Nowhere is it more difficult than in the general area of predator–prey interaction where the predator is a herbivore and the prey a plant. Indeed, this seemed almost intractable (Harper 1968) until recently when Noy-Meir (1975) and Caughley (1976) identified the features of predator–prey models which are applicable to herbivore–plant interactions. Rather surprisingly, in view of earlier misgivings, Caughley was able to remark that the fundamental generalizations of the Lotka–Volterra model are, in fact, even more applicable to these trophic levels than they are to those to which they have been traditionally applied. This is so because the logistic model is theoretically suitable for describing populations which cannot influence the supply of resources to the next generation. Models so far developed consider the growth of the plant in terms of increasing photosynthetic capacity as leaf area increases, followed by a plateau and subsequent decline as intraspecific competition takes place (Donald 1961; Noy-Meir 1975). Thus far, the models do not depart much from those derived for animal–animal interactions because it is assumed that intraspecific effects in the prey or branched food chains with competing predators will be the most important modifications to the basic Iota linkage.

Most authors have not considered the effect of competition among prey species or between a single prey species and a non-prey competitor taking place at the same time as predation or grazing. It has, however, been clearly established by Paine (1966) that removal of a predator may have a significant effect on the structure of a community by influencing the competitive relationships of the prey species. The theoretical background to this is explored by Hassell in this symposium. In the case of plants as 'prey', interspecific competition between prey is likely to be an even more common event (de Wit 1960; Harper 1967) in which the main 'driving force' determining plant distribution and abundance may be the competition, but the direction which this takes could be determined by the predator or herbivore. That is to say, the values of $\alpha$ and $\beta$ in the Lotka–Volterra competition equations may be determined by the plants' response to grazing as well as by their inhibitory effects on each other arising from their innate ability to compete. In such circumstances it is not germane to my general argument whether the effect of grazing is to damage the plant or to enhance its performance since either, although marginal in itself, may be the determinant of the competitive interaction.

There is, of course, absolutely no reason why this same effect should not be attributed to any other variable which affects the performance of the plants differentially. Indeed, Bennett & Runeckles (1977) have recently demonstrated just such an effect of low levels of ozone modifying the competitive abilities of crimson clover and annual ryegrass. An even more complex interaction of competition and grazing involving pollutants is discussed by Miller (1973). Competition between white fir (*Abies concolor* Lindl. & Gord) and Ponderosa pine (*Pinus ponderosa* Laws.) in California is thought to be tipped in favour of white fir by extensive mortality of the pine arising from bark beetle attack. This is enhanced, in turn, by oxidant air pollution weakening the trees.

The field observations and experiments which I have described do, I believe, suggest that the outcome of competition between plant species may in some circumstances be fully explicable only in terms of invertebrate grazing occurring at the same time as competition. To what extent this is a determinant of vegetation structure in the field can only be a matter for speculation at the present time, and experiments of the kind conducted by Cantlon (1969) and Waloff & Richards (1977), in which the removal of invertebrates from the field resulted in a change in vegetation, are greatly needed.

If there is a general moral in what I have written here, it is surely that factors affecting the dynamics of plants and animals can be considered in isolation only when they are themselves a dominating influence. Darwin (1859) made us well aware that 'the forces [of nature] are so nicely balanced . . . that . . . the merest trifle would give the victory to one organic being over another', and grazing by invertebrates is by no means a trifle.

## ACKNOWLEDGMENTS

This paper owes much to the stimulating writings of J.L. Harper, C.B. Huffaker and P. Harris amongst many. My experimental work was carried out with support from the National Research Council of Canada, the University of Calgary and the Natural Environmental Research Council.

## REFERENCES

**Alcock M.B. (1964)** The physiological significances of defoliation and subsequent regrowth of grass–clover mixtures and cereals. *Grazing in Terrestrial and Marine Environments* (Ed. by D.J. Crisp). Symposium of the British Ecological Society, 4.

**Andres L.A. & Bennett F.D. (1975)** Biological control of aquatic weeds. *Annual Review of Entomology*, **20**, 31–46.

**Bennett J.P. & Runeckles V.C. (1977)** Effects of low levels of ozone on plant competition. *Journal of Applied Ecology*, **14**, 877–880.

**Bentley S. & Whittaker J.B. (1979)** Effects of grazing by a chrysomelid beetle, *Gastrophysa viridula*, on competition between *Rumex obtusifolius* and *Rumex crispus*. *Journal of Ecology*, **67** (in press).

**Brouwer R. (1962)** Nutritive influences on the distribution of dry matter in the plant. *Netherlands Journal of Agricultural Science*, **10**, 399–408.

**Bryant H.T. & Blaser R.E. (1961)** Yields and stands of orchardgrass compared under clipping and grazing intensities. *Agronomy Journal*, **53**, 9–11.

**Cantlon J.E. (1969)** The stability of natural populations and their sensitivity to technology. *Diversity and Stability in Ecological Systems* (Ed. by G.M. Woodwell & H.H. Smith) pp. 197–205. Brookhaven Symposia in Biology, 22.

**Caughley G. (1976)** Plant–herbivore systems. *Theoretical Ecology, Principles and Applications* (Ed. by R.M. May) pp. 94–113. Blackwell Scientific Publications, Oxford.

**Chater E.H. (1931)** A contribution to the study of the natural control of gorse. *Bulletin of Entomological Research*, **22**, 225–235.

**Conway G. (1976)** Man versus pests. *Theoretical Ecology, Principles and Applications* (Ed. by R.M. May) pp. 257–281. Blackwell Scientific Publications, Oxford.

**Coulson J.C. & Whittaker J.B. (1978)** Ecology of moorland animals. *The Ecology of Some British Moors and Montane Grasslands* (Ed. by O.W. Heal & D.F. Perkins) pp. 52–93. Springer-Verlag, Berlin.

**Cull D.C. & van Emden H.F. (1977)** The effect on *Aphis fabae* of diet changes in their food quality. *Physiological Entomology*, **2**, 109–115.

**Culley M.J., Campbell R.S. & Canfield R.H. (1933)** Values and limitations of clipped quadrats. *Ecology*, **31**, 488–489.

**Curtis O.F. & Clark D.G. (1950)** *An Introduction to Plant Physiology*. McGraw-Hill, New York.

**Cussans G.W. (1974)** The biological contribution to weed control. *Biology in Pest and Disease Control* (Ed. by D. Price Jones & M.E. Solomon) pp. 97–105. Symposium of the British Ecological Society, 13.

**Darwin C. (1859)** *The Origin of Species by Means of Natural Selection*, VIth edition. John Murray, London.

**Dixon A.F.G. (1971)** The role of aphids in wood formation. II. The effect of the lime aphid, *Eucallipterus tiliae* L. (Aphididae), on the growth of the lime, *Tilia x vulgaris* Hayne. *Journal of Applied Ecology*, **8**, 393–399.

**Donald C.M. (1961)** Competition for light in crops and pastures. *Mechanisms in Biological Competition* (Ed. by F.L. Milthorpe) pp. 283–313. Symposium of the Society for Experimental Biology, 15.

**Doutt R.L. (1960)** Natural enemies and insect speciation. *Pan-Pacific Entomologist*, **36**, 1–14.

**Dyer M.I. & Bokhari U.G. (1976)** Plant–animal interactions: studies of the effects of grasshopper grazing on blue grama grass. *Ecology*, **57**, 762–772.

**Ellison L. (1960)** Influence of grazing on plant succession of rangelands. *Botanical Review*, **26**, 1–78.

**van Emden H.F. & Way M.J. (1973)** Host plants in the population dynamics of insects. *Insect/Plant Relationships* (Ed. by H.F. van Emden) pp. 181–199. Symposium of the Royal Entomological Society of London, 6.

**Feeny P.P. (1968)** Effect of oak leaf tannins on larval growth of the winter moth *Operophtera brumata*. *Journal of Insect Physiology*, **14**, 805–817.

**Goeden R.D., Fleschner C.A. & Ricker D.W. (1967)** Biological control of prickly pear cacti on Santa Cruz Island, California. *Hilgardia*, **38**, 579–606.

**Grime J.P. (1973)** Competitive exclusion in herbaceous vegetation. *Nature, London*, **242**, 344–347.

**Hairston N.G., Smith F.E. & Slobodkin L.B. (1960)** Community structure, population control, and competition. *American Naturalist*, **94**, 421–425.

**Harper J.L. (1967)** A Darwinian approach to plant ecology. *Journal of Ecology*, **55**, 247–270.

**Harper J.L. (1968)** The regulation of numbers and mass in plant populations. *Population Biology and Evolution* (Ed. by R.C. Lewontin) pp. 139–158. Syracuse University Press, New York.

**Harper J.L. (1977)** *Population Biology of Plants*. Academic, London.

**Harris P. (1973a)** The selection of effective agents for the biological control of weeds. *Canadian Entomologist*, **105**, 1495–1503.

**Harris P. (1973b)** Insects in the population dynamics of plants. *Insect/Plant Relationships* (Ed. by H.F. van Emden) pp. 201–209. Symposium of the Royal Entomological Society of London, 6.

**Harris P. (1974)** A possible explanation of plant yield increases following insect damage. *Agro-Ecosystems*, **1**, 219–225.

**Heady H.F. (1961)** Continuous vs. specialized grazing systems: a review and application to the California annual type. *Journal of Range Management*, **14**, 182–193.

**Hollaway J.K. & Huffaker C.B. (1952)** Insects to control a weed. *Yearbook of Agriculture*, **1952**, 135–140.

**Holm L. (1971)** Chemical interactions between plants on agricultural lands. *Biochemical Interactions Among Plants*. Proceedings of the National Academy of Science Symposium, Washington, D.C., 1971, 95–101.

**Horsfield D. (1977)** Relationships between feeding of *Philaenus spumarius* (L.) and the amino acid concentration in the xylem sap. *Ecological Entomology*, **2**, 259–266.

**Huffaker C.B. (1952)** Quantitative Studies on the Biological Control of St. John's Wort (Klamath Weed) in California. *Proceedings of the 7th Pacific Science Congress*, Vol. 4, pp. 303–313.

**Huffaker C.B. (1962)** Some concepts on the ecological basis of biological control of weeds. *Canadian Entomologist*, **94**, 507–514.

**Jameson D.A. (1963)** Responses of individual plants to harvesting. *Botanical Review*, **29**, 532–594.

**Janzen D.H. (1970)** Herbivores and the number of tree species in tropical forests. *American Naturalist*, **104**, 501–528.

**Janzen D.H. (1974)** The role of the seed predator guild in a tropical deciduous forest, with some reflections on tropical biological control. *Biology in Pest and Disease Control* (Ed. by D. Price Jones & M.E. Solomon) pp. 3–14. Symposium of the British Ecological Society, 13.

**Johnson W.M. (1956)** The effect of grazing intensity of plant composition, vigor, and growth of pine-bunchgrass ranges in central Colorado. *Ecology*, **37**, 790–798.

**Kincade R.T., Laster M.L. & Brazzel J.R. (1970)** Effect on cotton yield of various levels of simulated *Heliothis* damage to squares and bolls. *Journal of Economic Entomology*, **63**, 613–615.

**Kulman H.M. (1971)** Effects of insect defoliation on growth and mortality of trees. *Annual Review of Entomology*, **16**, 289–324.

**McNeill S. (1973)** The dynamics of a population of *Leptoterna dolabrata* (Heteroptera: Miridae) in relation to its food resources. *Journal of Animal Ecology*, **42**, 495–507.

**Miller P.I. (1973)** Oxidant-induced community change in a mixed conifer forest. *Air Pollution Damage to Vegetation* (Ed. by J.A. Naegele) pp. 101–117. American Chemical Society.

**Mueggler W.F. (1967)** Response of mountain grassland vegetation to clipping in south-western Montana. *Ecology*, **48**, 942–949.

**Newman E.I. (1973)** Competition and diversity in herbaceous vegetation. *Nature, London*, **244**, 310.

**Noy-Meir I. (1975)** Stability of grazing systems: an application of predator–prey graphs. *Journal of Ecology*, **63**, 459–481.

**Paine R.T. (1966)** Food web complexity and species diversity. *American Naturalist*, **100**, 65–75.

**Parkinson D., Visser S. & Whittaker J.B. (1979)** Effects of collembolan grazing on fungal colonization of leaf litter. *Soil Biology and Biochemistry* (in press).

**Rovira A.D. (1969)** Plant root exudates. *Botanical Review*, **35**, 35–57.

**Sibma L., Kort J. & de Wit C.T. (1964)** Experiments on competition as a means of detecting possible damage by nematodes. *Jaarboek, Instituut voor biologischen scheikundig onderzoek van Landbouwgewassen*, **1964**, 119–124.

**Skuhravy V. (1968)** Einfluss der Entblätterung und des Kartoffelkäferfrasses auf die Kartoffelernte. *Anzeiger für Schädlingskunde*, **41**, 180–188.

**Tenow O. (1972)** The outbreaks of *Oporinia autumnata* Bkh. and *Operophtera* spp. (Lep. Geometridae) in the Scandinavian mountain chain and northern Finland 1862–1968. *Zoologiska bidrag fran Uppsala Suppl.*, **2**, 1–107.

**Troughton A. (1957)** The underground organs of herbage grasses. *Commonwealth Bureau Pastures and Field Crops Bulletin*, **44**, 163.

**Visser S. & Whittaker J.B. (1977)** Feeding preferences for certain litter fungi by *Onychiurus subtenuis* (Collembola). *Oikos*, **29**, 320–325.

**Waloff N. & Richards O.W. (1977)** The effect of insect fauna on growth, mortality and natality of broom, *Sarothamnus scoparius. Journal of Applied Ecology*, **14**, 787–798.

**Weinmann H. (1948)** Underground development and reserves of grasses. *Journal of the British Grassland Society*, **3**, 115–140.

**Weiss P.W. (1976)** Effect of defoliation on growth of spiny emex (*Emex australis*). *Australian CSIRO Div. Plant Ind. Field Stn. Rec.*, **15**, 27–33.

**Weldon L.W., Blackburn R.D. & Durdan W.C. (1973)** Evaluation of Agasicles n.sp. for biological control of alligatorweed. *Aquatic Plant Control Programme Tech. Rep.* **3**. Biological Control of Alligatorweed, U.S. Army Engineers Waterways Experimental Station, Bicksburg, Massachusetts, pp. D1–D54.

**Williams J.R. (1954)** The biological control of weeds. *Report of the Commonwealth Entomological Conference*, **6**, 95–98.

**Wilson F. (1943)** The entomological control of St. John's wort (*Hypericum perforatum* L.) with particular reference to the insect enemies of the weed in southern France. *Council for Scientific & Industrial Research Australia, Bulletin No.* **169**, 1–87.

**de Wit C.T. (1960)** On competition. *Verslagen van het landbouwkundig onderzoek in Nederland*, **66**, 1–82.

# 11. BETWEEN THE DEVIL AND THE DEEP BLUE SEA: ON THE PROBLEM OF BEING A HERBIVORE

J. H. LAWTON[1] AND S. McNEILL[2]

[1] *Department of Biology, University of York, Heslington, York YO1 5DD*

[2] *Imperial College Field Station, Silwood Park, Sunninghill, Ascot, Berkshire SL5 7PY*

## INTRODUCTION

In this paper we review some of the mechanisms determining the characteristic levels of abundance of plant-feeding insects, focusing first on two apparently separate facets of the problem, namely natural enemies and plant chemistry. We then show how these interact, emphasizing that the equilibrium sizes and dynamical behaviour of populations of many phytophagous insects are largely determined by the properties of the trophic levels above and below them: that plant-feeding insects live in a world dominated on the one hand by their natural enemies and on the other by a sea of food that, at best, is often nutritionally inadequate and, at worst, is simply poisonous.

Generalizations of this nature are fraught with difficulties. It has been estimated that at least 60% of the earth's eukaryotic species are angiosperms and the insects that feed on them (Gilbert 1977). Recall also that at least half the world's insects are either parasitoids or predators of other insects (Varley, Gradwell & Hassell 1973), and it becomes abundantly clear that we are dealing with more than our fair share of the world's biota. Not surprisingly, bizarre ecologies with very low probabilities of occurrence may, nevertheless, be represented by many hundreds of examples which confound the unwary theoretician. We can do no more than say what usually happens, and try to interpret exceptions in this light.

## SOME EXAMPLES

Figures 11.1 and 11.2 summarize population data for the adapted herbivorous insects on two species of plants with which we are particularly familiar, the grass *Holcus mollis* (McNeill & Southwood 1978; S. McNeill unpublished data) and bracken *Pteridium aquilinum* (Lawton 1976). The communities of insects

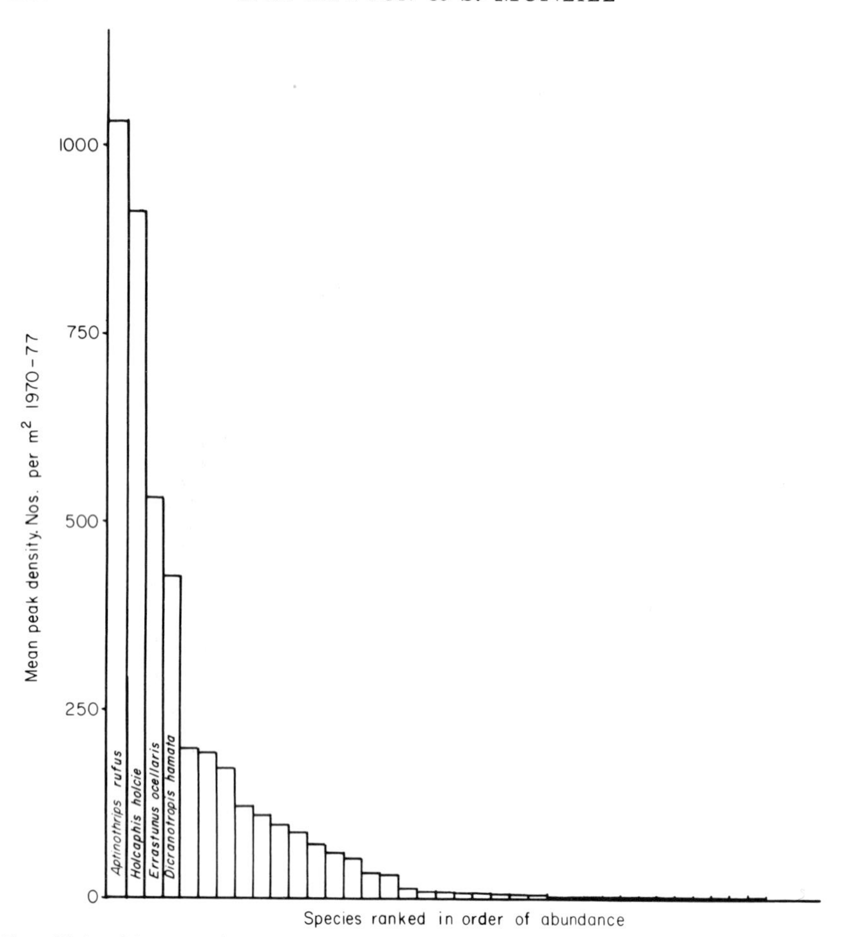


FIG. 11.1. Mean maximum densities of phytophagous insects on the grass *Holcus mollis* at Silwood Park, Berkshire, between 1970 and 1977.

on both plants share two features that are typical of many others (Tilden 1951; Menhinick 1964, 1967; Waloff 1968a; Feeny 1970; Denno 1977). First, a small proportion of the species may become temporarily very abundant and, when they do, cause serious damage to their host plant. *Aptinothrips* and *Holcaphis* on *Holcus* are good examples, reaching 2 851 $m^{-2}$ and 1 849 $m^{-2}$ in exceptional years. Certain of the Lepidoptera and sawflies on bracken behave in a similar way (Lawton 1976, and unpublished observations). However, by and large, these are rare events. If they were not, the world would not be green (Hairston, Smith & Slobodkin 1960). Second, the majority of species are never very abundant. Indeed, it is this feature of phytophagous insect communities, rather than the occasional spectacular outbreak, which makes them particularly interesting.

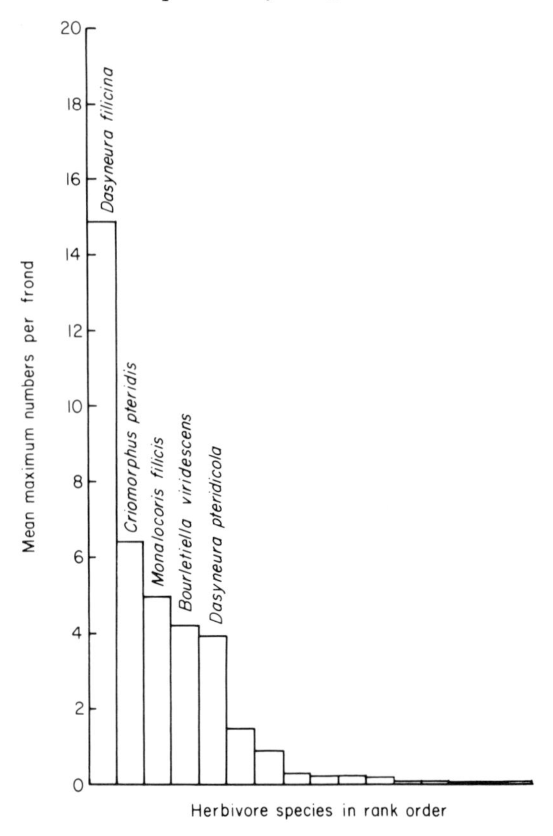


FIG. 11.2. Mean maximum abundance of phytophagous insects on bracken (*Pteridium aquilinum*) at Skipwith Common, Yorkshire, between 1972 and 1977 (excluding 1973, when very few samples were taken).

It is now well established that the abundance of phytophagous insects is influenced by a number of factors, including the size of the clump of plants which they are attacking (Cromartie 1975); plant density (Pimentel 1961; Root 1973); and the presence of other species of plants (Tahvanainen & Root 1972; Atsatt & O'Dowd 1976), although different species of insects may respond in opposite ways to similar changes in the environment (contrast Ralph 1977 with Thompson & Price 1977). Indeed, this is to be expected, because each of these factors is merely one of many different ways in which the demography, and hence abundance, of a plant-feeding insect can be altered. In this paper we have focused on just two factors that we believe are more important than many others; namely, natural enemies and plant chemistry, fully recognizing that their effects may be modified by other components of the insects' environment.

## THE IMPORTANCE OF NATURAL ENEMIES

Natural enemies play a key role in keeping many phytophagous insects rare. Indeed, the importance of insect parasitoids, in particular the Hymenoptera, has long been recognized (Marchal 1897; Howard & Fiske 1911).

The extent to which an insect parasitoid can depress the size of its host's population is shown dramatically in those cases of successful biological control where estimates of host abundance are available, first in the absence ($K$) and subsequently in the presence ($N^*$) of an introduced natural enemy (Fig. 11.3). The ratio ($N^*/K$) measures the impact of the parasitoid, and has been referred

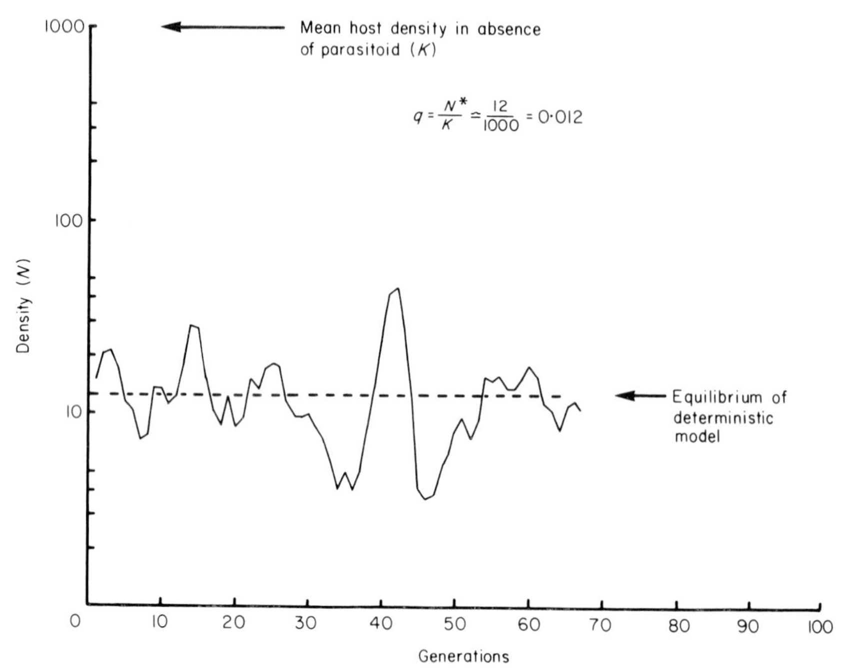


FIG. 11.3. A simulated host-parasitoid interaction of the form specified by eqn (11.1) incorporating pseudo-interference (eqn (11.4)). The mean rate of increase of the host, $\hat{r}$, and the mean attack rate of the parasitoid, $\hat{a}$, are independent, normally distributed random variables with a standard deviation/mean ratio of 0·5 and 0·3 respectively. This is comparable to the variation in $r$ shown by winter moth (Varley, Gradwell & Hassell 1973; Hassell, Lawton & May 1976) and in $a$ by a tachinid (Klomp 1959). In the absence of the parasitoid, the host population rises to $K$, set by food or space. The deterministic equations are $N_s = N_t \exp(-a'P_t)$, $N_{t+1} = N_s \exp(r[1 - N_s/K])$ and $P_{t+1} = N_t(1 - \exp[-a'P_t])$, with $a' = a(aP_t)^{-m}$. $N_s$ is the number of hosts surviving parasitism. The other terms are defined on pages 231 and 234. This particular example has $K = 1\,000$; $\hat{r} = 0{\cdot}5$; $\hat{a} = 0{\cdot}05$; $m = 0{\cdot}5$.

to elsewhere as $q$ (Beddington, Free & Lawton 1975, 1976, 1978). Some examples are shown in Figure 11.4. Obviously these values for $q$, obtained by averaging field data, are fairly crude, but they give an order of magnitude measure of the impact of the parasitoid.

Of the 120 attempts at biological control listed by DeBach (1974), approximately one-third (42) were described as completely successful. A further 48 were classified as 'substantial successes' and only 30 as 'partial'. In other words, dramatic depression of the host-population is common, but by no means inevitable (Huffaker 1974). Studies involving control by predators rather than parasitoids imply similar minimum $q$-values (Beddington, Free & Lawton 1978), although, again, some predators have little impact.

Whilst the data on successful biological control are useful, there are obvious dangers in extrapolating from what are, relatively speaking, a tiny

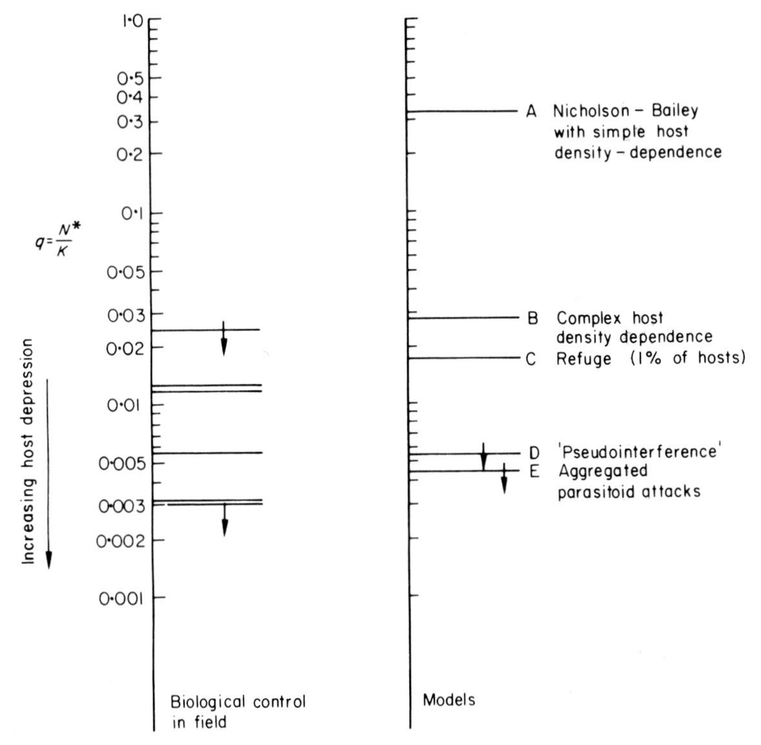


FIG. 11.4. On the left are field $q$-values for cases of successful biological control, where $q$ is the ratio of the host population in the presence ($N^*$) and absence ($K$) of the parasitoid. (From Beddington, Free & Lawton 1978.) On the right are shown minimum $q$-values generated by host-parasitoid models A–C (see text). The arrows on the results for Models D and E indicate that these are solutions comparable to those seen in the field for realistic parameter values, and that much lower $q$-values are possible.

number of studies involving a disproportionate number of species in ecosystems highly influenced by man to phytophagous insects and their enemies in general. Alternative evidence on the potential importance of natural enemies must, therefore, be sought from detailed field studies (e.g. Eickwort 1977) and particularly from an analysis of insect life tables.

Most species of phytophagous insects support a rich guild of their own parasitoids and predators (Zwölfer 1971; Lawton 1978), and if one or more of the complex acts as an effective controlling agent, models of host–parasitoid interactions suggest that this should be relatively easy to detect using *k*-factor analysis (Figs 11.3 and 11.5) (Varley, Gradwell & Hassell 1973). Podoler & Rogers (1975) and Stubbs (1977) summarize 18 life tables for 15 species of plant-feeding insects. Nine (50%) have parasitoids (or predators) acting in a directly density-dependent or, more rarely, delayed density-dependent manner, with similar effects suspected in two others. (All the species suffered measurable losses from parasitoids.) Only two of the life tables revealed no evidence of

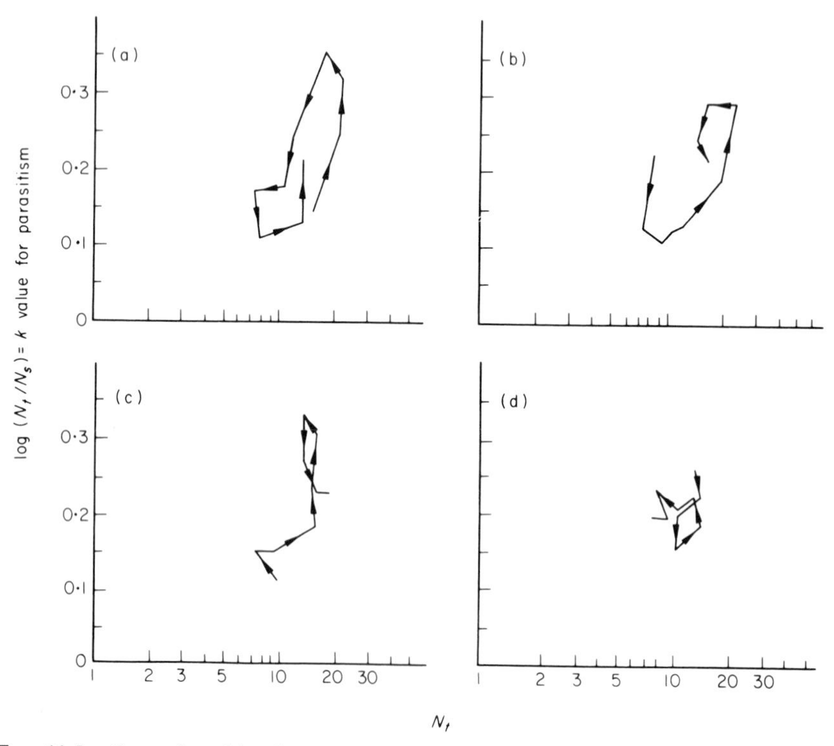


FIG. 11.5. Examples of *k*-values over ten generations for the host-parasitoid model illustrated in Figure 11.3. Most of the plots are either clearly density dependent (c) or some combination of delayed and direct density dependence (a, b). Only about 10% of the simulations were apparently random (d). (J.H. Lawton unpublished.)

density dependence (approximately 10%; cf. Fig. 11.5), and only two showed intraspecific competition for food to be the main regulatory factor. In the remaining three cases, 'competition' for an unknown resource (2) and disease (1) provided the density dependence.

In combination with the data on biological control, this leaves very little room for doubt about the importance of natural enemies, particularly parasitoids, in reducing the population sizes of phytophagous insects. Obviously not all species are controlled in this way, but on present evidence a majority appear to be.

## PLANT CHEMISTRY AND RESOURCE LIMITATION

Before analysing insect host–parasitoid interactions in detail, it is worthwhile pausing to say something briefly about intraspecific competition for food.

If predators or parasitoids hold the population at a low level (for example, $q \leqslant 0{\cdot}1$), then intraspecific competition will be negligible. However, a plant-feeding insect does not necessarily have to be very abundant to be food-limited,

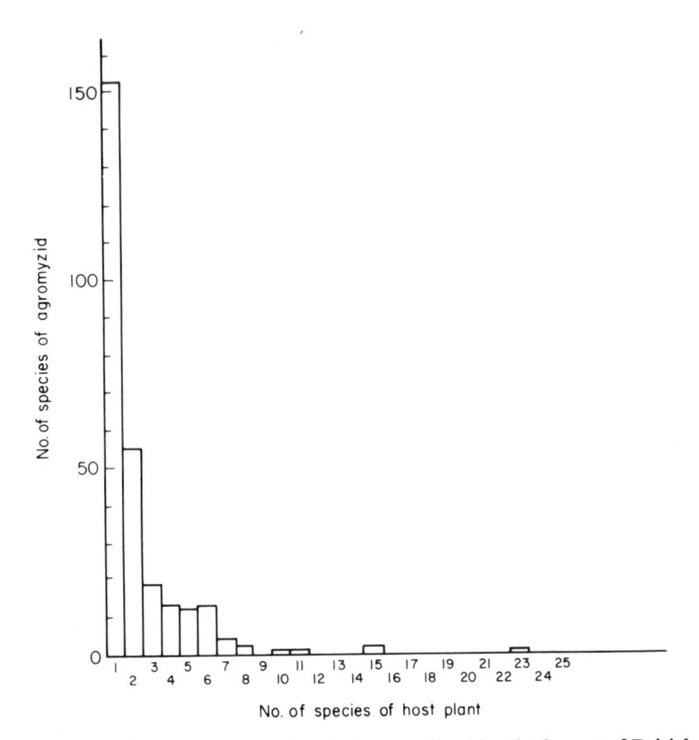


FIG. 11.6. The number of species of host-plants utilized by the larvae of British Agromyzidae. The vast majority are monophagous or oligophagous. (From Spencer 1972.)

because plant chemistry often severely limits the ability of herbivores to exploit plants as food (Feeny 1976; McNeill & Southwood 1978; Southwood 1973; van Emden & Way 1973). One simple manifestation is that most herbivorous insects are specialists, exploiting only a very restricted range of species (Fig. 11.6, from Spencer 1972). Similar results apply to grassland leaf-hoppers (Waloff & Solomon 1973), the herbivorous Cynipidae (Kinsey 1920, in Imms 1960) and many other groups.

Even on one host species most insects are confined to particular parts of the plant, often those with a high nutritional status (for a wide range of examples see Southwood (1973) and McNeill & Southwood (1978)), or are restricted to narrow windows of host suitability in time (Feeny 1970; McNeill 1973; Lawton 1976, 1978; Ikeda, Matsumura & Benjamin 1977; Thompson & Price 1977). We return to these developmental and seasonal changes in plant chemistry in a later section.

It is also interesting to speculate whether some insect populations are kept rare by the equivalent of an 'immune response' on the part of their host plant; we particularly have in mind the work of Ryan & Green (1974) showing that certain plants respond to local damage by the general production of 'wound-induced proteinase inhibitors'. A clear role in restricting phytophagous insects has yet to be demonstrated, but the whole problem deserves further study, not least because other plant defences are also thought to be inducible. Haukioja & Hakala (1975) provide a brief review.

## MODELS OF PHYTOPHAGOUS INSECT–PARASITOID INTERACTIONS

Although it is usual to treat plant chemistry and natural enemies separately, in reality they interact. To establish this, we first review some general phytophagous insect–parasitoid models in which both host and parasitoid have discrete (e.g. annual) generations. Elaboration of the models to describe a host with continuous reproduction (e.g. an idealized aphid) makes virtually no difference to the qualitative insights that we need before we move to the next section (Beddington, Free & Lawton 1978).

A general insect host–parasitoid model has the form:

$$\begin{aligned} N_{t+1} &= N_t f_1(N_t) f_2(N_t, P_t) \\ P_{t+1} &= P_t f_3(N_t, P_t) \end{aligned} \qquad (11.1)$$

where $N_t$ and $P_t$ are host and parasitoid densities. The function $f_1(N_t)$ describes the per capita rate of increase of the host as a function of its own density: $f_2(N_t, P_t)$ the proportional survival of $N_t$ hosts confronted by $P_t$ parasitoids; and $f_3(N_t, P_t)$ the per capita rate of increase of the parasitoid as a function of its own and its host's density.

Equation (11.1) permits a wide variety of models to be constructed with varying degrees of realism (Beddington, Free & Lawton 1975, 1976; Hassell 1978), of which one of the simplest is:

$$\begin{aligned} N_{t+1} &= N_t \exp\{r(1 - N_t/K) - aP_t\} \\ P_{t+1} &= N_t\{1 - \exp(-aP_t)\} \end{aligned} \tag{11.2}$$

where $a$ is the attack rate of the parasitoid and $r$ and $K$ the intrinsic rate of increase and equilibrium density (set by food or space) of the host in the absence of the parasitoid. The stability of this model has been analysed by Beddington, Free & Lawton (1975, 1976); Figure 11.7a shows the results. Two things should be noted. First, high host rates of increase tend to destabilize the interaction, changing a stable equilibrium at low $r$ into a stable cycle, the amplitude and complexity of which gets larger the larger the value of $r$. Second, within the locally stable region the host's equilibrium population size, $N^*$, in the presence of the parasitoid increases with $r$. An example is shown in Figure 11.8a, where $q$ is given by the solution of the equation:

$$r(1 - q) - aKq\{1 - \exp[-r(1 - q)]\} = 0 \tag{11.3}$$

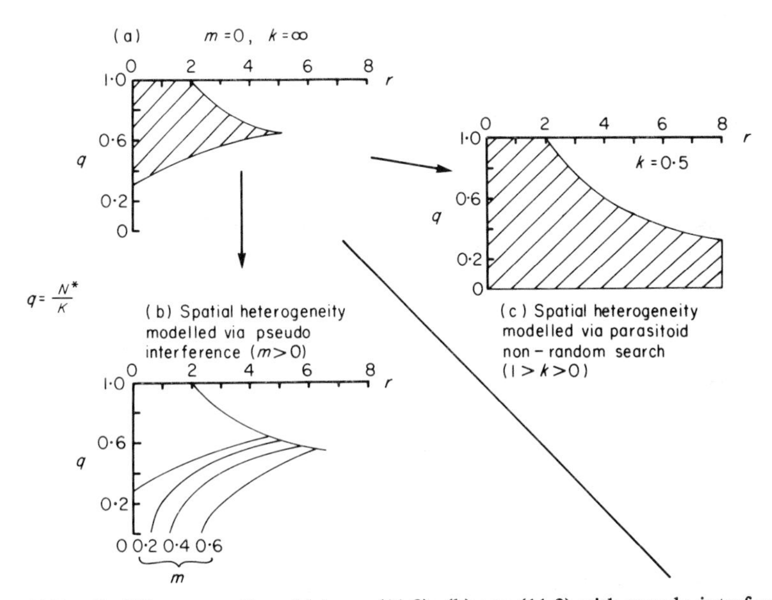


FIG. 11.7. Stability properties of (a) eqn (11.2); (b) eqn (11.2) with pseudo-interference (eqn (11.4)); (c) eqn (11.5). (b) and (c) are two alternative ways of describing a host–parasitoid interaction in a patchy environment. The shaded regions in (a) and (c) are locally stable; the locally stable region in (b) is similar to that in (a), but its lower bound is extended as $m$ increases.

This result is important because, other things being equal, plant chemistry has a profound effect on $r$.

Equation (11.2) has one important failing. If a locally stable equilibrium is to be preserved using this model, the biggest impact that a parasitoid can have is to depress the host population to one-third of the abundance which it maintained in the absence of the parasitoid ($q = 0{\cdot}33$) (Fig. 11.4, Model A, and Fig. 11.7). For locally unstable (i.e. oscillatory but persistent) solutions, the host population fluctuates round an average level which is higher than this. In other words, eqn (11.2) fails to reproduce the field $q$-values shown in Figure 11.4 by one or even two orders of magnitude (Beddington, Free & Lawton 1978).

An important difference between eqn (11.2) and the field is that the model assumes that host and parasitoid interact in a spatially homogeneous world. The effects of spatial heterogeneity can be incorporated explicitly by assigning some fraction of the hosts to a refuge (Model C, Fig. 11.4), or they can be included implicitly via the effects which patchily distributed hosts have on the rate at which parasitoids encounter them (Hassell & May 1974; Cowie & Krebs 1979). One such approach is the 'pseudo-interference' model of Free, Beddington & Lawton (1977), in which the differential exploitation of host patches leads to a decline in realized searching efficiency, $a'$, with parasitoid density which can be represented at equilibrium by:

$$a' = a(aP_t)^{-m} \tag{11.4}$$

(where $m$ measures the effects of both genuine interference and pseudo-interference). An alternative model which mimics aggregated search by parasitoids is that of May (1978) (see also Hassell 1979):

$$\begin{aligned} N_{t+1} &= N_t \exp\{r(1 - N_t/K)\}(1 + aP_t/k)^{-k} \\ P_{t+1} &= N_t\{1 - (1 + aP_t/k)^{-k}\} \end{aligned} \tag{11.5}$$

Here the random search assumed in eqn (11.2) is replaced by aggregated search, specified by the clumping parameter, $k$, of the negative binomial distribution.

Both methods of introducing spatial heterogeneity permit low stable $q$-values (Fig. 11.4, Models D and E, and Fig. 11.7), and both retain the important property that $N^*$ increases as $r$ increases (Fig. 11.8). Indeed, $N^*$ is, if anything, more sensitive to increasing $r$, particularly in the biologically interesting region $0 < r < 3$. (Rates of increase per generation, $r$, larger than $2{\cdot}5$ have been observed only rarely in insect populations in the field (Hassell, Lawton & May 1976).)

Interestingly, spatial heterogeneity may not be the only mechanism permitting low, stable $q$'s, although current evidence suggests that it is the most important (Beddington, Free & Lawton 1978). An alternative explanation is provided by the form of the host density-dependent function $f_1(N_t)$ (eqn (11.1)).

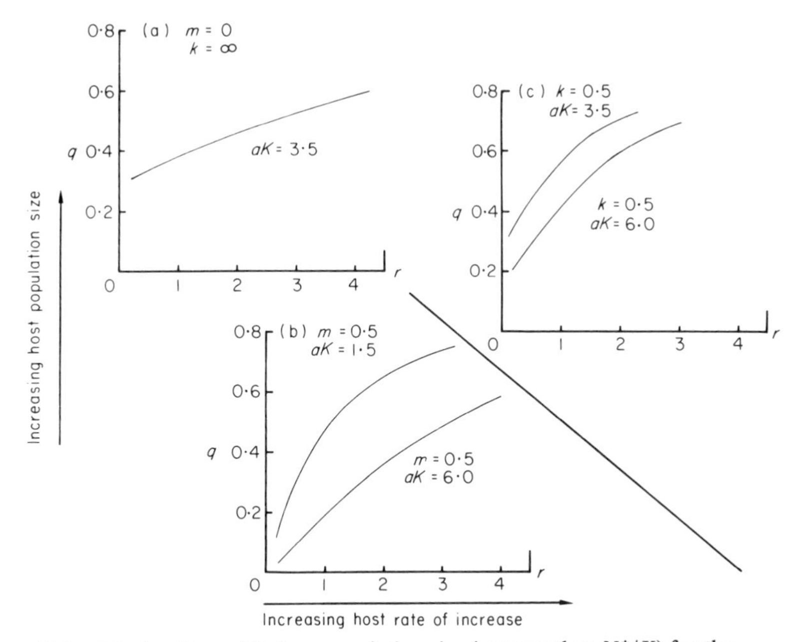


FIG. 11.8. The host's equilibrium population size (expressed as $N^*/K$) for the models shown in Figure 11.7.

If this is complex, it may give rise to two locally stable equilibria; an upper one where the host is resource-limited ($K$) and a lower one maintained by background mortality; for example, uncoupled polyphagous predators and parasitoids (Takahashi 1964; Southwood & Comins 1976; Peterman, Clark & Holling 1979). May (1977) provides a detailed review. The resulting lower equilibrium (Fig. 11.4, Model B) is not globally stable, so that any reduction in the efficiency of the background predators may cause the host to escape to the upper one ($K$) (Southwood 1975). However, populations of phytophagous insects that are kept rare by this mechanism are insensitive to increasing $r$ until the lower equilibrium becomes locally unstable. Then, when fluctuations become sufficiently large, the population outbreaks rather suddenly. Some aphids may escape control by background predators 'due in part to an increased reproductive rate' (Southwood & Comins 1976, p. 180).

Drawing the results of this section together, we see that models of closely coupled host–parasitoid interactions predict that the equilibrium population size of the host $N^*$ increases as $r$ increases, and that very high $r$-values may cause oscillations. If we replace the parasitoid by a coupled predator, these results are retained (Beddington, Free & Lawton 1976). Furthermore, models which capture the essence of spatially heterogeneous environments predict that small changes in $r$ make big differences to $N^*$. Only populations regulated by uncoupled background predators are relatively insensitive to small changes in $r$, but even these are sensitive to large ones.

Finally, the results in this section have mainly expressed the equilibrium population size of the host as the dimensionless ratio $q$: that is, $N^*/K$. If large parts of the host plant are inedible for the reasons discussed in the previous section ($K$ small), then the effect of parasitoids and predators will be to make the insect even rarer.

## THE EFFECTS OF PLANT CHEMISTRY ON RATES OF INCREASE

Phytochemists deal with a bewildering variety of compounds but, for ecological purposes at least, it is often possible to classify them into two main types: 'nutrients', for example the amino acid or sugar content of a plant, and 'secondary compounds' (toxins, poisons and antifeedants). Secondary compounds themselves can be further split into two categories which Feeny (1976) called 'qualitative' and 'quantitative' defences and Rhoades & Cates (1976) 'toxic' and 'digestibility reducing' respectively. Examples of the former include cyanide, glucosinolates and alkaloids; of the latter, tannins and resins.

In this section we are not concerned with the defensive role, already touched on, that secondary compounds play in preventing most species of phytophagous insects from feeding on most of the species of plants in their environment most of the time (Fig. 11.6). Indeed, this problem has received an almost disproportionate amount of attention, because in the final analysis plant chemistry makes very little difference to the total number of species of insects which eventually evolve to exploit a plant (Lawton 1978). Rather, we are concerned with the effects that plant chemistry has on the interactions between phytophagous insects and their normal food plants; that is, with adapted herbivores.

If we assume that more insects mean more damage to the plant (which does not seem unreasonable), then the results of the previous section suggest that it must be selectively advantageous for plants to reduce the rates of increase of their adapted herbivores as much as possible (e.g. Whittaker 1979). This can be achieved by reducing the growth rates and increasing the generation times of the insects; by influencing the number of generations a year; and by reducing their survival and fecundity (e.g. Istock, Zisfein & Vavra 1976).

The extent to which natural selection has achieved this for wild species of plants is difficult to judge. Crops, however, provide numerous examples of the sort of unfit phenotypes that natural selection has presumably eliminated in the wild. They are called susceptible varieties.

The reasons why particular varieties of crops are resistant to attack and others susceptible are numerous (Day 1972; Hill 1976; Jones 1977) and may involve more than just plant chemistry (Way & Murdie 1965; Hedin, Thompson & Gueldner 1976). The important point in the context of the present discussion, however, is that resistant varieties are rarely totally immune from attack; rather they support much lower populations of adapted species than do susceptible varieties. Furthermore, fecundity, growth and survival may all be markedly reduced on the resistant variety (Pathak 1975, and Fig. 11.9) and control by parasitoids greatly enhanced (Wyatt 1970; Starks, Muniappan & Eikenbary 1972 (Fig. 11.10); van Emden & Way 1973; Huffaker 1974, p. 207). These results are exactly what we would predict from the analysis presented in the previous section, although elaborations and complications are possible (van Emden 1978).

Presumably, in most plant populations, very susceptible varieties have been eliminated by natural selection. However, it should still be possible to show that plant chemistry has an effect on the rate of increase of adapted herbivores, even if the evidence is necessarily indirect.

We have already pointed out that an important consequence of the marked changes in plant chemistry that take place in all plants with season and with the age of their foliage (Feeny 1976; Rhoades & Cates 1976) is that many species of phytophagous insects are confined to narrow windows of host suitability in time. The number of generations that can be 'fitted in' during the year is therefore reduced and may often only be one (Slansky 1974).

Whether having more than one generation per year, and hence higher annual rates of increase, implies a higher $q$ in the models developed in the previous section depends upon details of the parasitoid's biology, and in particular whether it attacks every generation. If it does not (e.g. Waloff 1975),

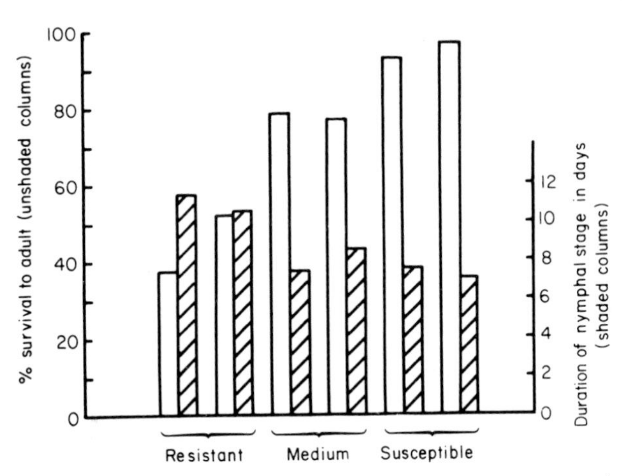


FIG. 11.9. Survival and development times in the cotton jassid *Amrasca devastans* on two resistant, two intermediate and two susceptible varieties of cotton (Agarwal & Krishnananda 1976).

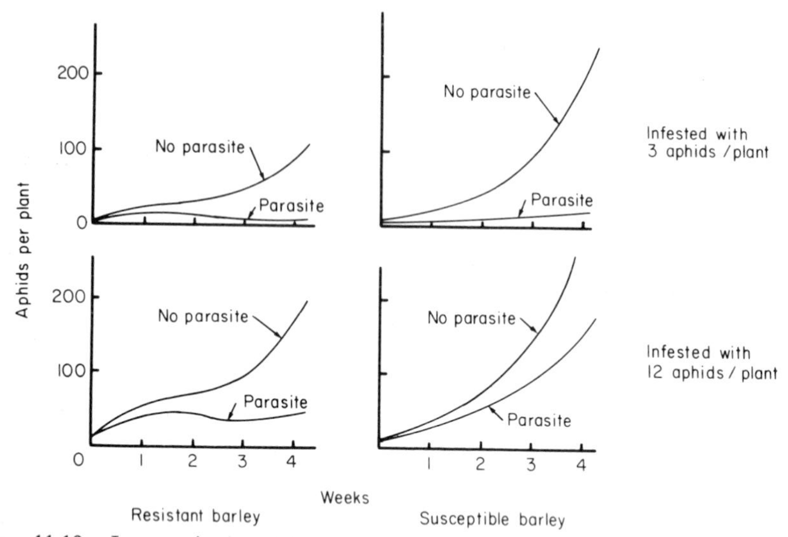


FIG. 11.10. Increase in the aphid *Schizaphis graminum* on resistant and susceptible varieties of barley in the presence and absence of the parasitoid, *Lysiphlebus testaceipes* (Starks, Muniappan & Eikenbary 1972).

then reducing the number of generations which the herbivore has in a year is always advantageous to the plant.

The effects of plant chemistry on the growth rates, fecundity and survival of plant-feeding insects have been comprehensively reviewed by van Emden & Way (1973), Beck & Reese (1976), Feeny (1976) and Rhoades & Cates (1976), amongst others. Southwood (1973) initially drew attention to the generally poor performance of plant-feeding insects compared with carnivorous or parasitic forms, and an extensive literature now suggests that plant nutrients, in particular nitrogen (the protein or amino acid content of the plant), are crucial in this respect (McNeill & Southwood 1978). The generally low levels of available nitrogen in plants are further aggravated by quantitative (digestability-reducing) secondary compounds. The consensus of opinion is that low nitrogen, and high levels of quantitative defences, singly or in combination, result in reduced feeding rates, slow growth rates, long development times (Strong & Wang 1977 provide a dramatic example), poor survival and reduced fecundity, even in adapted species. Poor nutrition also results in weakened individuals being more easily killed by external factors like weather or disease (McNeill 1973; Anderson 1979), and slow growth rates necessarily imply longer periods of exposure to predators and parasitoids.

These arguments ignore one important complication. In some insects (e.g. *Pieris rapae*; Slansky & Feeny 1977) falling nitrogen levels trigger a compensatory increase in feeding rate, so that growth and survival are maintained, but damage to the plant increases. Fox & Macauley (1977) provide a brief review. However, in most cases, increasing nitrogen levels in the foliage lead to an increase in herbivore populations (and hence damage) and not the reverse (Nickel 1973; Suski & Badowska 1975; Jones 1977; Onuf, Teal & Valiela 1977; Webb & Moran 1978).

*Eucalyptus* provides another intriguing exception. Eucalypts contain high levels of tannins and other phenolics, have very low nitrogen levels in the mature foliage and yet are also heavily damaged by phytophagous insects (Fox & Macauley 1977). We would predict that this is due to peculiarities in the parasitoid and predator complexes attacking *Eucalyptus*-feeding insects, reinforcing the view that it is plant chemistry and natural enemies acting together that are important in determining how much damage insects do to their host-plant, and not simply plant chemistry.

Whilst the effects of quantitative defences and nutrients are now well documented, the importance of qualitative (toxic) defences in reducing the rate of increase of adapted herbivores is much less clear. By definition, adapted insects detoxify or neutralize these compounds, and both detoxifying and sequestering may impose metabolic costs of unknown magnitude (Krieger, Feeny & Wilkinson 1971; Brower & Glazier 1975; Brattsten, Wilkinson & Eisner 1977). Presumably resources (e.g. nitrogen or energy) used for detoxification are not available for growth and reproduction.

However, although these arguments are sound in theory, in practice it has sometimes proved extremely difficult to detect any variation in the performance of insects faced with widely different concentrations of qualitative defences (toxins) in their normal food plants (Erickson 1973; Erickson & Feeny 1974; Feeny 1976). In other cases, effects have been easier to show (van Emden 1972), particularly in the case of oligophages or polyphages adapted to cope with more than one toxin (van Emden 1972; Erickson & Feeny 1974; Chew 1975). We therefore believe that the effects of qualitative defences on the population growth rates of adapted herbivores are probably small, but they may not be negligible.

We can summarize the results of this section as follows. Plant chemistry has been shown to have a marked effect on one or more of the components of $r$, even in adapted herbivores. The clearest examples are provided by plant nutrients (particularly nitrogen) and quantitative defences, but qualitative (toxic) defences may also play a part. In consequence, control by parasitoids and predators is facilitated and equilibrium population sizes are reduced.

## PREDICTING LEVELS OF ABUNDANCE

Although it is straightforward to argue that plants which are able to keep the rates of increase of their adapted insects as low as possible gain a selective advantage, it is much more difficult to decide what the characteristic levels of abundance of these insects will be. Indeed, an examination of the multiplicity of factors which control and disrupt populations of different species of insect on one species of plant (Waloff 1968b) imply that there are no easy answers. Nevertheless, some simple predictions suggest themselves.

Obviously we should compare like with like (e.g. Lepidoptera with Lepidoptera) on a common scale—the insects per unit weight or area of plant, for example. Within these constraints, and other things being equal, we would predict that insects attacking ephemeral, early successional herbs and weeds (non-apparent plants with largely qualitative defences (Feeny 1976)) will have higher $r$-values than those exploiting perennial long-lived plants (apparent species with considerable investment in quantitative defences). In parentheses, note that we could have arrived at this prediction via Southwood's 'habitat templet' (Southwood 1977); the fact that we are able to make it independently by focusing on plant chemistry is encouraging.

Again, other things being equal, high $r$-values imply large equilibrium population sizes and a greater propensity for the population to become unstable and to outbreak in populations controlled by predators and parasitoids (and, incidentally, in populations regulated only by intraspecific competition (May 1974; Hassell 1975; Hassell, Lawton & May 1976)). Hence we might expect the average population sizes of insects which attack non-apparent plants to be higher than those attacking apparent plants and to fluctuate more.

In reality, the situation will be more complex than this, because control by

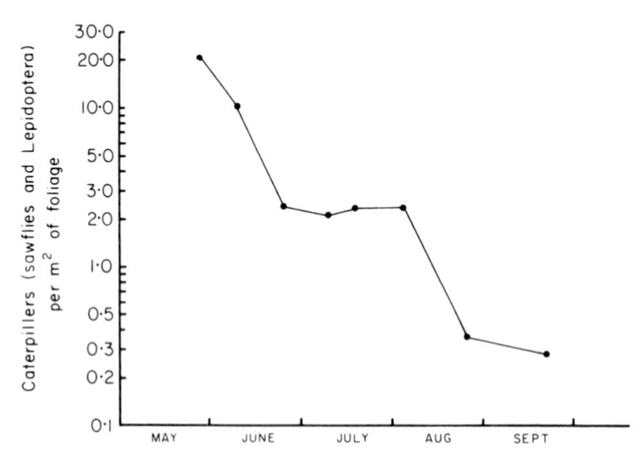


FIG. 11.11. The mean number of non-mining caterpillars per square metre of foliage at different times of the year on bracken (1972–77). The species change with season (Lawton 1976). Both Lepidoptera and sawflies are included in the analysis; and although sawflies are more abundant per frond later in the season (Lawton 1976), the fronds are also much larger (Lawton 1978) and hence numbers per unit area of foliage decline markedly. Protein levels decline and quantitative defences (tannins, silicate and lignin) increase in the pinnae during the growing season (Lawton 1976, 1978).

predation and parasitism is undoubtedly more difficult and uncertain on ephemeral plants (Southwood 1977; Stubbs 1977). However, this will merely serve to reinforce the differences that already exist between the two types of plants, favouring even larger and more variable populations on non-apparent species.

These predictions urgently need testing along a successional gradient over a comparatively small spatial scale.

Similar arguments might well apply to the young and old foliage of apparent plants (Feeny 1976; Rhoades & Cates 1976). Feeny's own work on the foliage-feeding Lepidoptera on oak clearly support this prediction (Feeny 1970), as do our own data for *Pteridium* (Fig. 11.11) and *Holcus* (Fig. 11.12). In general, species attacking the plant at times when nutrients are low and quantitative defences high have smaller average population sizes than species present at more favourable times of the growing season.

Finally, populations of gall-forming insects, which markedly alter the chemistry of their host (McNeill & Southwood 1978), may be relatively or indeed completely insensitive to these problems, and therefore have population sizes which are much less dependent on plant apparency.

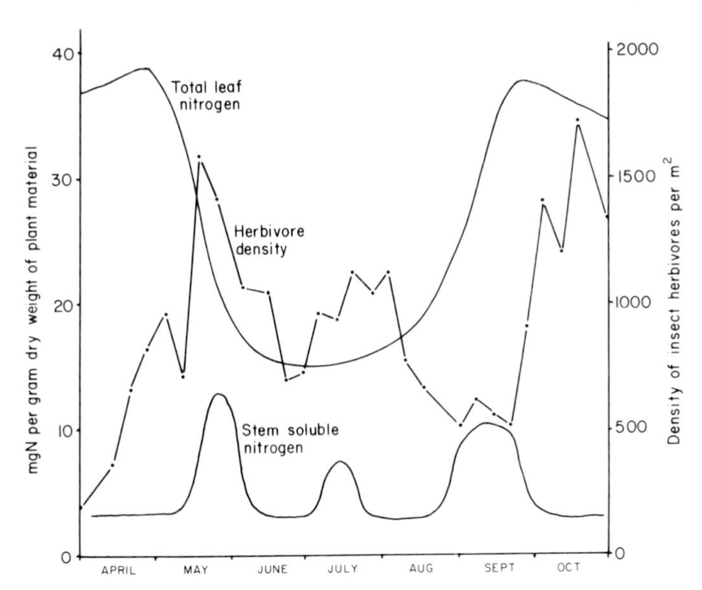


FIG. 11.12. Seasonal changes in the mean numbers of *Holcus* insects; as in Figure 11.11, the species change with season (McNeill & Southwood 1978). The peaks in density are related to peaks in food quality (measured as soluble N) in the leaves (upper solid curve) and stems (lower solid curve), the latter approximating phloem flows to the flowers and developing seeds in the middle of the summer. The nitrogen curves are idealized curves drawn from the mean times and concentrations of peaks and troughs in soluble N levels for the period 1970–76 (McNeill & Southwood 1978).

Of necessity, these predictions are crude, although most of them have been made before, at least in part by others, particularly Feeny, Rhoades and Cates and Southwood (loc. cit.). A really powerful general theory capable of predicting approximate or even relative levels of abundance of the different sorts of insects on various types of plants within a major biome is probably impossible to attain at the present time. Like the insects themselves, we are caught by opposing forces: between the devil of oversimplification on the one hand and a deep blue sea of endless unrelated facts on the other. The important thing is to ensure that neither gets the upper hand.

## ACKNOWLEDGMENTS

This paper draws extensively on ideas developed with John Beddington, Charles Free and Dick Southwood. We are extremely grateful to them, and to Valerie Brown for stimulating discussions. The work was supported in part by an NERC grant to J.H.L.

## REFERENCES

**Agarwal R.A. & Krishnananda N. (1976)** Preference to oviposition and antibiosis mechanism to jassids (*Amrasia devastans* Dist.) in cotton (*Glossypium* sp.). *The Host-plant in Relation to Insect Behaviour and Reproduction* (Ed. by T. Jermy) pp. 13–22. Plenum, New York & London.

**Anderson R.M. (1979)** The influence of parasitic infection on the dynamics of host population growth. *Population Dynamics* (Ed. by R.M. Anderson, B.D. Turner & L.R. Taylor) pp. 245–281. Blackwell Scientific Publications, Oxford.

**Atsatt P.R. & O'Dowd D.J. (1976)** Plant defense guilds. *Science*, **193**, 24–29.

**Beck S.D. & Reese J.C. (1976)** Insect–plant interactions: nutrition and metabolism. *Recent Advances in Phytochemistry*, **10**, 41–92.

**Beddington J.R., Free C.A. & Lawton J.H. (1975)** Dynamic complexity in predator–prey models framed in difference equations. *Nature, London*, **255**, 58–60.

**Beddington J.R., Free C.A. & Lawton J.H. (1976)** Concepts of stability and resilience in predator–prey models. *Journal of Animal Ecology*, **45**, 791–816.

**Beddington J.R., Free C.A. & Lawton J.H. (1978)** Characteristics of successful natural enemies in models of biological control of insect pests. *Nature, London*, **273**, 513–519.

**Brattsten L.B., Wilkinson C.F. & Eisner T. (1977)** Herbivore–plant interactions: mixed-function oxidases and secondary plant substances. *Science*, **196**, 1349–1352.

**Brower L.P. & Glazier S.C. (1975)** Localisation of heart poisons in the monarch butterfly. *Science*, **188**, 19–25.

**Chew F.S. (1975)** Coevolution of pierid butterflies and their cruciferous foodplants. I. The relative quality of available resources. *Oecologia*, **20**, 117–127.

**Cowie R.J. & Krebs J.R. (1979)** Optimal foraging in patchy environments. *Population Dynamics* (Ed. by R.M. Anderson, B.D. Turner & L.R. Taylor) pp. 183–205. Blackwell Scientific Publications, Oxford.

**Cromartie W.J. Jr (1975)** The effect of stand size and vegetational background on the colonization of cruciferous plants by herbivorous insects. *Journal of Applied Ecology*, **12**, 517–533.

**Day P.R. (1972)** Crop resistance to pests and pathogens. *Pest Control Strategies for the Future*, pp. 257–271. National Academy of Sciences, Washington, D.C.

**DeBach P. (1974)** *Biological Control by Natural Enemies*. Cambridge University Press.

**Denno R.F. (1977)** Comparison of the assemblages of sap-feeding insects (Homoptera-Hemiptera) inhabiting two structurally different salt marsh grasses in the genus *Spartina*. *Environmental Entomology*, **6**, 359–372.

**Eickwort K.R. (1977)** Population dynamics of a relatively rare species of milkweed beetle (*Labidomera*). *Ecology*, **58**, 527–538.

**Erickson J.M. (1973)** The utilization of various *Asclepias* species by larvae of the Monarch butterfly *Danaus plexippus*. *Psyche*, **80**, 230–244.

**Erickson J.M. & Feeny P. (1974)** Sinigrin: a chemical barrier to the black swallowtail butterfly, *Papilio polyxenes*. *Ecology*, **55**, 103–111.

**Feeny P. (1970)** Seasonal changes in oak leaf tannins and nutrients as a cause of spring feeding by winter moth caterpillars. *Ecology*, **51**, 565–581.

**Feeny P. (1976)** Plant apparency and chemical defense. *Recent Advances in Phytochemistry*, **10**, 1–40.

**Fox L.R. & Macauley B.J. (1977)** Insect grazing on *Eucalyptus* in response to variation in leaf tannins and nitrogen. *Oecologia*, **29**, 145–162.

**Free C.A., Beddington J.R. & Lawton J.H. (1977)** On the inadequacy of simple models of mutual interference for predation and parasitism. *Journal of Animal Ecology*, **46**, 543–554.

**Gilbert L.E. (1977)** Development of theory in the analysis of insect–plant interactions. *Analysis of Ecological Systems* (Ed. by D.J. Horn, R.D. Mitchell & G.R. Stairs) Ohio State University Press.

**Hairston N.G., Smith F.E. & Slobodkin L.B. (1960)** Community structure, population control and competition. *American Naturalist*, **94**, 421–425.

**Hassell M.P. (1975)** Density-dependence in single species populations. *Journal of Animal Ecology*, **44**, 283–295.

**Hassell M.P. (1978)** *The Dynamics of Arthropod Predator–Prey Systems*. Princeton University Press, Princeton.

**Hassell M.P. (1979)** The dynamics of predator–prey interactions: polyphagous predators, competing predators and hyperparasitoids. *Population Dynamics* (Ed. by R.M. Anderson, B.D. Turner & L.R. Taylor) pp. 283–306. Blackwell Scientific Publications, Oxford.

**Hassell M.P., Lawton J.H. & May R.M. (1976)** Patterns of dynamical behaviour in single-species populations. *Journal of Animal Ecology*, **45**, 471–486.

**Hassell M.P. & May R.M. (1974)** Aggregation of predators and insect parasites and its effect on stability. *Journal of Animal Ecology*, **43**, 567–594.

**Haukioja E. & Hakala T. (1975)** Herbivore cycles and periodic outbreaks. Formulation of a general hypothesis. *Report Kevo Subarctic Research Station*, **12**, 1–9.

**Hedin P.A., Thompson A.C. & Gueldner R.C. (1976)** Cotton plant and insect constituents that control boll weevil behaviour and development. *Recent Advances in Phytochemistry*, **10**, 271–350.

**Hill M.G. (1976)** The population and feeding ecology of five species of leaf hoppers (Homoptera) on *Holcus mollis*. *Unpublished Ph.D. thesis, University of London.*

**Howard L.O. & Fiske W.F. (1911)** The importation into the United States of the parasites of the gipsy-moth and the brown-tail moth. *Bulletin of the Bureau of Entomology, U.S. Department of Agriculture*, **91**, 1–312.

**Huffaker C.B. (Ed.) (1974)** *Biological Control*. Plenum, New York.

**Ikeda T., Matsumura F. & Benjamin D.M. (1977)** Chemical basis for feeding adaptation of pine sawflies *Neodiprion rugifrons* and *Neodiprion swainei*. *Science*, **197**, 497–499.

**Imms A.D. (1960)** *General Textbook of Entomology*. Methuen, London.

**Istock C.A., Zisfein J. & Vavra K.J. (1976)** Ecology and evolution of the pitcher-plant mosquito. 2. The substructure of fitness. *Evolution*, **30**, 535–547.

**Jones F.G.W. (1977)** Pests, resistance and fertilizers. *12th Colloquium of the International Potash Institute*, 111–135.

**Klomp H. (1959)** Infestations of forest insects and the role of parasites. *Proceedings of the 15th International Congress of Zoology, London, 1958*, 797–800.

**Krieger R.I., Feeny P.P. & Wilkinson C.F. (1971)** Detoxication enzymes in the guts of caterpillars: an evolutionary answer to plant defenses? *Science*, **172**, 579–581.

**Lawton J.H. (1976)** The structure of the arthropod community on bracken. *Botanical Journal of the Linnean Society*, **73**, 187–216.

**Lawton J.H. (1978)** Host-plant influences on insect diversity: the effects of space and time. *Symposia of the Royal Entomological Society of London. Diversity of Insect Faunas* (Ed. by L.A. Mound & N. Waloff) pp. 105–125. Blackwell Scientific Publications, Oxford.

**Marchal P. (1897)** L'équilibre numérique des espèces et ses relations avec les parasites chez les insectes. *C.R. Société de Biologie*, **49**, 129–130.

**May R.M. (1974)** Biological populations with non-overlapping generations: stable points, stable cycles and chaos. *Science*, **186**, 645–647.

**May R.M. (1977)** Thresholds and breakpoints in ecosystems with a multiplicity of stable states. *Nature, London*, **269**, 471–477.

**May R.M. (1978)** Host-parasitoid systems in patchy environments: a phenomenological model. *Journal of Animal Ecology*, **47**, 833–844.

**McNeill S. (1973)** The dynamics of a population of *Leptoterna dolobrata* (Heteroptera: Miridae) in relation to its food resources. *Journal of Animal Ecology*, **42**, 495–507.

**McNeill S. & Southwood T.R.E. (1978)** The role of nitrogen in the development of insect/plant relationships. *Biochemical Aspects of Plant and Animal Coevolution* (Ed. by J.B. Harborne) pp. 77–98. Academic Press, London & New York.

**Menhinick E.F. (1964)** A comparison of some species diversity indices applied to samples of field insects. *Ecology*, **45**, 859–861.

**Menhinick E.F. (1967)** Structure, stability, and energy flow in plants and arthropods in a sericea lespedeza stand. *Ecological Monographs*, **37**, 255–272.

**Nickel J.L. (1973)** Pest situation in changing agricultural systems—a review. *Bulletin of the Entomological Society of America*, **19**, 136–142.

**Onuf C.P., Teal J.M. & Valiela I. (1977)** Interactions of nutrients, plant growth and herbivory in a mangrove ecosystem. *Ecology*, **58**, 514–526.

**Pathak M. (1975)** Utilization of insect–plant interactions in pest control. *Insects, Science and Society* (Ed. by D. Pimentel) pp. 121–148. Academic Press, New York & London.

**Peterman R.M., Clark W.C. & Holling C.S. (1979)** The dynamics of resilience: shifting stability domains in fish and insect systems. *Population Dynamics* (Ed. by R.M. Anderson, B.D. Turner & L.R. Taylor) pp. 321–341. Blackwell Scientific Publications, Oxford.

**Pimentel D. (1961)** Species diversity and insect population outbreaks. *Annals of the Entomological Society of America*, **54**, 76–86.

**Podoler H. & Rogers D. (1975)** A new method for the identification of key factors from life-table data. *Journal of Animal Ecology*, **44**, 85–114.

**Ralph C.P. (1977)** Effect of host plant density on populations of a specialised seed-sucking bug, *Oncopeltus fasciatus*. *Ecology*, **58**, 799–809.

**Rhoades D.F. & Cates R.G. (1976)** Toward a general theory of plant antiherbivore chemistry. *Recent Advances in Phytochemistry*, **10**, 168–213.

**Root R.B. (1973)** Organization of a plant–arthropod association in simple and diverse habitats: the fauna of collards (*Brassica oleracea*). *Ecological Monographs*, **43**, 95–124.

**Ryan C.A. & Green T.R. (1974)** Proteinase inhibitors in natural plant protection. *Recent Advances in Phytochemistry*, **8**, 123–140.

**Slansky F. Jr (1974)** Relationship of larval food-plants and voltinism patterns in temperate butterflies. *Psyche*, **81**, 243–253.

**Slansky F. Jr & Feeny P. (1977)** Stabilization of the rate of nitrogen accumulation by larvae of the cabbage butterfly on wild and cultivated food plants. *Ecological Monographs*, **47**, 209–228.

**Southwood T.R.E. (1973)** The insect/plant relationship—an evolutionary perspective. *Symposia of the Royal Entomological Society:* **6.** *Insect/Plant Relationships* (Ed. by H.F. van Emden) pp. 3–30. Blackwell Scientific Publications, Oxford.

**Southwood T.R.E. (1975)** The dynamics of insect populations. *Insects, Science and Society* (Ed. by D. Pimentel) pp. 151–199. Academic Press, New York & London.

**Southwood T.R.E. (1977)** Habitat, the templet for ecological strategies? *Journal of Animal Ecology*, **46**, 337–365.

**Southwood T.R.E. & Comins H.N. (1976)** A synoptic population model. *Journal of Animal Ecology*, **45**, 949–965.

**Spencer K.A. (1972)** *Handbooks for the Identification of British Insects. Diptera, Agromyzidae.* Royal Entomological Society, London.

**Starks K.J., Muniappan R. & Eikenbary R.D. (1972)** Interaction between plant resistance and parasitism against the greenbug on barley and sorghum. *Annals of the Entomological Society of America*, **65**, 650–655.

**Strong D.R. & Wang M.D. (1977)** Evolution of insect life histories and host plant chemistry: hispine beetles on *Heliconia*. *Evolution*, **31**, 854–862.

**Stubbs M. (1977)** Density dependence in the life-cycles of animals and its importance in *K*- and *r*-strategies. *Journal of Animal Ecology*, **46**, 677–688.

**Suski Z.W. & Badowska T. (1975)** Effect of the host plant nutrition on the population of the two spotted spider mite, *Tetranychus urticae* Koch (Acarina: Tetranychidae). *Ekologia Polska*, **23**, 185–209.

**Tahvanainen J.O. & Root R.B. (1972)** The influence of vegetational diversity on the population ecology of a specialized herbivore, *Phyllotreta cruciferae* (Coleoptera: Chrysomelidae). *Oecologia*, **10**, 321–346.

**Takahashi F. (1964)** Reproductive curve with two equilibrium points: a consideration on the fluctuation of insect population. *Researches on Population Ecology*, **6**, 28–36.

**Thompson J.N. & Price P.W. (1977)** Plant plasticity, phenology, and herbivore dispersion: wild parsnip and the parsnip webworm. *Ecology*, **58**, 1112–1119.

**Tilden J.W. (1951)** The insect associates of *Baccharis pilularis* De Candolle. *Microentomology*, **16**, 149–185.

**van Emden H.F. (1972)** Aphids as phytochemists. *Phytochemical Ecology* (Ed. by J.B. Harborne) pp. 25–43. Academic Press, London & New York.

**van Emden H.F. (1978)** Insects and secondary plant substances—an alternative viewpoint with special reference to aphids. *Biochemical Aspects of Plant and Animal Coevolution* (Ed. by J.B. Harborne) pp. 309–323. Academic Press, London & New York.

**van Emden H.F. & Way M.J. (1973)** Host plants in the population dynamics of insects. *Symposia of the Royal Entomological Society of London:* **6.** *Insect/Plant Relationships* (Ed. by H.F. van Emden) pp. 181–199. Blackwell Scientific Publications, Oxford.

**Varley G.C., Gradwell G.R. & Hassell M.P. (1973)** *Insect Population Ecology, An Analytical Approach.* Blackwell Scientific Publications, Oxford.

**Waloff N. (1968a)** Studies on the insect fauna on Scotch broom *Sarothamnus scoparius* (L.) Wimmer. *Advances in Ecological Research*, **5**, 87–208.

**Waloff N. (1968b)** A comparison of factors affecting different insect species on the same host plant. *Symposia of the Royal Entomological Society of London:* **4**. *Insect Abundance* (Ed. by T.R.E. Southwood) pp. 76–87. Blackwell Scientific Publications, Oxford.

**Waloff N. (1975)** The parasitoids of the nymphal and adult stages of leaf hoppers (Auchenorrhyncha: Homoptera) of acidic grassland. *Transactions of the Royal Entomological Society of London*, **126**, 637–686.

**Waloff N. & Solomon M.G. (1973)** Leafhoppers (Auchenorrhyncha: Homoptera) of acid grasslands. *Journal of Applied Ecology*, **10**, 189–212.

**Way M.J. & Murdie G. (1965)** An example of varietal variations in resistance of Brussels sprouts. *Annals of Applied Biology*, **56**, 326–328.

**Webb J.W. & Moran V.C. (1978)** The influence of the host on the population dynamics of *Acizzia russellae* (Homoptera: Psyllidae). *Ecological Entomology*, **3**, 313–321.

**Whittaker J.B. (1979)** Invertebrate grazing, competition and plant dynamics. *Population Dynamics* (Ed. by R.M. Anderson, B.D. Turner & L.R. Taylor) pp. 207–222. Blackwell Scientific Publications, Oxford.

**Wyatt I.J. (1970)** The distribution of *Myzus persicae* (Sulz.) on year-round chrysanthemums. II. Winter season: the effect of parasitism by *Aphidius matricariae* Hal. *Annals of Applied Biology*, **65**, 31–41.

**Zwölfer H. (1971)** The structure and effect of parasite complexes attacking phytophagous host insects. *Proceedings Advanced Study Institute: Dynamics of Numbers in Populations* (Oosterbeek, 1970) (Ed. by P.J. den Boer & G.R. Gradwell) pp. 405–418. PUDOC, Wageningen, The Netherlands.

# 12. THE INFLUENCE OF PARASITIC INFECTION ON THE DYNAMICS OF HOST POPULATION GROWTH

ROY M. ANDERSON

*Zoology Department, Imperial College, London University, Prince Consort Road, London SW7*

## INTRODUCTION

The toll of life exacted by an epidemic outbreak of disease is a spectacle both fascinating and repellent. To an ecologist the fascination lies in witnessing a natural regulatory mechanism in action; epidemics invariably occur when host population densities are high.

There has been a tendency among population biologists to think of disease principally in terms of epidemics which suddenly arise, sweep through a host population and then disappear, as if by magic. The literature contains many accounts of such phenomena within both human and animal populations (Bird & Elgee 1957; Stiven 1962; Tanada 1964; Herman 1969; Vaughan & Vaughan 1969; Vizoso 1969; Duggan 1970; Sinnecker 1976). The concept of an epidemic, however, perhaps as a direct result of its dramatic nature, gives rise to a false impression of the interaction between host and parasite populations. Broadly speaking, examination of the available long-term studies of host–parasite associations reveals patterns of stable co-existence. This point is clearly portrayed by the examples presented in Figures 12.1 and 12.2. Here we see populations of various species of parasites, ranging from viruses to helminths, exhibiting a remarkable degree of temporal persistence.

It is of interest to note that wide-scale population fluctuations are more commonly observed in the associations involving microparasitic organisms, such as viruses or bacteria (Figs 12.1b,c). In contrast, metazoan parasites such as helminths often demonstrate a high degree of temporal population stability (Fig. 12.3). These patterns are in part determined by the relationship between the reproductive attributes of the parasite and its host; in particular, the rates of reproduction and generation times. They are also influenced, however, by the nature of the host's immunological response to the parasite, which may confer either transient or lasting protection against reinfection.

The observed stability of host–parasite interactions within natural communities suggests that, in addition to thinking in terms of the impact of

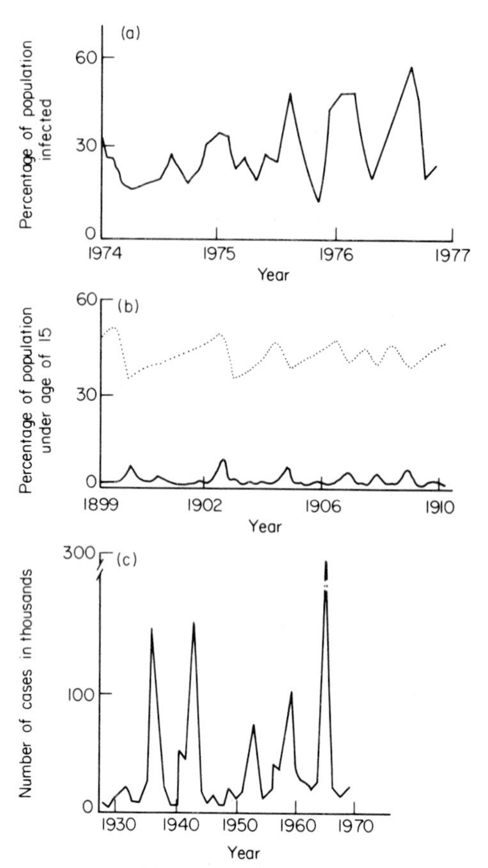


FIG. 12.1. Some examples of long-term trends of fluctuations in parasite populations: parasites of man.
(a) *Entamoeba histolytica* (Protozoa) in Gambia (Bray & Harris 1977).
(b) Measles (virus) in Baltimore, U.S.A. Infections in human population under 15 years of age; dotted line—susceptibles, solid line—infected (Hedrich 1933).
(c) Rubella (virus) in U.S.A. (Krugman 1973).

infrequent outbreaks of disease, we should also consider the long-term effects of stable co-existence between host and parasite populations. In this latter context we may envisage parasites as agents which depress host populations from the equilibrium levels that they would have achieved in the absence of infection. The validity of such a conceptual view of the mode of action of parasitic organisms is supported by a variety of laboratory studies (Park 1948; Finlayson 1949) (Fig. 12.4). In natural communities parasites are likely, therefore, to play an analogous, and often complimentary, role to that of predators or resource limitation in constraining the growth of animal populations (Lack

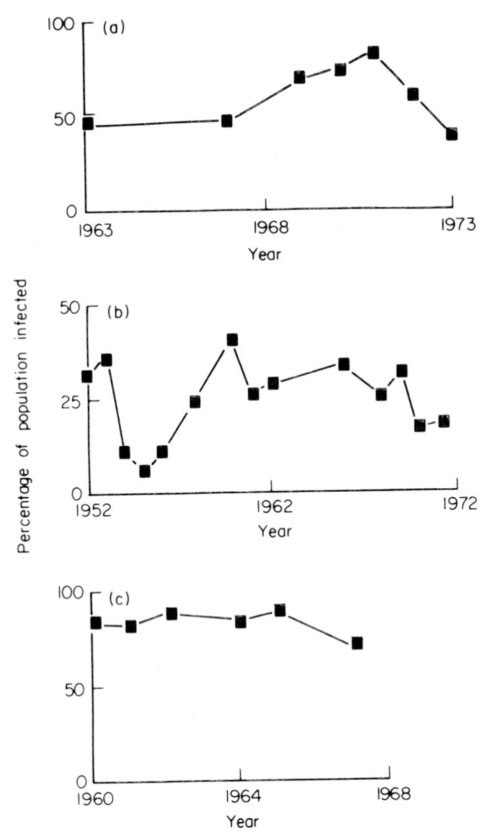


FIG. 12.2. Some examples of long-term trends of fluctuations in parasite populations: parasites of animals and man.
(a) *Trypanosoma* (Protozoa) of *Glossina morsitans submorsitans* in Nigeria (Riordan 1977).
(b) *Eubothrium salvelini* (Cestoda) in *Oncorhynchus nerka* (Smith 1973).
(c) *Schistosoma haematobium* (Trematoda) in man in Iran (Rosenfield, Smith & Wolman 1977).

1954). This regulatory potential, however, has received very little attention in the ecological literature.

The principal aims of this paper are two-fold. The first is to identify the biological attributes of parasitic organisms which determine the degree to which a parasite is able to depress host population growth. The second aim is to seek causal explanations of epidemic outbreaks of disease in the light of observed long-term co-existence of host and parasite. Population models will be employed to examine both of these problems.

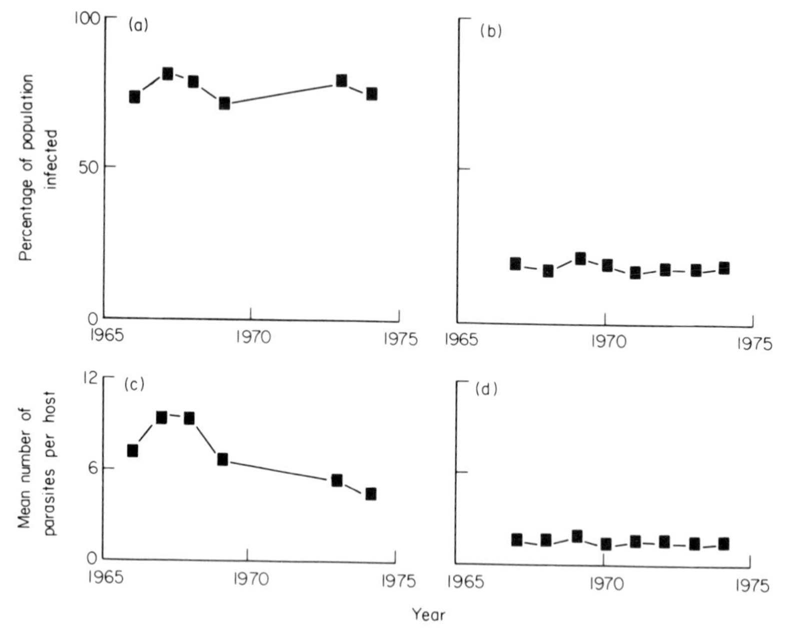


FIG. 12.3. Long-term population trends of *Pomphorhynchus laevis* (Acanthocephala) infections in its definitive fish host (*Leuciscus leuciscus*) and its arthropod intermediate host (*Gammarus pulex*) (data from Kennedy & Rumpus 1977).
(a) Percentage infection in final host.
(b) Percentage infection in intermediate host.
(c) Mean parasite burden per final host.
(d) Mean parasite burden per intermediate host.

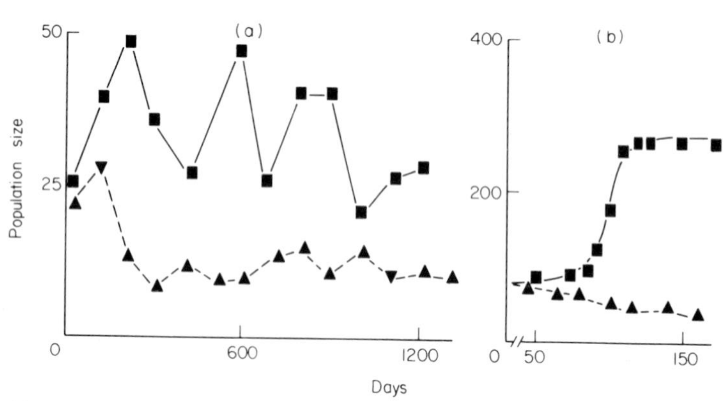


FIG. 12.4. Laboratory examples of the depression of insect population growth by protozoan parasites.
(a) *Tribolium castaneum* infected with *Adelina triboli* (data from Park 1948).
(b) *Laemophloeus minutus* infected with *Mattesia dispora* (data from Finlayson 1950).
Dotted line infected population; solid line uninfected population.

The paper is divided into three major sections. The first part considers the influence of parasites on host survival and reproduction with respect to the direct effects of infection, increased susceptibility to predation and reduced competitive fitness.

The second section explores the dynamical properties of two distinct patterns of infection: *persistent infections*, in which the parasite does not induce a lasting degree of host immunity to reinfection, and *transient infections*, where the recognition of an immune category of hosts is essential.

The final section examines the relationship between host 'stress' and parasitic infection. Specifically, we explore the impact of parasites in high-density host populations where competition for resources is severe.

## THE EFFECTS OF PARASITIC INFECTION ON HOST SURVIVAL AND REPRODUCTION

In general, we regard a species as parasitic if it exhibits, during part or nearly all of its life cycle, a degree of habitat and nutritional dependency on one or more host species. These conditions are necessary but not sufficient. Sufficiency is created if the organism in question causes 'harm' or 'damage' to its host. In population terms we can measure 'harm' by reference to the parasites' influence on the natural intrinsic growth rate of the host population. Such measurement provides a template for assessing the potential of a parasitic organism to regulate or suppress host population growth (Anderson & May 1978; May & Anderson 1978). We can usefully consider this potential under three general headings: direct effects on host survival and reproduction, increased susceptibility to predation and reduced competitive fitness.

### *Direct effects*

By the very nature of their chosen life style, parasites exhibit direct effect on host survival and reproduction within host populations not subject to predatory and competitive pressures. The severity of such effects is invariably associated with the number, or burden, of parasites harboured by a given host (Anderson & May 1978). This point is clearly demonstrated by a variety of laboratory studies as portrayed in Figure 12.5. These examples illustrate the influence of parasite burden on host survival and traverse a broad spectrum of associations, ranging from protozoan parasites of coelenterates to arthropod parasites of mammals. It is clear that the rate of parasite-induced host mortality [$\hat{\alpha}(i)$] is some function of parasite burden ($i$). (Further quantitative examples are documented in Anderson & May 1978.) An important consequence of such relationships is that the net rate of host mortality due to parasitic infection ($D$) is

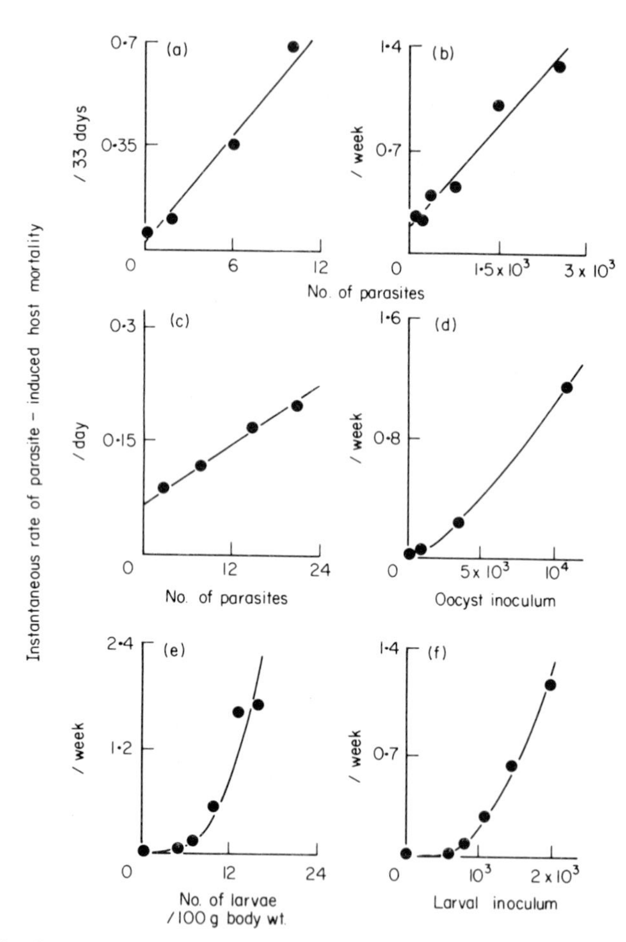


FIG. 12.5. Some examples of the influence of parasite burden on host survival.
(a) *Hydrometra myrae* infected with the mite *Hydryphantes tenuabilis* (Lanciani 1975).
(b) Sheep infected with the fluke *Fasciola hepatica* (Boray 1969).
(c) *Chlorohydra viridissima* infected with the protozoan *Hydromoeba hydroxena* (Stiven 1962).
(d) Turkey poults infected with the protozoan *Eimeria adenoides* (Clarkson 1958).
(e) Guinea pigs infected with the arthropod *Callitroga hominivorax* (Stone 1964).
(d) Rats infected with the nematode *Nippostrongylus muris* (Hunter & Leigh 1961).

critically dependent on the statistical distribution of parasite numbers within a host population. More formally, we can define this loss rate as

$$D = H \sum_{i=0}^{\infty} \hat{\alpha}(i) p(i) \quad (12.1)$$

where $p(i)$ is the probability that a given host harbours $i$ parasites and $H$ is host population size. The function $\hat{\alpha}(i)$ is approximately linear in three of the examples portrayed in Figure 12.5 (Fig. 12.5a,b,c), where

$$\hat{\alpha}(i) = b + \alpha i \quad (12.2)$$

The parameter $b$ represents the instantaneous death rate of the host in the absence of parasitic infection ($i = 0$). In the remaining examples, $\hat{\alpha}(i)$ is a more complex non-linear function of parasite burden (Fig. 12.5d,e,f).

Parasites may also influence the reproductive potential of their host (May & Anderson 1978; Anderson 1978). Here again, the degree of depression of host reproduction appears to be closely related to parasite numbers. Within a population of hosts the reduction in the net rate of reproduction due to infection ($B$) may be formally defined as

$$B = H \sum_{i=0}^{\infty} \hat{c}(i) p(i) \quad (12.3)$$

where $\hat{c}(i)$ denotes the functional relationship between reproduction and parasite burden ($i$). In a few isolated instances parasitic castration may occur, with a subsequent improvement in host survival as a direct result of infection. This arises because reproduction by healthy hosts reduces energy reserves which could otherwise be deployed to enhance survival. The parasite, therefore, by inhibiting reproduction, improves host survival. The net effect of infection, however, is always detrimental to the intrinsic growth rate of the host population.

The direct effects of parasitic infection are invariably linked with a number of other factors, in addition to parasite burden. The age of the host, for example, is often of utmost importance, young and old animals being particularly susceptible to certain infectious diseases. Past experiences may also be relevant, since a degree of resistance (immunologically-based) is often acquired by those hosts that survive their first exposure to infection by a specific parasite. Such mechanisms are particularly important in the case of viral, bacterial and protozoan diseases (Jackson, Herman & Singer 1969; Mimms 1977). Immunity is usually specific in nature, providing protection against one type of organism (homologous immunity) (Soulsby 1972). In certain cases, however, experiences of one parasite may confer resistance to a range of other organisms (heterologous immunity) (Yoeli, Becker & Bernkopf 1955; Schultz, Huang & Gordon 1968; Cox 1975). A contrasting situation sometimes arises where infection with one disease results in increased susceptibility to other parasites (immunodepression) (Cox, Wedderburn & Salaman 1974; Roitt 1976).

The severity of the parasite's influence on host survival and reproduction is also often affected by environmental factors such as temperature. Adverse environmental conditions invariably place a degree of 'strain' on those hosts that harbour infections, as indicated by the laboratory studies on both poikilothermic and homoiothermic hosts by Steinhaus (1958), Sheppe & Adams (1957) and Kolodny (1940).

### *Increased susceptibility to predation*

In the case of human infections, heavy parasite burdens are often said to lead to morbidity rather than mortality. More generally, within the animal kingdom, the 'morbid' host tends to suffer proportionately more from other mortality factors such as predation. Empirical evidence of increased susceptibility to predation as a result of parasitic infection is mainly of an anecdotal nature. Comparatively few studies have demonstrated unequivocably that predation mortality is linked with parasite burden.

A wide variety of field studies indicate that predators tend to select the more vulnerable prey individuals (Rudebeck 1950; Hirst 1965; Hornocker 1970). Vulnerability is often associated with the age of the prey, but in certain cases the presence or absence of parasitic infection is of major significance (Murie 1944; Crisler 1956; Borg 1962; Fuller 1962; Jenkins, Watson & Miller 1963; Mech 1966).

A few laboratory and field studies demonstrate clearly that the presence or absence of parasitic infection in prey species influences their chances of capture by predators (Fig. 12.6). Van Dobben (1952), for example, reported that cormorants (*Phalacrocorax carbo sinensis*) capture a disproportionately large number of roach (*Rutilus rutilus*) infected with the tapeworm *Ligula intestinalis* when compared with the prevalence of infection within the fish population as a whole (Fig. 12.6a). Similarly, Holmes & Bethel (1972) demonstrated that mallards consumed *Gammarus pulex* infected with the larval acanthocephalan *Polymorphus minutus* much more often than uninfected crustaceans (Fig. 12.6b). An intriguing study by Weiser (1958) showed that populations of the arthropod *Tyrophagus noxius* protected from predators contained large numbers of individuals infected with the protozoan parasite *Nosema steinhaus*. Populations subject to predation contained comparatively low numbers of infected insects. The author concluded that infected insects were more susceptible to predation and hence the proportion of infected prey was suppressed by predator activity (Fig. 12.6c).

A qualitative indication of the influence of parasite burden on predation is provided by the work of Vaughan & Coble (1975) on predation of the yellow perch *Perca flavescens* infected with metacercariae of *Crassiphiala bulboglossa*

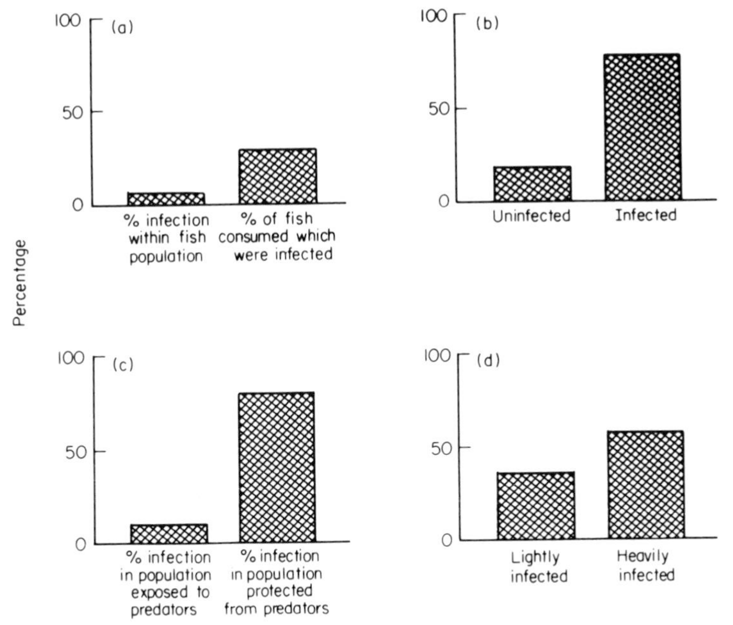


FIG. 12.6. The influence of parasitic infection on predation (see text for further details).
(a) *Rutilus rutilus* infected with *Ligula intestinalis* subject to predation by the Cormorant *Phalacrocorax carbosinensis* (Van Dobben 1952).
(b) *Gammarus pulex* infected with *Polymorphus paradoxus* subject to predation by the Mallard (*Anas platyrhynchos*) (Holmes & Bethel 1972).
(c) Prevalence of infection with *Nosema steinhaus* in populations of *Tyrophagus noxius* either subject to predation by insect species or protected from predation (see text for further details) (Weiser 1958).
(d) *Perca flavescens* infected with metacercariae of *Crassiphiala bulboglossa* subject to predation by *Micropterus salmoides* (Vaughan & Coble 1975).

by the large-mouth bass *Micropterus salmoides*. Heavily infected fish were more susceptible to predation when compared with lightly infected hosts (Fig. 12.6d). Studies of this nature are rare, and hence it is not possible to correlate predation mortality closely with parasite burdens. Intuitively, however, it appears highly probable that a close relationship exists in natural communities. If this hypothesis is correct, predation pressures will increase the severity of the functional relationships, portrayed in Figure 12.5, of the direct effects of parasite numbers on host survival.

In many instances, where the parasite utilizes two or more host species in its life cycle, passage from one host to the next is achieved by means of a predator–prey link within a community food web. Increased host susceptibility to predation may therefore enhance the ability of the parasite to complete its life cycle, provided the predator is a suitable host for the next developmental stage of the parasite (Holmes & Bethel 1972). In the majority of cases, however, predation of infected prey individuals leads to the death of the parasites contained within. Predation may therefore have two very different effects on the population dynamics of parasitic species.

### *Reduced competitive fitness*

The 'strain' of parasitic infection will often result in host mortality due to a reduction in the ability of the infected animal to successfully compete for available resources. For example, certain studies suggest the existence of a correlation between the ability of an animal to gain a territory and the level of parasitic infection (Jenkins, Watson & Miller 1963).

One of the most elegant studies of the interaction between parasitism and competitive ability is that of Park (1948), who demonstrated experimentally that the sporozoan parasite *Adelina triboli* is able to reverse the outcome of competition between two species of flour beetle, *Tribolium confusum* and *T. castaneum*. In the absence of infection, *T. castaneum* has the competitive advantage over *T. confusum* due to its higher reproductive capacity. When the sporozoan is present, however, *T. castaneum* suffers higher mortalities than *T. confusum* due to its greater susceptibility to infection, and hence the competitive advantage is often reversed.

The inability of a host to obtain adequate nutrients often increases the pathogenic effects of the parasite, transforming a comparatively innocuous infection into a lethal disease (Chandler 1953). For example, nutritional stress in insect populations may so weaken the hosts as to make them fatally susceptible to micro-organisms ordinarily of limited pathogenicity (Steinhaus 1958). Infectious diseases within human populations are invariably of greater significance in areas where malnutrition is prevalent (Scrimshaw, Taylor & Gordon 1968; Cole & Parkin 1977).

A number of laboratory studies illustrate clearly the interaction between host nutrition and mortality due to parasitic infection (Fig. 12.7). Sheppe & Adam (1957), for instance, demonstrated that infections of the blood protozoan *Trypanosoma duttoni* in laboratory mice were more often fatal in hosts maintained on half rations when compared with fully-fed mice (Fig. 12.7a). Similarly, Brooke (1945) showed that the blood parasite *Plasmodium relictum* increased mortality of birds maintained on an inadequate diet (Fig. 12.7b). Helminth infections may also be of greater significance in malnourished hosts. For example, sheep fed on a low protein diet and infected with the nematode *Trichostrongylus colubriformis* have much higher mortality rates than those fed on a high plane diet (Gordon 1963) (Fig. 12.7c).

In the case of vertebrates, available evidence suggests that poorly nourished hosts are unable to mount a fully effective immunological response to parasitic invasion (Roitt 1976; Mimms 1977). Depressed immunological competence

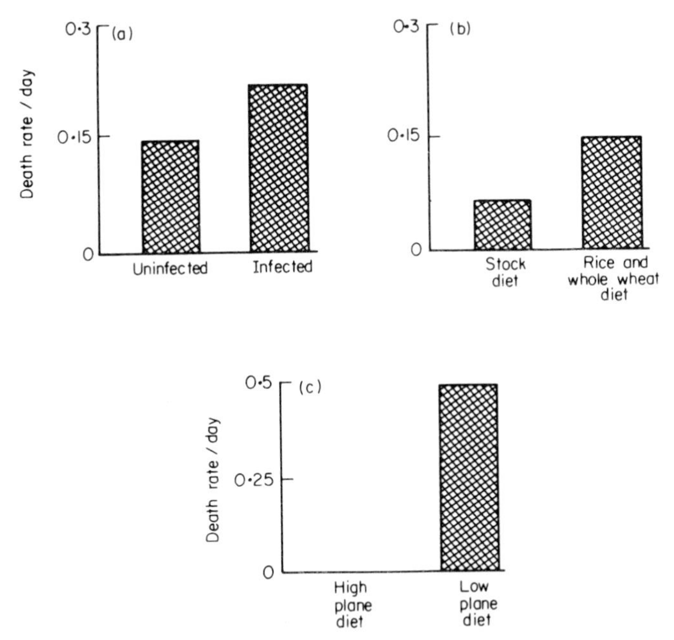


FIG. 12.7. Host diet and pathogenicity of parasitic species.
(a) Mortality of mice infected with *Trypanosoma duttoni* and fed on half rations (Sheppe & Adams 1957).
(b) Mortality of pigeons infected with *Plasmodium relictum* and fed on either a good stock diet or a poor diet of rice and whole wheat (Brooke 1945).
(c) Mortality of sheep infected with *Trichostrongylus colubriformis* and fed on either a high plane or low plane diet (Gordon 1963).

tends to enhance parasite survival and reproduction. This point is illustrated clearly by the studies of Brooke (1945) on bird malaria. The growth of blood protozoan populations was enhanced in birds fed an inadequate diet. Similar evidence is reported by Gordon (1963) for nematode infections in sheep; the parasites showing enhanced survival and reproduction in malnourished hosts.

Care must be taken, however, in making broad generalizations on the interaction between host nutrition and parasite population growth. In certain instances poor host diet may also suppress parasite reproduction and survival. Egg production by certain intestinal helminths is often inhibited in hosts maintained on a low protein diet (Bawden 1969). Similarly, the blood protozoan *Plasmodium cynomogi* shows poor population growth in monkeys maintained on an inadequate diet (Garnham 1966).

On the one hand, malnutrition within the host population may increase the rate of parasite-induced mortality while, on the other hand, it may suppress parasite population growth. In this latter case the rate of parasite-induced host mortality may decrease, provided the reduction in parasite burden is *not* accompanied by increased pathogenicity of the parasite due to the weakened state of the malnourished host. All the same, the available empirical evidence suggests that poor host nutrition is more commonly associated with reduced survival of infected animals (Fig. 12.7).

Parasites may not only increase mortality in undernourished hosts but may, in addition, be the cause of malnutrition by reducing the ability of infected hosts to compete for available resources. We may thus expect parasite-induced host losses to be greater when competition for finite resources is severe at high host population densities.

## POPULATION MODELS

The impact of parasitic infection on the dynamics of host population growth can best be examined within a model framework. Before delineating this, however, it is necessary to make certain distinctions between different types of parasitic infection.

Broadly speaking, we can distinguish two patterns of population association within host–parasite interactions. The majority of helminth and arthropod parasites, and certain protozoan species, do not tend to induce permanent host immunity to reinfection. Hosts may therefore harbour parasite populations for long periods of time due to continual reinfection. It is well established that hosts are able to mount immunological responses to the invasion of metazoan parasites (Soulsby 1972) but such reactions tend to act in a density-dependent manner, restricting but not entirely prohibiting reinfection. If a degree of immunity is induced, it tends to be of short duration in relation to the expected life span of the host. We will term models of such associations *persistent infection models.*

Many microparasitic organisms, the viruses, bacteria and, to a lesser extent, the protozoa, induce lasting immunity to reinfection if the host survives the initial onslaught of the parasite. Population models of such interactions must therefore identify three distinct types of host: susceptible (uninfected), infected and immune. We will refer to such models as *transient infection models.*

A further distinction between transient and persistent infections can be based on the reproductive characteristics of the parasite. Parasitic organisms exhibit two distinct types of reproduction: 'normal births' (*direct reproduction*) which result in instant population growth and births which give rise to transmission stages (*transmission reproduction*). The viruses, bacteria and protozoa invariably reproduce within their host, directly contributing to population growth. This type of reproduction is distinct from the production of transmission stages such as eggs, spores or cysts which, as a developmental necessity, pass outside the host (either into the external habitat or into a vector) before becoming infective to other members of the host population. In general, microparasitic organisms and protozoan species exhibit both direct and transmission reproduction.

In contrast, the vast majority of helminth and arthropod parasites simply display transmission reproduction capabilities. It is important to note, however, that certain exceptions exist; for example, the digenean flukes reproduce directly within their intermediate molluscan hosts.

Transient parasitic infections are generated by organisms which exhibit both direct and transmission reproduction. The development of effective immunological protection to such parasites is undoubtedly linked, in evolutionary terms, with the parasites' ability to display direct population growth within the host. If unconstrained, such growth would inevitably lead to host death. Persistent infections are more commonly associated with organisms such as tapeworms and nematodes, which are capable only of transmission reproduction.

This distinction, however, is not absolute since certain protozoan parasites create persistent infections and exhibit both direct and transmission reproduction.

We will examine the dynamical properties of persistent and transient infection within separate population models. The recurrent aim will be to identify (a) the important general features of parasitism that promote the stable co-existence of the association between host and parasite populations and (b) the biological attributes of parasitic organisms which determine the degree to which a parasite is able to depress a host population from the equilibrium state which would have been achieved in the absence of infection.

As a first step, we will pursue these aims by examining the direct effects of parasitic infection on host survival and reproduction, coupled with the influence of infection on increased susceptibility to predation.

## DYNAMICS OF PERSISTENT INFECTIONS

The question of whether or not persistent infections can regulate host population growth in the absence of other constraints will concern us only briefly. Recent work has explored this problem in depth, yielding the general conclusion that such parasites are very capable of regulating host population growth in a stable manner (Anderson & May 1978; May & Anderson 1978). More specifically, the stable co-existence of host–parasite associations is facilitated by (i) aggregated distributions of parasite numbers per host, (ii) density-dependent parasite mortality or reproduction within individual hosts, (iii) rates of parasite-induced host mortality that increase faster than linearly with parasite burden (see Fig. 12.5d,e,f), and (iv) parasite-induced host mortality rates that are more severe at high host densities. Conversely, a number of processes have a destabilizing effect: (i) parasite-induced reduction in host reproduction, (ii) direct parasite reproduction within their hosts, (iii) time delays in the development of infective stages from the point of parasite reproduction, and (iv) random distributions of parasite numbers per host. Observed host–parasite associations exhibit all the above effects to a greater or lesser extent and, as such, they are in tension between stabilizing and destabilizing elements (May & Anderson 1978).

In this section we shall build on these theoretical insights and explore the factors which determine the degree of depression of the host population induced by parasitic infection. The first step is to have at hand a suitable host–parasite model. In this we shall follow Anderson & May (1978) in adopting a differential equation framework for a coupled one host one parasite interaction. We shall assume that in the absence of infection the host population exhibits logistic growth to an equilibrium population level $K$ representing the carrying capacity of the host's environment. For convenience the reproductive rate ($a$) is assumed to be constant and density-dependent constraints are placed on the host mortality rate $\hat{b}$ such that

$$\hat{b}(H) = b + \hat{\beta}H$$

The parasite is assumed to influence host population growth by reducing host survival in direct proportion to parasite burden (see Fig. 12.5a,b,c and eqns (12.1) and (12.2)). The linear relationship between host mortality and parasite burden (of slope $\alpha$) will be assumed to reflect both the direct effects of the parasite and increased host susceptibility to predation, albeit in a most detail-independent manner.

We will start by considering metazoan parasites which exhibit transmission reproduction (at a rate $\lambda$ per parasite) and no phase of direct reproduction. We will assume that density-dependent constraints (either due to intraspecific competition or immunological attack) act on parasite survival within the host.

We may formally represent the rates of change of the host [$H(t)$] and parasite [$P(t)$] populations with respect to time as follows:

$$dH/dt = (a - b - \hat{\beta}H)H - \alpha H \sum_{i=0}^{\infty} ip(i) \qquad (12.4)$$

$$dP/dt = \lambda HP/(H + H_0) - bP - \hat{\beta}HP - (\mu + \alpha)H \sum_{i=0}^{\infty} i^2 p(i) \qquad (12.5)$$

The first term in eqn (12.4) depicts logistic growth in the absence of parasitic infection [$(a - b - \hat{\beta}H)H$], while the second term [as defined in eqn (12.1)] portrays the net rate of host mortality due to the influence of the parasite [$\alpha H \sum_{i=0}^{\infty} ip(i)$]. The net rate of parasite-induced host mortality is, as discussed previously, dependent on the probability distribution of parasite numbers per host [$p(i)$], where $\alpha$ measures the severity of the parasites' influence.

The first term in the second equation [eqn (12.5)] represents the net rate of parasite transmission reproduction ($\lambda P$) modified by the success of these stages to 'contact' new hosts [$H/(H + H_0)$]. Transmission success is proportional to host density but is dependent on the parameter $H_0$. This parameter varies inversely with transmission efficiency and is equivalent to the death rate of the infective stages divided by their instantaneous rate of infection (May & Anderson 1978). The next two terms portray parasite losses resulting from natural host deaths ($bP$ and $\hat{\beta}HP$). The final term in eqn (12.5) depicts the net rate of loss of parasites due to parasite-induced host mortalities [$\alpha H \sum_{i=0}^{\infty} i^2 p(i)$] and density-dependent natural parasite deaths [$\mu H \sum_{i=0}^{\infty} i^2 p(i)$]. Both of these terms are dependent on the statistical distribution of parasite numbers per host.

This distribution is invariably aggregated or overdispersed in form within natural host populations (Crofton 1971; Anderson 1978; Anderson & May 1978). Such patterns may be conveniently represented by the negative binomial probability model; a distribution defined by two parameters, the mean parasite burden per host ($P/H$) and a parameter $k$ which varies inversely with the degree of aggregation or contagion.

If we make the realistic assumption that the probability terms ($p(i)$) follow the negative binomial pattern, our model becomes

$$dH/dt = (a - b - \hat{\beta}H)H - \alpha P \qquad (12.6)$$

$$dP/dt = P\{\lambda H/(H + H_0) - b - \hat{\beta}H - (\alpha + \mu)[P(k + 1)/(kH) + 1]\} \qquad (12.7)$$

The equilibrium population levels ($H^*$ and $P^*$) are easily obtained from these equations; positive equilibria where $H^*$ is less than or equal to $K$, and $P^* > 0$ are locally stable.

Dependent on the precise parameter values, the model possesses two distinct patterns of dynamical behaviour: (1) the parasite and host may co-exist in a stable association, in which case the host population equilibria $H^*$ is depressed below $K$ (the value it would have reached in the absence of the parasite where $K = (a - b)/\hat{\beta}$); (2) the parasite fails to establish (as a result of low transmission efficiency or low transmission reproduction) and the host population rises to the carrying capacity $K$ as a result of parasite extinction. With these two dynamical patterns in mind, we need to address the questions of how the different population parameters determine population behaviour and, in the case of stable co-existence, how they influence the degree of depression of the host population.

### *Pathenogenicity of the parasite* ($\alpha$)

As portrayed in Figure 12.8, the maximum degree of depression (minimum $H^*$) is achieved by moderate degrees of pathogenicity as measured by the parameter $\alpha$. Parasites at the mutualistic end of the spectrum cause little

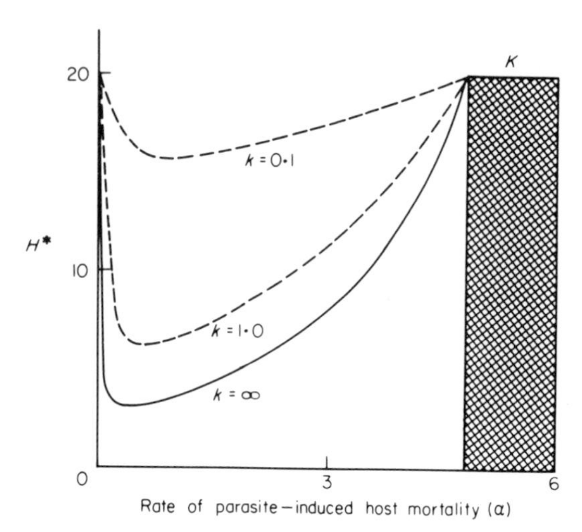


FIG. 12.8. *Persistent infection model.* The relationship between the host population equilibrium ($H^*$) and the parameter $\alpha$ which measures the severity of the parasites' influence on host survival. In the shaded region the parasite is extinct and the host population equilibrium value is $K$. The solid line depicts the relationship for random distributions ($k \rightarrow \infty$) of parasite numbers per host, while the dotted lines depict the relationship for two patterns of over-dispersion (negative binomial distribution where $k = 0{\cdot}1$ and $k = 1{\cdot}0$). ($\lambda = 10$, $a = 3$, $b = 2$, $\mu = 0{\cdot}1$, $H_0 = 5$, $\hat{\beta} = 0{\cdot}1$, $K = 20$.)

suppression of $H^*$, but as $\alpha$ increases $H^*$ falls to a minimum value and then gradually rises to $K$, at which point the parasite fails to co-exist and becomes extinct. This pattern can be explained on intuitive grounds. Host deaths result in the loss of parasites and those hosts that die most rapidly contain on average the larger burdens of parasites. As $\alpha$ increases in severity, the deaths induced by the parasite result in too many parasite losses and hence the total parasite population begins to decline in size. Consequently, the host population is able to recover and the degree of suppression induced by the parasite gradually declines as $\alpha$ becomes large. If $\alpha$ is too high, infected hosts die extremely quickly, preventing effective parasite transmission within the host population and hence the parasite becomes extinct. The mean parasite burden per host ($P^*/H^*$) declines gradually as $\alpha$ rises until it reaches zero at the point of parasite extinction.

Theory therefore suggests that the prevalence of infection (within a host population) and the average parasite load will, to a large extent, reflect parasite pathogenicity in stable interactions within the real world. Highly pathogenic species will be characterized by very low levels of infection within their host population, both in the proportion of hosts infected and the average burden.

One of the most interesting points to emerge from Figure 12.8 pertains to the use of parasites as biological control agents of a pest host species. *The lesson for biological control is clear: parasites of moderate to low pathogenicity are the most effective suppressors of host population equilibria.* Highly pathogenic species will cause little depression and may even fail to persist in a stable association with the host species.

### *Distribution of parasite numbers per host*

Random distributions of parasites within the host population ($k \rightarrow \infty$) tend to have a destabilizing influence on the stability of the host–parasite association. This influence, however, may be counteracted by strong density-dependent constraints on the parasite death rate ($\mu$). In such cases random patterns lead to maximum depression of the host population (Fig. 12.8). As the parasites become more aggregated ($k \rightarrow 0$), fewer and fewer hosts harbour the major proportion of the parasite population. It is precisely these hosts that suffer the highest mortality rates (as a direct consequence of infection) and hence the total parasite population decreases in size and the degree of suppression lessens (Fig. 12.8). The statistical distribution of parasite numbers per host within natural host population will therefore reflect, to some extent, the severity of the parasites' influence on the equilibrium host population. In the case of biological control of a pest host species, the most desirable parasite will be one that exhibits low levels of overdispersion.

### *Parasite reproduction and mortality* ($\lambda$ and $\mu$)

High rates of parasite transmission reproduction ($\lambda$ large) and high transmission efficiency ($H_0$ small) result in low host population equilibria (high depression) and heavy parasite burdens per host. Alternatively, severe density-dependent parasite mortality, due either to intraspecific competition for finite resources such as food and space or immunological attack by the host, lessen the impact of the parasite, leading to low levels of host population suppression or even parasite extinction.

### *Host reproduction* ($a$)

Hosts with a high reproductive potential ($a$) can, to a large extent, offset the impact of parasitic infection. For a fixed value of $\alpha$ increased host reproduction ($a$) leads to a smaller degree of host population depression. If the host reproductive rate is too high, the parasite may even fail to persist in a stable manner within the host population. The pattern of host mortality will therefore change with increased reproductive potential. For a fixed rate of parasite reproduction low levels of host reproduction are associated with high parasite-induced host mortality and consequently very low levels of intraspecific competition within the host population (low $H^*$). For high levels of host reproduction a very small proportion of deaths is due to infection, the majority being a direct result of intraspecific competition for finite resources ($H^*$ high). The impact of parasitic infection is therefore closely correlated with the reproductive potential of the host. If this potential is high, parasites remove only those hosts that would most probably have succumbed anyway due to severe levels of competition. The parasite, however, may be important in such cases if its reproductive potential is also extremely high. Put another way, on the $r$–$K$ continuum (Southwood 1976) populations of $r$ strategist hosts will only be significantly depressed by infection if the parasite itself lies at the very extreme of the $r$ end of the spectrum.

### *Direct parasite reproduction*

The impact of direct parasite reproduction within the host, on the stable co-existence of the association, can be examined by modifying our population model.

This can simply be achieved by adding an extra term, $rP$, to the right-hand side of eqn (12.7). This term depicts direct reproduction at an instantaneous rate, $r$, per parasite. This framework mirrors the dynamics of persistent infections which do not confer on the host a lasting degree of resistance to reinfection. Many protozoan infections may be appropriately discussed within this framework since the immunological response they induce often fails to eliminate the parasite population, simply suppressing it at a low level within the host. In a few instances the parasite may seek refuge within the host (immunologically privileged site) such as the central nervous system, where it can escape the onslaught of antibody and cell-mediated attack. Such patterns of response bring to mind a prey species seeking refuge from a predator.

Three types of dynamical behaviour are possible: (1) parasite extinction and host equilibrium at $K$, (2) stable co-existence of host and parasite, and (3) extinction of host and parasite. The rate of direct reproduction $r$ determines, to a large extent, the boundaries between the three patterns of behaviour (Anderson in prep.). If $r$ is too high, the immunological responses of the host are unable to cope and host and parasite populations rapidly become extinct. High rates of direct reproduction can, however, be counteracted by very efficient immunological attack, leading to high rates of parasite mortality ($\mu$). Such patterns of behaviour are commonly observed in associations between protozoan parasites and mammalian hosts (Clark & Allison 1974), the evolution of a sophisticated immunological response promoting the stable co-existence of host and parasite populations.

Generally speaking, parasites with the ability to reproduce directly within their host are more able to depress host population equilibria severely. This is particularly noticeable amongst invertebrate hosts with their comparatively unsophisticated immunological defences. The studies of Park (1948) and

Finlayson (1949) on protozoan parasites of insect hosts demonstrate this potential clearly (Fig. 12.4).

## DYNAMICS OF TRANSIENT INFECTIONS

The viruses, bacteria and, to a large extent, the protozoan parasites are characterized by their small size, short generation times and extremely high rates of direct reproduction. Such organisms exhibit very rapid population growth within the host and are transmitted from host to host by direct contact, the utilization of an arthropod vector, by means of a transmission stage which may persist outside of the host for short periods of time or, in certain circumstances, by cannibalism or transovarial infection (Mimms 1977).

In order to survive the onslaught of the parasite, the host responds by mounting an immunological response and may be successful in limiting parasite population growth. The success or failure of this response determines the outcome of infection. Infection with certain types of viruses and bacteria invariably leads to host death while others may only marginally decrease the chances of host survival or reproduction during the course of an infection which will eventually be overcome by the immunological responses of the host (Mimms 1977).

The members of the host population that survive their first experience of infection usually exhibit varying degrees of resistance to reinfection. Our model framework must therefore encompass three types of hosts, namely susceptible (uninfected), infected and immune individuals. We will represent these categories by the three population variables $x(t)$, $y(t)$ and $z(t)$ respectively. As portrayed in Figure 12.9, members of the host population will pass from one category to the next as determined by rates of flow such as the rate of infection ($\beta$), rate of recovery from infection ($v$) and rate of loss of immunity ($\gamma$).

The mathematical theory of infectious diseases has been based, to a large extent, on the compartmentalization of the host population into categories representing susceptible, infected and immune individuals (Bailey 1975). This large body of theory is extremely sophisticated in certain respects, such as the stochastic treatment of the growth and decay of epidemics (Bartlett 1956, 1960, 1964; Bailey 1964). In other respects, however, it is far too simplistic. For example, such theory is based almost entirely on the assumption that the host population is of constant size and unaffected by the presence of parasitic infection. Such a framework is inadequate for the exploration of the impact of disease within natural animal populations and, furthermore, for human populations in specific regions of the world where infectious diseases are still a major cause of mortality.

This shortcoming can easily be remedied within a deterministic framework by making the host population a dynamic variable with its own birth and death rates. Such a framework is portrayed in Figure 12.9, where the three categories

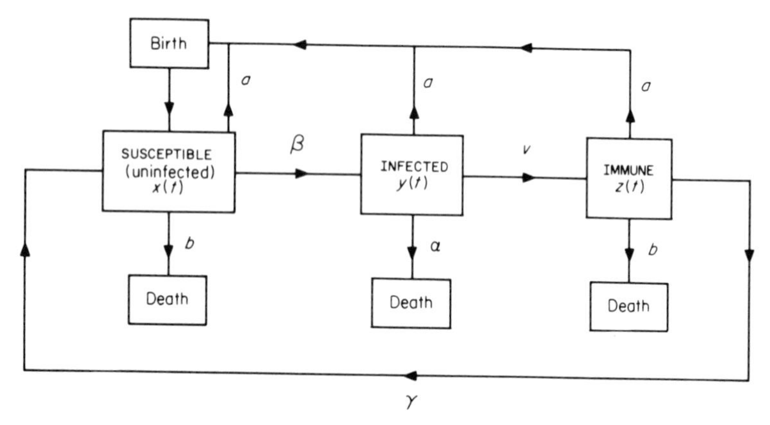


FIG. 12.9. Diagrammatic flow chart of the dynamics of transient infections within three categories of hosts: susceptible [$x(t)$], infected [$y(t)$] and immune [$z(t)$]. The instantaneous rates of population flow are defined as follows: $a$—host birth rate, $b$—uninfected host death rate, $\alpha$—infected host death rate, $v$—rate of recovery from infection, $\gamma$—rate of loss of immunity.

of hosts are assumed to reproduce at a rate $a$ per individual and die at rates determined by their current status. For example, the flow diagram displayed in Figure 12.9 assumes that infected hosts die at a rate $\alpha$, while uninfected and immune individuals die at a rate $b$, where $b$ is assumed to be less than $\alpha$ in value.

This framework is very general since, by altering specific parameter values, we can cater for a wide range of parasitic infections. For instance, if $\gamma$, the rate of loss of immunity, is zero, hosts that recover from infection remain immune for life. Similarly, if $v$, the rate of recovery from infection is zero, all hosts die from infection. Conversely, the duration of infection can be altered by changing the value of $v$ since the mean expected duration is $1/v$. We can also alter the impact of infection on host reproduction since the infected individuals may reproduce at a lower rate than the susceptible and immune classes.

The framework is, however, inadequate in one particular sense; it fails to relate host mortality to parasite burden. This feature of host–parasite associations (see Fig. 12.5) was faithfully mirrored in the persistent infection models of the previous section. Luckily, this shortcoming is of limited significance in the case of transient infections since the duration of infection is usually short in relation to the life span of the host. We can therefore realistically consider infection as a temporal entity which induces a change in host mortality and/or reproduction. In a practical sense, it is also rather unrealistic to attempt to measure the number of viral particles, bacteria or protozoa within an individual host and to relate this quantity to host mortality.

We can translate the flow diagram of Figure 12.9 into three coupled differential equations (where the rates of change are as defined in the figure legend) for the populations of susceptible $[x(t)]$, infected $[y(t)]$ and immune $[z(t)]$ individuals as follows:

$$dx/dt = a(x + y + z) - bx - \beta xy + \gamma z \tag{12.8}$$

$$dy/dt = \beta xy - \alpha y - vy \tag{12.9}$$

$$dz/dt = vy - bz - \gamma z \tag{12.10}$$

We have assumed that the parasite influences host survival such that the death $\alpha$ of infected individuals is greater than the death rate of susceptible and immune hosts $b$. Reproduction is assumed to be unaffected by infection. The net rate of infection ($\beta xy$) is assumed to be directly proportional to the product of the number of susceptible and infected hosts (see Bailey (1975) for a fuller discussion of this point).

We will use this model to examine one particular question: are transient infections able to regulate host population growth stably in the absence of other constraints?

### *Regulation of host population growth*

The model depicted by eqns (12.8), (12.9) and (12.10) readily yields the equilibrium population densities $x^*$, $y^*$ and $z^*$, where

$$x^* = (\alpha + v)/\beta \tag{12.11}$$

$$y^* = [(a - b)(\alpha + v)(b + \gamma)]/\beta[(\alpha + v)(b + \gamma) - a(b + \gamma) - v(a + \gamma)] \tag{12.12}$$

$$z^* = vy^*/(b + \gamma) \tag{12.13}$$

The total host population size at equilibrium $N^*$ is given by

$$N^* = x^* + y^* + z^* \tag{12.14}$$

As long as the host population growth rate in the absence of infection is positive ($a > b$), eqn (12.12) reveals that the parasite is able to regulate host population growth in the absence of other constraints, provided that

$$(\alpha - a) > v(a - b)/(b + \gamma) \tag{12.15}$$

In other words, the death rate of infected hosts must be greater than their reproductive rate by a factor of $v(a - b)/(b + \gamma)$. If this condition is not satisfied, the host population grows exponentially ($y^*$ and $z^*$ also grow exponentially, but the proportion of host infected, $x^* \rightarrow (a + v)/\beta$ until other factors such as resource limitation constrain growth. When the parasite and host co-exist [eqn (12.15) satisfied], positive equilibria are locally stable (Anderson in prep.).

The condition defined in eqn (12.15) for the stable co-existence of host and parasite can be used to explore the biological characteristics which enhance the regulatory potential (creation of locally stable equilibria) of transient infections. The following attributes enable such infections to control host population growth.

1 Highly pathogenic parasites ($\alpha$ large) are able to regulate growth even when the duration of infection is short ($v$ large) or immunity to reinfection long lasting ($\gamma$ small).

2 Moderately pathogenic species may still act in a regulatory manner provided the duration of infection is long ($v$ small) or immunity to reinfection of limited duration ($\gamma$ large). If lasting immunity is conferred on those hosts that recover from infection ($\gamma \to 0$), the immune category of the host population acts as a refuge from infection (analogous to prey species escaping from the attention of predators) and the persistence of the parasite population is difficult to achieve unless the duration of infection is long ($v$ small).

3 Comparatively innocuous infections ($\alpha$ small) will rarely regulate host population growth. In certain rather specific instances, however, the stable co-existence of host and parasite populations may occur. In such cases the following conditions must be satisfied:

(a) If the duration of infection is extremely long and immunity to reinfection virtually non-existent, regulation may be achieved.

(b) The inability of the infected hosts to reproduce enhances the ability of the parasite to co-exist within the host population. Specifically, if $a$ is zero for the infected category of hosts, the condition expressed in eqn (12.15) becomes

$$\alpha > v(a - b)/(b + \gamma) \tag{12.16}$$

which is more easily satisfied for small values of $\alpha$.

The general conclusion to emerge from these points is that transient infections, created by viruses, bacteria or protozoa, may effectively regulate host population growth provided their influence on host survival or reproduction is sufficiently severe. The severity of their influence (in the cases of stable interactions between host and parasite populations) will be closely related to certain biological characteristics of their association with the host, such as duration of both infection and host immunity to reinfection.

One further point of interest concerns the influence of the pathogenicity of the parasite (measured by $\alpha$) on the size of the host population equilibria ($N^*$) when stable co-existence occurs.

Intriguingly, the picture which emerges is very similar to that found for the persistent infections in the previous section, even though the mathematical framework of our model is very different. These patterns are portrayed in Figure 12.10 and should be compared with Figure 12.8a (persistent infections). Concentrating on Figure 12.10a, we see that moderate degrees of pathogenicity lead to the lowest host population equilibrium ($N^*$), as in Figure 12.8a.

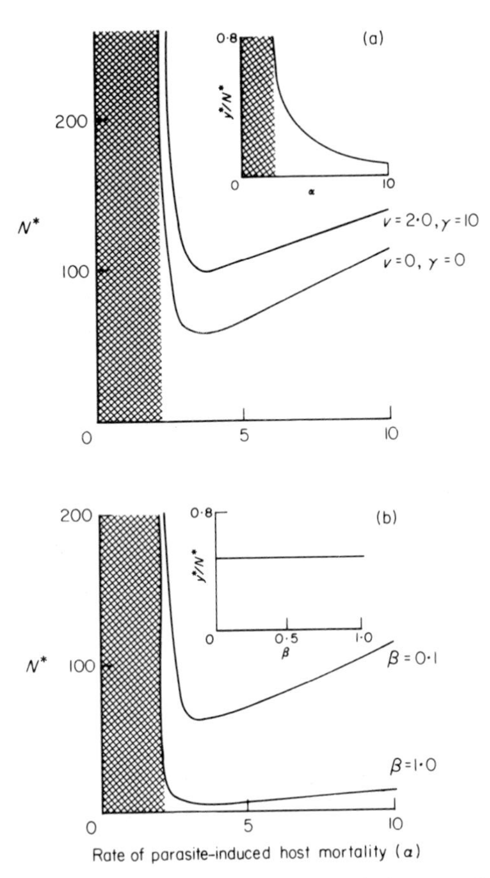


FIG. 12.10. *Transient infection models.* The relationship between the host equilibrium population level ($N^*$) and the pathogenicity of the parasite ($\alpha$). In the shaded regions the parasite fails to regulate host population growth.
(a) *Main graph:* The size of $N^*$ for an infection of short duration and inducing a long period of immunity to reinfection (line (1), $v = 2{\cdot}0$, $\gamma = 10{\cdot}0$). Line (2) represents $N^*$ for a highly pathogenic infection from which the infected hosts do not recover ($v = 0{\cdot}0$, $\gamma = 0{\cdot}0$). Other parameter values: $a = 2$, $b = 1$, $\beta = 0{\cdot}1$. *Inset graph:* Relationship between the proportion of infected hosts within the population ($y^*/N^*$) and $\alpha$. ($v = 0$, $\gamma = 0$.)
(b) *Main graph:* The size of $N^*$ for two different levels of parasite transmission: $\beta = 0{\cdot}1$ (line (1)); $\beta = 1{\cdot}0$ (line (2)). ($a = 2$, $b = 1$, $v = 0{\cdot}1$, $\gamma = 10$.) *Inset graph:* The relationship between the proportion of infected hosts ($y^*/N^*$) and $\beta$, the rate of infection ($\alpha = 3{\cdot}0$).

Small values of $\alpha$ are insufficient to lead to regulation, while high values lead to the death of large numbers of infected hosts, reducing transmission and permitting high host population equilibria. Infections of short duration ($v$ large), which create lasting immunity to reinfection ($\gamma$ small), lessen the regulatory effect of the parasite. The proportion of infected hosts within the population ($y^*/N^*$) declines as the parasites' pathogenicity increases. As in the case of persistent infections, high rates of transmission efficiency ($\beta$ large) decrease the equilibrium level of the host population (Fig. 12.10b). However, it is interesting to note that for fixed values of $\alpha$, $v$ and $\gamma$ the rate of infection $\beta$ has little effect on the proportion of infected hosts within the population ($y^*/N^*$) (Fig. 12.10b). The pathogenicity of the parasite ($\alpha$) is the principal determinant of this proportion.

The model framework outlined above is convenient for exploring the regulatory potential of transient infections. However, the factors which determine the degree of depression of the host population caused by infection are best explored within a framework which assumes logistic growth in the absence of the parasite, since this provides a point of reference for comparison.

### *Depression of host population equilibria*

We will assume that in the absence of infection the host population grows in a logistic manner with an equilibrium state at $K$, the carrying capacity of the environment. We will further assume that the parasite is highly pathogenic such that infected hosts do not recover from the disease. The immune category $[z(t)]$ is therefore excluded from our model.

This framework more closely mirrors viral infections of invertebrate hosts such as insect species, where the infected animal is unable to mount an effective immunological response to parasitic invasion and invariably dies as a direct result.

Our model is therefore of the form:

$$dx/dt = a(1 - (x + y)/K)(x + y) - \beta xy \tag{12.17}$$

$$dy/dt = \beta xy - \alpha y \tag{12.18}$$

The first term in eqn (12.17) represents logistic growth of the host population in the absence of infection to the carrying capacity $K$, while the remaining terms in both eqn (12.17) and eqn (12.18) are as defined for the previous model, eqns (12.8), (12.9) and (12.10).

When the disease is present ($y^* > 0$), the equilibrium value of $x^*$ is obtained from eqn (12.20), where

$$x^* = \alpha/\beta \tag{12.19}$$

The equilibrium population of infected hosts $y^*$ is obtained from the quadratic equation

$$ay^{*2} + y^*[K(a - \alpha)] + (\alpha/\beta)(\alpha/\beta - K) = 0 \tag{12.20}$$

A single positive and realistic value of $y^*$ exists provided

$$\beta K > \alpha \qquad (12.21)$$

If this condition is not satisfied, the parasite becomes extinct and the host population settles to the equilibrium value $K$. Where host and parasite populations co-exist (eqn (12.21) satisfied), it can be shown that the equilibrium populations are locally stable.

The biological characteristics of the infection which determine the degree of suppression of the host population (from the level $K$) are very similar to those discussed for persistent infections and for the transient infection model in which density-dependent constraints on host population growth were excluded. These characteristics are portrayed graphically in Figure 12.11 and may be summarized as follows.

1 Moderate degrees of pathogenicity (represented by $\alpha$) lead to the maximum degree of host population depression (Fig. 12.11a).

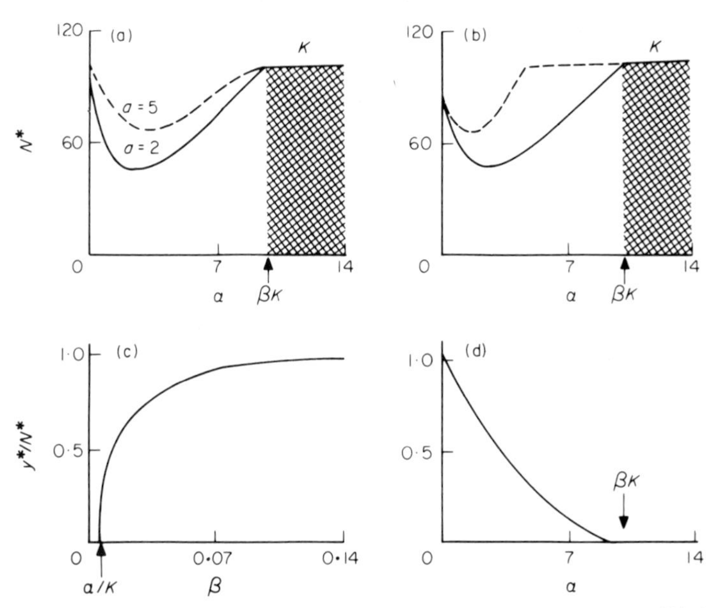


FIG. 12.11. *Transient infection models.* The degree of host population equilibrium depression induced by a highly pathogenic parasite; infected hosts do not recover from infection. In the shade regions the parasite becomes extinct and the host population equilibrium levels is $K$.

(a) The relationship between $N^*$ and $\alpha$ for two different rates of host reproduction $a$ ($a = 2$, $a = 5$). ($b = 1$, $\beta = 0{\cdot}1$, $K = 100$.)

(b) The relationship between $N^*$ and $\alpha$ for two different rates of infection $\beta$ ($\beta = 0{\cdot}1$, $\beta = 0{\cdot}05$). ($b = 1$, $a = 2$, $K = 100$.)

(c) The relationship between $y^*/N^*$ and $\beta$ ($\alpha = 0{\cdot}5$, $a = 2$, $b = 1$, $K = 100$).

(d) The relationship between $y^*/N^*$ and $\alpha$ ($\beta = 0{\cdot}1$, $a = 2$, $b = 1$, $K = 100$).

2 Severe parasite pathogenicity will lead to parasite extinction (Fig. 12.11a).
3 Hosts with a high reproductive potential ($a$) are more able to withstand the impact of infection and will therefore exhibit low degrees of depression (Fig. 12.11a).
4 High rates of parasite reproduction usually reflect high transmission potential and hence high rates of infection ($\beta$). Such characteristics will enhance the suppression of the host population equilibria (Fig. 12.11b). If the rate of infection is too low, the parasite will become extinct (Fig. 12.11c).
5 The proportion of infected hosts in the population ($y^*/N^*$) declines as parasite pathogenicity ($\alpha$) rises (Fig. 12.11d) but will increase as the rate of infection ($\beta$) rises (Fig. 12.11c).

The preceding analysis and discussion of the dynamics of both persistent and transient infections indicates that host and parasite populations may co-exist in a stable manner provided certain biological criteria are met.

So far, however, the models do not exhibit dynamical patterns of behaviour, which explain sudden outbreaks of disease where a state of high host population density and low average parasite burden dramatically changes to a state of low host density and high prevalence and intensity of infection.

In an attempt to seek causal explanations of such events, we now proceed to examine the relationship between the pathogenicity of the parasite and host nutritional status.

## HOST NUTRITIONAL STATUS AND THE IMPACT OF INFECTION

Within a competitive framework the inability of a host to obtain adequate nutrients often increases the impact of parasitic infection on host survival or reproduction. Reduced competitive fitness is often a direct result of parasitic infection and hence the parasite may be responsible for both *causing malnutrition* and *increasing mortality* in undernourished hosts. The influence of parasitic infection is therefore likely to be more pronounced at high host population densities when competition for available resources is severe.

We may explore the influence of such effects within the framework of either the persistent or transient infection models. The persistent infection model, however, captures the impact of infection on host mortality in a more detailed manner and will therefore serve as our model framework (see eqn (12.1) and Fig. 12.5).

We will assume that high host population densities (near or at the carrying capacity of the host environment $K$) lead to increased parasite pathogenicity as a result of *either* the reduced competitive fitness of infected hosts (inability to obtain adequate nutrients for survival) *or* the additional strain placed on infected hosts with low levels of food intake (reduced immunological competence in malnourished hosts). The net rate of parasite-induced host mortality ($D$) may therefore be conveniently portrayed as a function of both parasite burden ($i$) and host population density ($H$) as follows:

$$D = H \sum_{i=0}^{\infty} \hat{\alpha}(i, H)p(i) \tag{12.22}$$

Quantitative information on the nature of the function $\hat{\alpha}(i, H)$ is unavailable at present, but we may deduce some general patterns from a knowledge of the biology of host–parasite associations. For simplicity we will assume that for a constant value of $H$ the relationship between host mortality and parasite burden ($i$) is of linear form, as portrayed in the empirical examples documented in Figure 12.5a,b,c. That parasite-induced host mortality increases in poorly nourished hosts is not in doubt (see Fig. 12.7), but a number of possible forms of $\alpha(i, H)$ may exist in natural populations. We will consider two types of functional relationship as follows.

1 $\hat{\alpha}(i, H) = \alpha Hi$: The rate of host losses is directly proportional to host population density.

2 $\hat{\alpha}(i, H) = \alpha H^2 i/(H^2 + g)$: The rate of host losses is non-linearly related to host population density such that the rate is small at low population density but rises to a plateau at high densities (this function is analogous to the type III functional responses of predators to prey densities; Holling 1965). The biological interpretation of such patterns is that in well-nourished hosts (low population density) the impact of the parasite is minimal due to the effective immunological defences of the host. As the average level of food intake decreases (high host densities), the host's ability to cope with infection declines due to a decreased ability to mount an effective immunological response. Eventually, at very high host densities, the parasites' pathogenicity reaches a plateau ($\alpha$) as the host's immunological defences crumble. The other constant in the function, $g$, determines the rate at which $\alpha(i, H)$ approaches the plateau $\alpha$.

### *Parasite pathogenicity directly proportional to host density*

Following the procedures used earlier in the formulation of a persistent infection model, we employ a framework based on two coupled differential equations representing the dynamics of the host and parasite populations. Our only departure is to assume that the probability distribution of parasite numbers per host is of Poisson form and that the rate of parasite-induced host mortality is linearly related to host population density. We consider first parasitic species which exhibit only transmission reproduction.

Our new model takes the form

$$dH/dt = (a - b - \beta H)H - \alpha HP \tag{12.23}$$

$$dP/dt = P[\lambda H/(H + H_0) - b - \beta H - \mu - \mu P/H - \alpha P - \alpha H] \tag{12.24}$$

[The equilibrium populations $H^*$ and $P^*$ and the stability properties of this model are discussed more fully in Anderson in prep.] The mathematical complexity of this model makes local stability analysis difficult, but we may gain a qualitative feel for the properties of the equations by numerical and graphical analysis.

The model possesses two patterns of dynamical behaviour.

1 For certain parameter values a unique stable state exists which *either* represents stable co-existence of host and parasite populations ($P^* > 0$, $H^* < K$) *or* parasite extinction and host population equilibrium at the carrying capacity of the environment ($P^* = 0$, $H^* = K$).

2 For other parameter values two stable states occur. This type of pattern is displayed in Figure 12.12a, where the relationship between $H^*$ and $\alpha$ is depicted. At low levels of $\alpha$ (low parasite pathogenicity) the host and parasite co-exist in a stable association ($P^* > 0$, $H^* < K$). High values of $\alpha$ lead to parasite extinction ($P^* = 0$, $H^* = K$). However, intermediary values of $\alpha$ (i.e. $\alpha = 1.5$ in Fig. 12.12a) may generate two stable states, the dynamical landscape between them, divided by an unstable equilibrium point (dotted line in Fig. 12.12a). The two states, for a fixed set of parameter values, are stable co-existence ($P^* > 0$, $H^* < K$) or parasite extinction ($P^* = 0$, $H^* = K$). These notions are expressed in a different manner in Figure 12.12b,c, where the relationships between

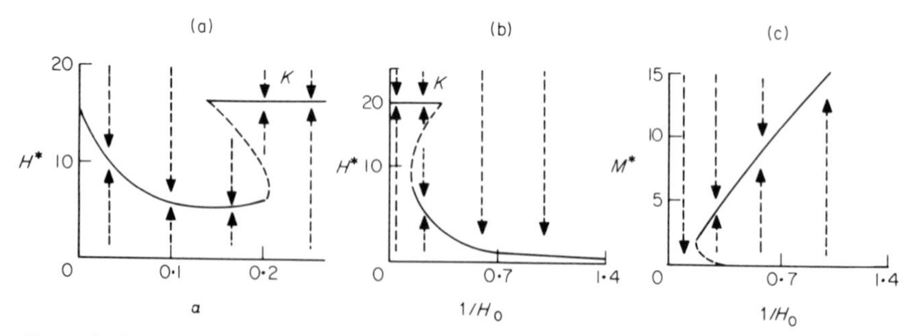


FIG. 12.12. *Persistent infection models.* Impact of increased parasite pathogenicity as host density rises. The solid line denotes a stable host equilibrium, while the dashed line denotes unstable states ($a = 3$, $b = 1$, $\mu = 0{\cdot}1$, $\lambda = 10$, $H_0 = 5$, $\beta = 0{\cdot}1$).
(a) The relationship between $H^*$ and $\alpha$.
(b) The relationship between $H^*$ and transmission efficiency ($1/H_0$) ($\alpha = 0{\cdot}15$).
(c) The relationship between the equilibrium mean parasite burden per host, $M^*$ and $1/H_0$ ($\alpha = 0{\cdot}15$).

parasite transmission efficiency ($1/H_0$) and both the host population equilibrium ($H^*$) and the mean parasite burden ($M^* = P^*/H^*$) are displayed. Certain levels of transmission efficiency will lead to two stable equilibrium states, one of which is the host population at $K$ and the parasite extinct while the other is stable co-existence of host and parasite.

These patterns of dynamical behaviour are of considerable interest. They suggest that the introduction of a disease (with given population characteristics) into a host population at or near its carrying capacity may result in a dramatic decrease in host population size with a concomitant rise in the mean parasite burden per host. Similarly, a stable association between host and parasite may suddenly be disrupted, leading to parasite extinction by a slight change in one population parameter such as parasite transmission efficiency perhaps induced by climatic change (dynamical properties of this type may be interpreted in the terminology of catastrophe theory; Ludwig, Jones & Holling 1978; Thom 1975). The behaviour of this model is reminiscent of many documented accounts of disease outbreaks in animal populations (Lovat 1911; Murie 1930; Bardach 1951; Cheatum 1951; Mykytowycz 1962; Tanada 1964; Herman 1969; Vizoso 1969).

### *Parasite pathogenicity non-linearly related to host density*

In many host–parasite associations it is probable that the pathogenicity of the parasite will rise to a plateau as host density increases. At this plateau level the immunological barriers of the weaker hosts (due to severe intraspecific competition for resources) will be lowered and the parasite will be able to live life to the full unhindered by cell-mediated or antibody attack. In this state the parasite will have its maximum impact on the survival of its host.

The following function captures the qualitative features of such effects, albeit in a most detail-independent manner:

$$\hat{\alpha}(i, H) = \alpha H^2 i/(H^2 + g) \qquad (12.25)$$

The new model incorporating eqn (12.25) may exhibit four distinct dynamical patterns depending on the values of the population parameters.

1 A unique stable state may exist in which the parasite is extinct and the host population equilibrium is at $K$.

2 A second type of unique stable state may occur in which the host and parasite co-exist with a low mean parasite burden per host and a high host population equilibrium (but less than $K$).

3 A third type of unique stable state can occur, similar to (2) above, but with a high mean parasite burden and low host population equilibrium.

4 Two stable states may exist, in both of which the host and parasite co-exist. One state has high host population equilibria (but less than $K$) and low average

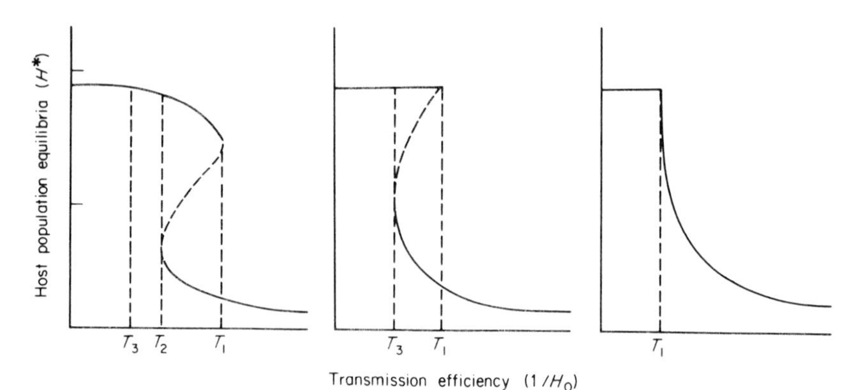


FIG. 12.13. *Persistent infection models.* Summary of the dynamical effects of a relationship between parasite pathogenicity and host density. The relationship between $H^*$ and transmission efficiency ($1/H_0$). Solid lines stable states, dashed lines unstable states ($a = 3$, $b = 1$, $\beta = 0{\cdot}1$, $\mu = 0{\cdot}1$, $\lambda = 6{\cdot}1$).
(a) $\alpha(H, i) = \alpha H^2/(H^2 + g)$.
(b) $\alpha(H, i) = \alpha i H$.
(c) $\alpha(H, i) = \alpha i$.
See text for an explanation of the points $T_1$, $T_2$ and $T_3$.

parasite burdens while the other has high average parasite burdens and low host densities.

These patterns of behaviour are displayed in Figures 12.13a and 12.14a, in which the host population equilibrium ($H^*$, Fig. 12.13a) and the equilibrium mean parasite burden ($M^*$, Fig. 12.14a) are depicted for various levels of parasite transmission efficiency (defined as $1/H_0$). Focusing our attention on the horizontal axis in these figures, levels of transmission between 0–$T_3$ give rise to pattern (1), levels between $T_3$–$T_2$ to pattern (2), levels greater than $T_1$ to pattern (3) and levels between $T_2$–$T_1$ pattern (4).

Of major interest is the behaviour of the model for levels of transmission

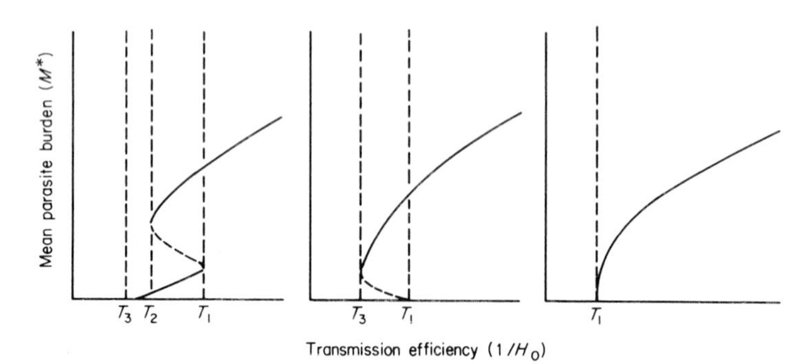


FIG. 12.14. Identical to Figure 12.13, except it depicts the relationship between the mean parasite burden $M^*$ and $1/H_0$.

between $T_2$ and $T_1$. In this region the association possesses two stable states separated by an unstable equilibrium point. The state to which the system settles will depend on the size of any perturbation from each equilibrium. For example, systems at the stable state of high host density and low parasite burdens may collapse to a stable state of low host density and high parasite burdens if the host population is depressed by a large downward perturbation (Fig. 12.13a).

The intuitive explanation of such behaviour lies in the fact that the parasite is highly pathogenic at high host densities (due to host nutritional stress), causing many parasite deaths by the host mortalities it induces. The net result is small parasite populations and low degrees of host population depression. At low host densities the parasite exhibits low to moderate pathogenicity, *but*, as we have already seen (Fig. 12.8), it is able to severely depress the host population since the deaths it induces remove only a small fraction of the total parasite population. Consequently, host equilibria are low and parasite burdens high.

These patterns of behaviour suggest that small changes in parasite transmission efficiency, perhaps due to climatic change (enhance infective stage survival in the free-living habitat), may induce 'epidemic' outbreaks of disease. The stable co-existence of high host densities and low prevalences of infection may suddenly collapse to a state of severe depression of the host population concomitant with a dramatic rise in the prevalence and intensity of infection.

A general qualitative summary of the patterns of dynamical behaviour shown by the persistent infection models discussed in this paper is displayed in Figures 12.13 and 12.14. If host losses due to infection are unrelated to changes in host density (eqns (12.4) and (12.5)), a unique stable equilibrium will occur which will represent either parasite and host co-existence or parasite extinction (Figs 12.13c and 12.14c). When the pathogenicity of the parasite is linearly related to host density (eqns (12.23) and (12.24)), either one or two stable states may exist. The two stable state situations represent stable co-existence or parasite extinction (Figs 12.13b and 12.14b). Complex non-linear relationships between pathogenicity and host density may also generate two stable states, but in certain instances both represent stable co-existence, one leading to high host densities and low parasite burdens and the other to low host densities and high parasite burdens (Figs 12.13a and 12.14a).

Finally, the existence of multiple stable points in host–parasite associations has important implications for the control of parasitic diseases. Encouragingly, Figures 12.13b and 12.14b suggest that lowering the transmission efficiency of the parasite may result in the eradication of a parasitic disease if the pathogenicity of the parasite is linearly dependent on host density. Alternatively, Figure 12.14b indicates that a decrease in the mean parasite burden (by the application of chemotherapeutic agents) may drive the parasite to extinction if a 'break-point' is crossed. These points have been discussed in a different context by

MacDonald (1965) and May (1977a) in relation to models of the human disease schistosomiasis. Where non-linear relationships occur between parasite pathogenicity and host density, our models suggest that parasite extinction may not necessarily be a feasible objective. In such cases we may be able to reduce parasite transmission such that the association switches from one characterized by high parasite burdens to a state of low levels of infection (Fig. 12.14a).

## CONCLUSIONS

Perhaps one of the most intriguing aspects of theoretical study in ecology is its use in seeking causal explanations of discontinuous patterns of population behaviour (May 1977b). Epidemic outbreaks of disease have not previously been examined in this context. Theory certainly supports the widely held opinion that epidemics are essentially density-dependent phenomena. The reason for this dependency, however, is probably associated with the increased pathogenicity of many parasitic organisms within poorly nourished or 'stressed' hosts rather than increased transmission efficiency at high host densities.

Host–parasite associations are likely to possess a number of stable equilibrium states. Changes in certain population parameter values, or perturbation of population equilibria, may switch the association from a stable state of high host density and low prevalence of infection to another stable state of low host density and high levels of parasitic infection. Such changes may occur on a regular basis due, perhaps, to the influence of seasonal climatic variation. On the other hand, cyclic changes may be driven by other variables such as limit cycle behaviour in the abundance of food for a herbivore population (Caughley 1976). The general point, however, is that the occurrence of multiple stable states of co-existence between host and parasite may be of central importance in explaining sudden epidemic outbreaks of disease within animal populations.

Broadly speaking, it is true to say that ecologists have in the past tended to underestimate the long-term impact (in contrast to epidemic phenomena) of parasitic infection within natural communities. A clear illustration of this point is provided by the extremely limited coverage given to the role of infectious diseases in determining the dynamics of animal populations by the numerous ecology texts published in the last ten years. This is in sharp contrast to the extensive coverage of predator–prey interactions. The importance of devoting more attention to the role of parasitism is highlighted by the numerous parasitological surveys of natural communities, which clearly indicate that the vast majority of animal populations harbour a large number of parasitic species (Elton 1942; Anderson 1962; Dogiel 1964; Welch 1965; Herman 1969; Avery 1969; Baker 1969; Young 1970; Kennedy 1972). Such studies, however, tend to underestimate the overall pattern of parasitic infection since they do not, in general, consider viral and bacterial diseases. The difficulties of identification and detection of microparasitic organisms invariably mean that their presence is noticed only when epidemic outbreaks of disease occur.

Empirical evidence of the relative importance of disease in the natural regulation of animal numbers is scarce (Lack 1954; Dunsmore 1971; Anderson 1976). The impact of parasitism is, in general, obscured from view due to the small size of parasitic organisms, their endoparasitic mode of life in many cases and, of course, the enormous practical difficulties associated with observing mortality in natural communities and ascribing cause of death.

It is hoped, however, that theoretical studies which underline the regulatory potential of parasitic species will provide a stimulus to overcome such practical difficulties. Support for, or falsification of, the hypothesis that this regulatory potential is frequently realized in natural communities will be dependent on an increased awareness of the need to include the study of infectious diseases beneath the umbrella of ecology.

## ACKNOWLEDGMENTS

I am very grateful to M.P. Hassell and R.M. May for their helpful comments and advice.

## REFERENCES

**Anderson R.C. (1962)** The parasites (helminths and arthropods) of white-tailed deer. *Proceedings of the First National Deer Disease Symposium, Georgia, U.S.A.*, 162–173.

**Anderson R.C. (1976)** Helminths. *Wildlife Diseases* (Ed. by L.A. Page) pp. 35–43. Plenum Publishing Corporation, New York.

**Anderson R.M. (1978)** The regulation of host population growth by parasitic species. *Parasitology*, **76**, 119–158.

**Anderson R.M. & May R.M. (1978)** Regulation and stability of host-parasite population interactions. I. Regulatory processes. *Journal of Animal Ecology*, **47**, 219–249.

**Avery R.A. (1969)** The ecology of tapeworm parasites in wildfowl. *Wildfowl*, **20**, 59–68.

**Bailey N.T.J. (1964)** Some stochastic models for small epidemics in large populations. *Applied Statistics*, **13**, 9–19.

**Bailey N.T.J. (1975)** *The Mathematical Theory of Infectious Diseases and its Application.* Griffin, London.

**Baker J.R. (1969)** Trypanosomes of wild mammals in the neighbourhood of the Serengeti National Park. *Diseases in Free-living Wild Animals* (Ed. by A. McDiamid) pp. 147–158. Zoological Society of London Symposium 24.

**Bardach J.E. (1951)** Changes in the yellow perch population of Lake Mendota, Wisconsin, between 1916–1948. *Ecology*, **32**, 719–728.

**Bartlett M.S. (1956)** Deterministic and stochastic models for recurrent epidemics. *Proceedings of the Third Berkeley Symposium on Mathematical Statistics and Probability*, **4**, 81–109. University of California Press, Berkeley.

**Bartlett M.S. (1960)** *Stochastic Population Models in Ecology and Epidemiology.* Methuen, London.

**Bartlett M.S. (1964)** The relevance of stochastic models for large-scale epidemiological phenomena. *Applied Statistics*, **13**, 2–8.

**Bawden R.J. (1969)** Some effects of the diet of mice on *Nematospiroides dubius* (Nematoda). *Parasitology*, **59**, 203–213.

**Bird F.T. & Elgee D.E. (1957)** A virus disease and introduced parasites as factors controlling the European spruce sawfly, *Diprion hercyniae* (Htg), in Central New Brunswick. *The Canadian Entomologist*, **139**, 371–378.

**Boray J.C. (1969)** Experimental Fascioliasis in Australia. *Advances in Parasitology* (Ed. by B. Dawes) pp. 95–209. Academic Press, London.

**Borg K. (1962)** Predation on roe deer in Sweden. *Journal of Wildlife Management*, **26**, 133–136.

**Bray R.S. & Harris W.G. (1977)** The epidemiology of infection with *Entamoeba histolytica* in Gambia, West Africa. *Transactions of the Royal Society of Tropical Medicine and Hygiene*, **71**, 401–407.

**Brooke M.M. (1945)** Effect of Dietary changes upon avian Malaria. *American Journal of Hygiene*, **41**, 81–108.

**Caughley G. (1976)** Plant–herbivore systems. *Theoretical Ecology: Principles and Applications* (Ed. by R.M. May) pp. 94–113. Blackwell Scientific Publications, Oxford.

**Chandler A.C. (1953)** The relation of nutrition to parasitism. *Journal of Egyptian Medical Association*, **36**, 533–552.

**Cheatum E.L. (1951)** Disease in relation to winter mortality of deer in New York. *Journal of Wildlife Management*, **15**, 216–220.

**Clark I.A. & Allison A.C. (1974)** *Babesia microti* and *Plasmodium berghei yoelii* infections in nude mice. *Nature, London*, **252**, 328.

**Clarkson M.J. (1958)** Life history and pathogenicity of *Eimeria adenoeides* Moore and Brown, 1951, in the turkey poult. *Parasitology*, **48**, 70–88.

**Cole T.J. & Parkin J.M. (1977)** Infection and its effect on the growth of young children: a comparison of Gambia and Uganda. *Transactions of the Royal Society of Tropical Medicine and Hygiene*, **71**, 196–198.

**Cox F.E.G. (1975)** Factors affecting infections of mammals with intra-erythrocytic protozoa. *Symbiosis* (Ed. by D.H. Jennings & D.L. Lee) pp. 429–451. Cambridge University Press, London.

**Cox F.E.G., Wedderburn N. & Salaman M.H. (1974)** The effect of Rowson-Parr virus on the severity of malaria in mice. *Journal of General Microbiology*, **85**, 358–364.

**Crisler L. (1956)** Observations of wolves hunting caribou. *Journal of Mammalogy*, **37**, 337–346.

**Crofton H.D. (1971)** A quantitative approach to parasitism. *Parasitology*, **62**, 179–194.

**Dogiel V.A. (1964)** *General Parasitology.* Oliver & Boyd, Edinburgh.

**Duggan A.J. (1970)** An historical perspective. *The African Trypanosomiases* (Ed. by H.W. Mulligan) pp. 41–88. Allen & Unwin, London.

**Dunsmore J.D. (1971)** A study of the biology of the wild rabbit in climatically different regions in eastern Australia. IV. The rabbit in the south coastal region of New South Wales, an area in which parasites appear to exert a population-regulating effect. *Australian Journal of Zoology*, **19**, 355–370.

**Elton C. (1942)** *Voles, Mice and Lemmings—Problems in Population Dynamics.* Oxford University Press, Oxford.

**Finlayson L.H. (1949)** Mortality of *Laemophloeus* (Coleoptera, Cucujidae) infected with *Mattesia dispora* Naville (Protozoa, Schizogregarinaria). *Parasitology*, **40**, 261–264.

**Fuller W.A. (1962)** The biology and management of the bison of Wood Buffalo National Park. *Wildlife Management Bulletin, Ottawa Series*, **1** (16).

**Garnham P.C.C. (1966)** *Malarial Parasites and Other Haemosporidia*. Blackwell Scientific Publications, Oxford.

**Gordon M.H. (1963)** Nutrition and helminthosis in sheep. *Proceedings of the Australian Society of Animal Production*, **3**, 93–104.

**Hedrich A.W. (1933)** Monthly estimates of the child population susceptible to measles, 1900–1931, Baltimore, Md. *American Journal of Hygiene*, **17**, 626.

**Herman C.M. (1969)** Blood protozoa of free-living birds. *Diseases in Free-living Wild Animals* (Ed. by A. McDiarmid) pp. 177–195. *Zoological Society of London Symposium*, **24**.

**Hirst S.M. (1965)** Ecological aspects of big game predation. *Fauna Flora Pretoria*, **16**, 3–15.

**Holling C.S. (1965)** The functional response of predators to prey density and its role in mimicry and population regulation. *Memoirs of the Entomological Society of Canada*, **45**, 3–60.

**Holmes J.C. & Bethel W.M. (1972)** Modification of intermediate host behaviour by parasites. *Behavioural Aspects of Parasite Transmission* (Ed. by E.U. Canning & C.A. Wright) pp. 123–149. Academic Press, London.

**Hornocker M.G. (1970)** An analysis of mountain lion predation upon mule deer and elk in the Idaho primitive area. *Wildlife Monographs*, **21**, 1–39.

**Hunter G.C. & Leigh L.C. (1961)** Studies on the resistance of rats to the nematode *Nippostrongylus moris* (Yokogawa, 1920). I. Dosage–mortality relationship. *Parasitology*, **51**, 347–351.

**Jackson G.J., Herman R. & Singer I. (1969)** *Immunity to Parasitic Animals*, Vols 1 and 2. North-Holland Publishing Co., Appleton-Century-Crofts, U.S.A.

**Jenkins P., Watson A. & Miller G.R. (1963)** Population studies on red grouse, *Lagopus lagopus scoticus* (Lath.), in north-east Scotland. *Journal of Animal Ecology*, **32**, 317–376.

**Kennedy C.R. (1972)** Parasite communities in freshwater ecosystems. *Essays in Hydrobiology* (Ed. by R.B. Clark & R.J. Wooton) pp. 124–138. University of Exeter Press, Exeter.

**Kennedy C.R. & Rumpus A. (1977)** Long-term changes in the size of the *Pomphorynchus laevis* (Acanthocephala) population in the River Avon. *Journal of Fish Biology*, **10**, 35–42.

**Kolodny M.H. (1940)** The effect of temperature upon experimental Trypanosomiasis (*T. cruzi*) of rats. *American Journal of Hygiene*, **32**, 21–23.

**Krugman S. (1973)** Newer vaccines (measles, mumps, rubella): potential and problems. *International Symposium on Vaccination Against Communicable Diseases* (Ed. by F.T. Perkins) pp. 55–63. *Symposia Series in Immunobiological Standardization*. S. Karger, London.

**Lack D. (1954)** *The Natural Regulation of Animal Numbers*. Oxford University Press, Oxford.

**Lanciani C.A. (1975)** Parasite-induced alterations in host reproduction and survival. *Ecology*, **56**, 689–695.

**Lovat Lord (Ed.) (1911)** *The Grouse in Health and in Disease*. Lovat, London.

**Ludwig D., Jones D.D. & Holling C.S. (1978)** Qualitative analysis of insect outbreak systems: the spruce budworm and forest. *Journal of Animal Ecology*, **47**, 315–332.

**Macdonald G. (1965)** The dynamics of helminth infections, with special reference to schistosomes. *Transactions of the Royal Society of Tropical Medicine and Hygiene*, **59**, 489–506.

**May R.M. (1977a)** Togetherness among schistosomes: its effects on the dynamics of the infection. *Mathematical Biosciences*, **35**, 301–343.

**May R.M. (1977b)** Thresholds and breakpoints in ecosystems with a multiplicity of stable states. *Nature, London*, **269**, 471–477.

**May R.M. & Anderson R.M. (1978)** Regulation and stability of host-parasite population interactions. II. Destabilizing processes. *Journal of Animal Ecology*, **47**, 249–268.

**Mech L.D. (1966)** The wolves of Isle Royale. *Fauna of the National Parks of the U.S., Fauna Series 7*. U.S. Government Printing Office, Washington.

**Mimms C.A. (1977)** *The Pathogenesis of Infectious Diseases*. Academic Press, London.

**Murie A. (1944)** The wolves of Mount McKinley. *U.S. National Park Service Fauna Series*, **5**, 1–238.

**Murie O.J. (1930)** An epizootic disease of elk. *Journal of Mammology*, **11**, 214–222.

**Mykytowycz R. (1962)** Epidemiology of coccidiosis (*Eimeria* spp.) in an experimental population of the Australian wild rabbit, *Oryctolagus cuniculus* (L.). *Parasitology*, **52**, 375–385.

**Park T. (1948)** Experimental studies of interspecies competition. 1. Competition between populations of the flour beetles *Tribolium confusum* Duval and *Tribolium castaneum* Herbst. *Ecological Monographs*, **18**, 267–307.

**Riordan K. (1977)** Long-term variations in trypanosome infection rates in highly infected tsetse flies on a cattle route in western Nigeria. *Annals of Tropical Medicine and Parasitology*, **71**, 11–20.

**Roitt I.M. (1976)** *Essential Immunology*, 3rd ed. Blackwell Scientific Publications, Oxford.

**Rosenfield P.L., Smith R.A. & Wolman M.G. (1977)** Development and verification of a schistosomiasis transmission model. *American Journal of Tropical Medicine and Hygiene*, **26**, 505–516.

**Rudebeck G. (1950)** The choice of prey and modes of hunting of predatory birds with special reference to their selective effect. *Oikos*, **2**, 65–88.

**Scrimshaw N.S., Taylor C.E. & Gordon J.E. (1968)** Interactions of nutrition and infection. *World Health Organization Monograph Series*, **57**, 328 pp.

**Schultz W.W., Huang K.Y. & Gordon F.B. (1968)** Role of interferon in experimental mouse malaria. *Nature, London*, **220**, 709–710.

**Sheppe W.A. & Adams J.R. (1957)** The pathogenic effect of *Trypanosoma duttoni* in hosts under stress conditions. *Journal of Parasitology*, **57**, 55–59.

**Sinnecker H. (1976)** *General Epidemiology*. John Wiley & Sons, London.

**Smith H.D. (1973)** Observations on the Cestode *Eubothrium salvelini* in Juvenile Sockeye Salmon (*Oncorhynchus nerka*) at Babine Lake, British Columbia. *Journal of the Fisheries Research Board of Canada*, **30**, 947–964.

**Soulsby E.J.L. (Ed.) (1972)** *Immunity to Animal Parasites*. Academic Press, New York.

**Southwood T.R.E. (1976)** Bionomic strategies and population parameters. *Theoretical Ecology* (Ed. by R.M. May) pp. 26–48. Blackwell Scientific Publications, Oxford.

**Steinhaus E.A. (1958)** Stress as a factor in insect disease. *Proceedings of the 10th International Congress of Entomology*, **4**, 725–730.

**Stiven A.E. (1962)** Experimental studies on the epidemiology of the host-parasite system hydra and *Hydramoeba hydroxena* (Entz.). I. The effect of the parasite on the individual host. *Physiological Zoology*, **35**, 166–178.

**Stone W.M. (1964)** Rate of survival in guinea pigs following infestation by screw-worm larvae. *Journal of Parasitology*, **50**, 152–154.

**Tanada Y. (1964)** Epizootiology of insect diseases. *Biological Control of Insect Pests and Weeds* (Ed. by P. DeBach) pp. 548–578. Chapman and Hall, London.

**Thom R. (1975)** *Structural Stability and Morphogenesis* (English translation by D.H. Fowler). Benjamin, Reading, Massachusetts.

**Van Dobben W.H. (1952)** The food of the cormorants in the Netherlands. *Ardea*, **40**, 1–63.

**Vaughan G.E. & Coble P.W. (1975)** Sublethal effects of three ectoparasites on fish. *Journal of Fish Biology*, **7**, 283–294.

**Vaughan H.E.N. & Vaughan J.A. (1969)** Some aspects of the epizootiology of myxomatosis. *Diseases in Free-living Wild Animals* (Ed. by A. McDiarmid) pp. 289–309. *Zoological Society of London Symposium*, **24**.

**Vizoso A.D. (1969)** A red squirrel disease. *Diseases in Free-living Wild Animals* (Ed. by A. McDiarmid) pp. 29–38. *Zoological Society of London Symposium*, **24**.

**Weiser J. (1958)** Protozoan diseases and insect control. *Proceedings of the 10th International Congress of Entomology*, **4**, 681–685.

**Welch H.E. (1965)** Entomophilic nematodes. *Annual Review of Entomology*, **10**, 275–302.

**Yoeli M., Becker Y. & Bernkopf H. (1955)** The effect of West Nile virus on experimental malaria infections (*Plasmodium berghei*) in mice. *Horefuah, Jerusalem*, **49**, 116–119.

**Young A.S. (1970)** Investigations of the epidemiology of blood parasites of small mammals with special reference to piroplasms. *Unpublished Ph.D. thesis, University of London.*

# 13. THE DYNAMICS OF PREDATOR–PREY INTERACTIONS: POLYPHAGOUS PREDATORS, COMPETING PREDATORS AND HYPERPARASITOIDS

M. P. HASSELL

*Department of Zoology and Applied Entomology, Imperial College, London SW7 2AZ*

## INTRODUCTION

Theoretical ecologists have been much devoted to models of coupled one predator–one prey interactions, some of which are now of considerable sophistication, incorporating, for example, complex functional responses (Hassell & Comins 1978) and optimal foraging in a patchy environment (Comins & Hassell 1979). Such two-species systems are best seen within the confines of a laboratory experiment, but also correspond well to many of the insect parasitoid–host interactions arising from successful biological control programmes. Indeed, one result of the development of these models is a renewed confidence that they can provide some theoretical basis for classical biological control practices (Huffaker, Luck & Messenger 1977; Beddington, Free & Lawton 1978; Hassell 1978).

At the same time, new insights into the dynamics of such simple interactions are becoming harder to extract. One avenue for future studies will be to make the models more appropriate to true predators rather than to the simplified life cycle of insect parasitoids. Parasitoids have only a single searching stage (the adult female) and, rather than eating the prey that they encounter, oviposit in or on them. A true predator–prey model should, therefore, include provision for age-dependent differences in searching efficiency and a net predator reproductive rate that is a complex function of the number of prey eaten. First steps in exploring the consequences of such reproduction have been taken by Beddington, Free & Lawton (1976).

An alternative avenue is to enlarge upon the one predator–one prey system in favour of those with three (or more) interacting species. This introduces the kinds of trophic configurations shown in Figure 13.1, each of which is to be examined in this paper. The recurrent aim will be to identify from each system the important general features of predation that promote the co-existence of

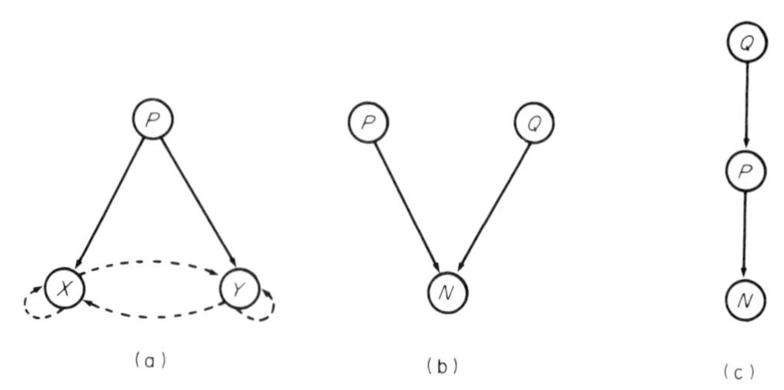


FIG. 13.1. Diagrams illustrating the three-species systems discussed in this paper. (a) A single predator species ($P$) attacking two prey species ($X$ and $Y$). The broken lines indicate intra- and inter-specific competitive interactions between the prey. (b) Two parasitoid species ($P$ and $Q$) attacking a single host ($N$). (c) A hyperparasitoid ($Q$)–parasitoid ($P$)–host ($N$) interaction.

the community as a whole. The models will all be couched in difference equations rather than in their differential counterparts, and hence will be particularly applicable to interactions with more or less discrete generations. As with the one predator–one prey systems, they will also be most appropriate to parasitoids rather than predators in that they neglect complications due to age structure and have the parasitoid rate of increase defined solely by the host mortality. It still remains convenient, however, to refer primarily to predators in general and only make the distinction between predators and parasitoids where some difference is to be stressed.

The common element linking the three types of model will be the form of the function [$f(P_t)$] describing the fraction of prey surviving predation. In this we shall follow May (1978) in assuming a negative binomial distribution of predator attacks amongst prey:

$$f(P_t) = \left[1 + \frac{aP_t}{k}\right]^{-k} \qquad (13.1)$$

where $P_t$ is the number of predators in generation $t$, $a$ is the searching efficiency per predator and $k$ is one of the parameters of the negative binomial distribution. Handling time has been omitted for the sake of simplicity and because its effects are unlikely to be important to the outcome (Hassell & May 1973). The parameter $k$ therefore reflects the extent to which the distribution of attacks on prey are clumped, with clumping increasing as $k$ gets smaller. In doing so it captures, albeit only in a phenomenological way, the ability of the predators to aggregate in local patches of high prey density. May has reinforced this conclusion by showing analytically that the negative binomial provides a good overall description of the distribution of attacks from a system with $n$ patches of prey and an aggregated predator search pattern between patches ($k$ is now related to the coefficient of variance of the number of predators per patch). Thus, in adopting this simple means of modelling non-random search, we are, in effect, looking at the end-product of predators tending to aggregate where prey are most numerous, a behaviour that is known to be generally widespread amongst predators (see Hassell & May 1974 for a range of examples). The aggregation is strongest when $k \rightarrow 0$ and weakest as $k \rightarrow \infty$ when the attack distribution becomes Poisson and the familiar Nicholson & Bailey (1935) function for predation is reobtained:

$$f(P_t) = \exp(-aP_t) \tag{13.2}$$

Equation (13.1), whose provenance lies in parasitology (Crofton 1971a,b; May 1977a; Bradley & May 1978; Anderson & May 1978), represents an important advance over eqn (13.2) with its assumed random search. It provides a simple means of incorporating non-random search into predator–prey models without the parameter proliferation characteristic of its predecessors (e.g. Hassell & May 1973, 1974), the effect of which can be to contribute markedly to stability. Indeed, the one predator–one prey interaction

$$N_{t+1} = \lambda N_t\left[1 + \frac{aP_t}{k}\right]^{-k} \qquad P_{t+1} = N_t\left[1 - \left\{1 + \frac{aP_t}{k}\right\}^{-k}\right] \tag{13.3}$$

where $N_t$, $N_{t+1}$, $P_t$ and $P_{t+1}$ are the prey and predator populations in successive generations and $\lambda$ is the prey rate of increase, proves to be stable for all $k < 1$ (May 1978), in contrast to the Nicholson & Bailey (1935) model which is always unstable. We are thus able to include in the multispecies models to follow a form of predation that can promote the stability of the individual predator–prey linkages.

## ONE PREDATOR AND TWO PREY

An important ecological problem centres on how, if at all, polyphagous predators can affect the co-existence of competing prey species, a problem that has received plentiful theoretical attention (e.g. Parrish & Saila 1970; Cramer & May 1972; Steele 1974; van Valen 1974; Murdoch & Oaten 1975; Roughgarden & Feldman 1975; Comins & Hassell 1976; Fujii 1977). Several experimental studies have also been addressed to this question (see Connell 1975 for a review), one of the best known being that of Paine (1966, 1974). He found that, following the removal of the starfish *Pisaster ochraceus* from an area of Washington shoreline, competition between the prey (limpets, chitons, mussels and barnacles) soon led to extinctions. After one year six species had been replaced, and the mussel *Mytilus* and the goose barnacle *Mitella* were now dominating the system. Clearly, *Pisaster* as a 'top predator' is playing a crucial role in maintaining the prey species diversity. But what are the essential ingredients for a predator to play this role? That it must be polyphagous is probably a primary requirement. What is less clear is the relative importance of random or non-random search for a given prey, of any preferences for certain prey species and of the ability to 'switch' from one prey species to another as their relative abundance changes.

The first step in examining these questions is to have at hand a suitable competition model for the prey species alone, one in which predation can then be included. In this we shall follow Hassell & Comins (1976) and Comins & Hassell (1976) in adopting a difference equation equivalent of the Lotka–Volterra model:

$$\begin{aligned} X_{t+1} &= X_t \exp\left[r - \frac{r}{K}(X_t + \alpha Y_t)\right] \\ Y_{t+1} &= Y_t \exp\left[r' - \frac{r'}{K'}(Y_t + \beta X_t)\right] \end{aligned} \tag{13.4}$$

where $X_t$, $X_{t+1}$, $Y_t$ and $Y_{t+1}$ are the numbers of species $X$ and $Y$ in successive generations, $r$ and $r'$ are the intrinsic rates of increase, $r = \ln \lambda$ in eqn (13.3), $\alpha$ and $\beta$ are the usual interspecific competition coefficients and $K$ and $K'$ are the equilibrium levels or 'carrying capacities'.

Since our aim is to determine how predation can enhance prey co-existence, we should first be quite clear about the conditions for an equilibrium between $X$ and $Y$ in the absence of predation. Equation (13.4) retains the linear zero-growth isoclines that are a familiar feature of the Lotka–Volterra model, and so also permits the three kinds of isocline configuration illustrated in Figure 13.2. If the isoclines intersect, they may do so as in Figure 13.2a to give a potentially stable equilibrium or as in Figure 13.2c, in which case the equilibrium is always unstable. Alternatively, the isoclines may fail to intersect as in Figure 13.2b, in which case one of the species ($X$ in this example) is clearly the superior competitor and replaces the other. In relation to the degree of

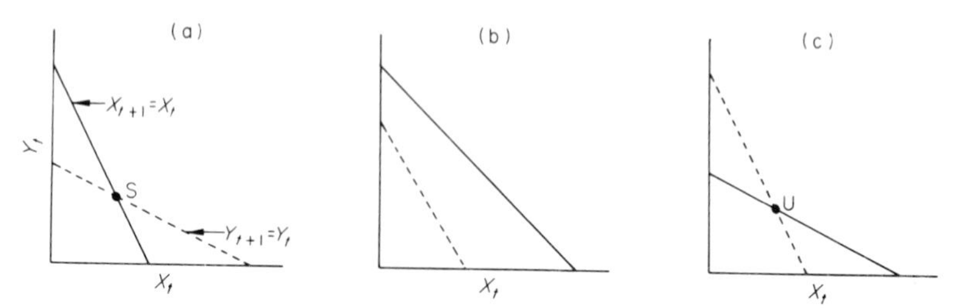


FIG. 13.2. Possible configurations for the zero growth isoclines separating the regions of positive and negative population growth for species $X$ and $Y$ in eqn (13.4). The solid line is for species $X$ and the broken line for species $Y$. (a) A locally stable equilibrium ($S$) is feasible. (b) The species whose isocline lies consistently below the other ($Y$ in this case) inevitably becomes extinct. (c) The equilibrium is unstable ($U$).

niche overlap, the configurations in Figures 13.2a and 13.2b are feasible if there is some niche separation ($\alpha\beta < 1$), that in Figure 13.2b, but with parallel slopes, occurs if there is complete niche overlap with $\alpha\beta = 1$ and both Figures 13.2b and 13.2c can occur if $\alpha\beta > 1$. The one important departure from the Lotka–Volterra model lies in the very different local stability properties possible when a stable equilibrium is feasible as in Figure 13.2a. Instead of the populations always approaching this equilibrium monotonically, the inherent time delays of one generation now also permit oscillatory damping, stable limit cycles and even apparent 'chaos' as shown in Figures 13.3 and 13.4 (May 1974, 1975; Hassell & Comins 1976).

With Figure 13.2 in mind, the central questions are three-fold. (1) Can predation stabilize a locally unstable equilibrium of the kind in Figure 13.2a, where limit cycles or higher order behaviour are occurring? (2) Can predation 'neutralize' the more able competitor ($X$ in Fig. 13.2b) and so permit a stable equilibrium (i.e. convert Fig. 13.2b to 13.2a)? (3) Can predation reverse the isoclines in Figure 13.2c to produce the potentially stable configuration of Figure 13.2a? To answer these questions we shall investigate three types of predation: (i) 'equivalent' predation, (ii) 'preference' and (iii) 'switching'.

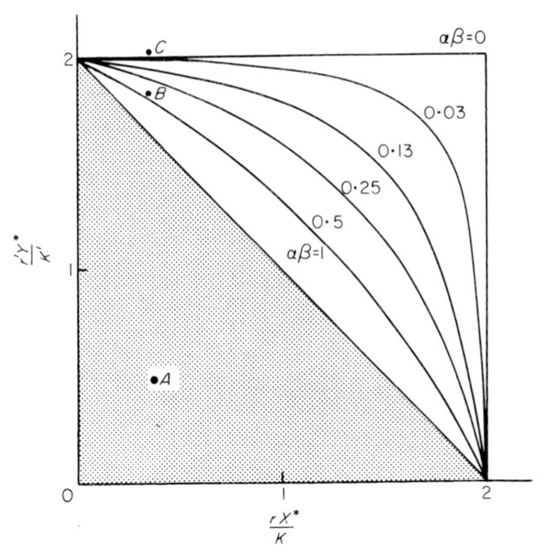


FIG. 13.3. Local stability boundaries from eqn (13.4) in terms of the equilibrium populations, $X^*$ and $Y^*$. Equilibrium points falling within the relevant $\alpha\beta$ boundary (from 0 to 1) are stable; points outside the boundary correspond to limit cycles or chaotic behaviour. The shaded area indicates the region of monotonic stability. No stable equilibrium is possible for $\alpha\beta > 1$. Points $A$, $B$ and $C$ correspond to simulations (a), (b) and (c) in Figure 13.4. (After Hassell & Comins 1976.)

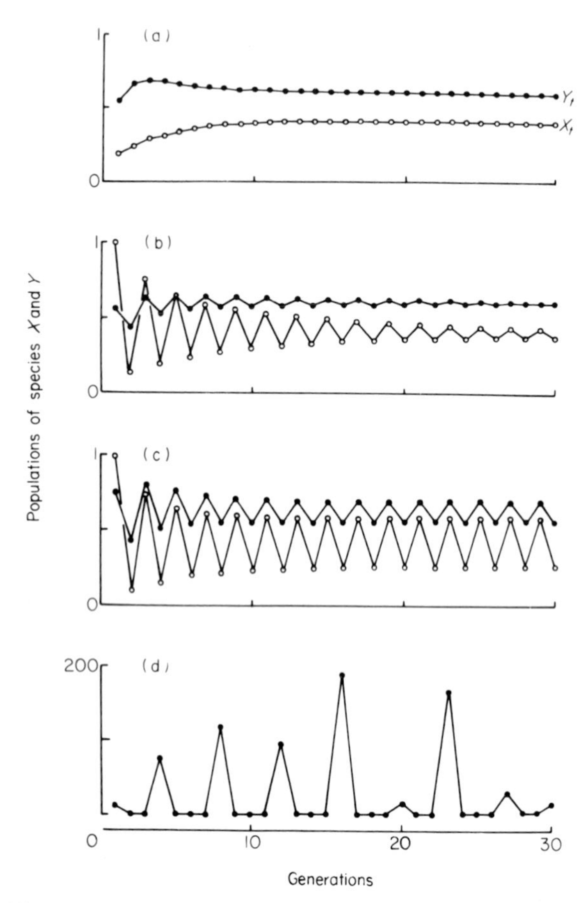


FIG. 13.4. Numerical simulations from eqn (13.4). The parameters used in (a), (b) and (c) correspond to points $A$, $B$ and $C$ in Figure 13.3 and show the effect of increasing the level of competition in species $Y$. (d) shows the apparently chaotic behaviour arising when competition is further increased. Note that species $X$ has been omitted since its fluctuations are not visible on this compressed scale. (After Hassell & Comins 1976.)

### *'Equivalent' predation*

Let us consider the general model

$$
\begin{aligned}
X_{t+1} &= X_t \exp\left[r - \frac{r}{K}(X_t + \alpha Y_t)\right] f(P_t) \\
Y_{t+1} &= Y_t \exp\left[r' - \frac{r'}{K'}(Y_t + \beta X_t)\right] f'(P_t) \qquad (13.5) \\
P_{t+1} &= X_t[1 - f(P_t)] + Y_t[1 - f'(P_t)]
\end{aligned}
$$

where $f$ and $f'$ take the form of eqn (13.1), namely

$$f(P_t) = \left[1 + \frac{aP_t}{k}\right]^{-k} \qquad f'(P_t) = \left[1 + \frac{a'P_t}{k'}\right]^{-k'} \tag{13.6}$$

and $a$ and $a'$ are the searching efficiencies and $k$ and $k'$ are the parameters of the negative binomial distribution that capture the degree of aggregated search for species $X$ and $Y$ respectively. The customary use of the term 'equivalent predation' implies not only that the predators have no predilection for either prey type (i.e. $a = a'$ and $k = k'$ from eqn (13.6)), but also that the prey are equivalent in having equal growth rates, $r = r'$. The conventional wisdom is that such predation can make no difference to competitive co-existence among the prey (van Valen 1974; May 1977b). Thus, the isocline configurations in Figure 13.2b, c cannot be converted to the potentially stable form of Figure 13.2a. This is certainly true but, in addition, in systems with time delays rather than those based on the differential Lotka–Volterra models, equivalent predation *can* have an affect on the local stability of the equilibrium in Figure 13.2a. It is thus possible to create a locally stable equilibrium where the two prey alone could exhibit only limit cycles or higher order behaviour, as shown by the simulation in Figure 13.5a. This effect is enhanced by small values of $k_1$ and $k_2$ making the individual predator–prey links more stable. At the same time, for other parameter combinations, the reverse is possible: a locally stable two-prey system may become a locally unstable three-species system on the introduction of a predator, as illustrated by the example in Figure 13.5b. (A detailed stability analysis of eqn (13.5), but with $k = k' \rightarrow \infty$, is given by Comins & Hassell (1976).)

At most, therefore, equivalent predation can affect the local stability only where a potentially stable equilibrium already exists.

## *Preference*

Equation (13.5) can also be used to illustrate the effects of predator preference for a particular prey type. Preference is normally measured in terms of the deviation of the proportion of a prey type attacked from the proportion available in the environment. Such a simple definition belies the complex behaviour upon which it may depend. For example, preference may result from differential searching rates, from active rejection of some prey types following their encounter, from differing abilities of prey to escape and from any combination of these factors (Hassell 1978).

The most straightforward means of including preference in population models is to assume unequal searching parameters [i.e. $a \neq a'$ and/or $k \neq k'$ in eqn (13.6)]. With such preference the interesting situation can arise where predation is maintaining a three-species equilibrium where the two-prey species

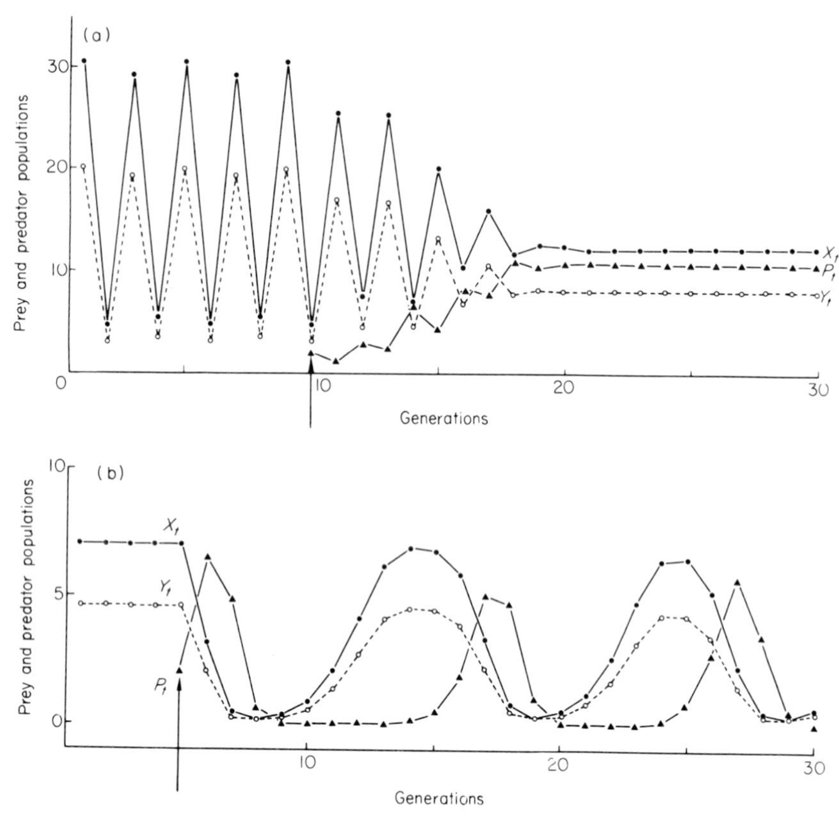


FIG. 13.5. Numerical simulations from eqn (13.5). (a) A locally stable three-species equilibrium maintained by the predator. $a = a' = 0{\cdot}07$, $k = k' \rightarrow \infty$, $r = r' = 2{\cdot}5$, $K = K' = 25$, $\alpha = 0{\cdot}65$ and $\beta = 0{\cdot}77$. (b) Three-species limit cycles arising from the introduction of the predator. $a = a' = 0{\cdot}4$, $k = k' \rightarrow \infty$, $r = r' = 1{\cdot}0$, $K = K' = 10{\cdot}0$, $\alpha = 0{\cdot}65$ and $\beta = 0{\cdot}77$. The arrows indicate the point of introduction of the predators.

alone could exhibit no form of equilibrium (i.e. where the isoclines are as in Fig. 13.2b, with no intersection in the positive quadrant). The effect of unequal predation is now to shift these isoclines downwards by differing amounts without changing their slopes. If preference is for the 'dominant' competitor, the upper isocline is shifted more than the lower one and a potentially stable equilibrium becomes possible, as shown in Figure 13.6. A simulation of such a case, where species $X$ would normally displace $Y$ but is prevented from doing so by predator preference for $X$, is shown in Figure 13.7.

There is, unfortunately, a difficulty in viewing the phenomenon shown in Figure 13.6 solely in terms of predator behaviour. Exactly the same effect may be achieved with no behavioural predator preference but unequal prey growth

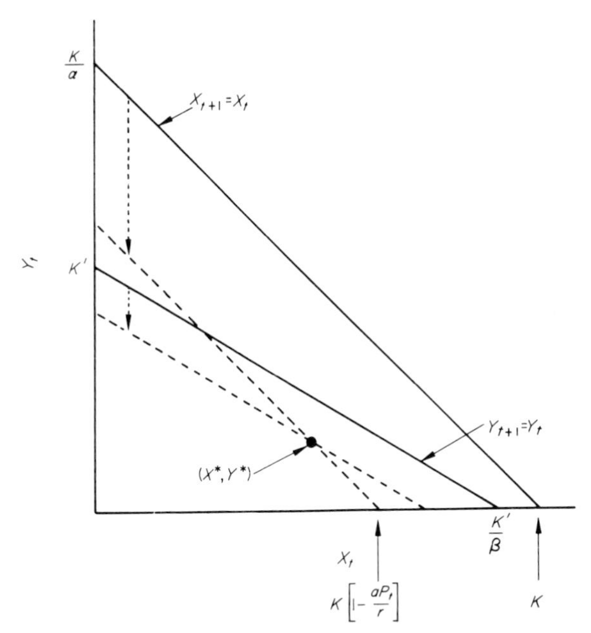


FIG. 13.6. Zero growth isoclines for the prey species $X$ and $Y$. In the absence of predation the isoclines do not intersect (solid lines). Predation with preference for the superior competitor (species $X$) depresses its isocline more than that for other species (broken lines) and a potentially stable equilibrium may result.

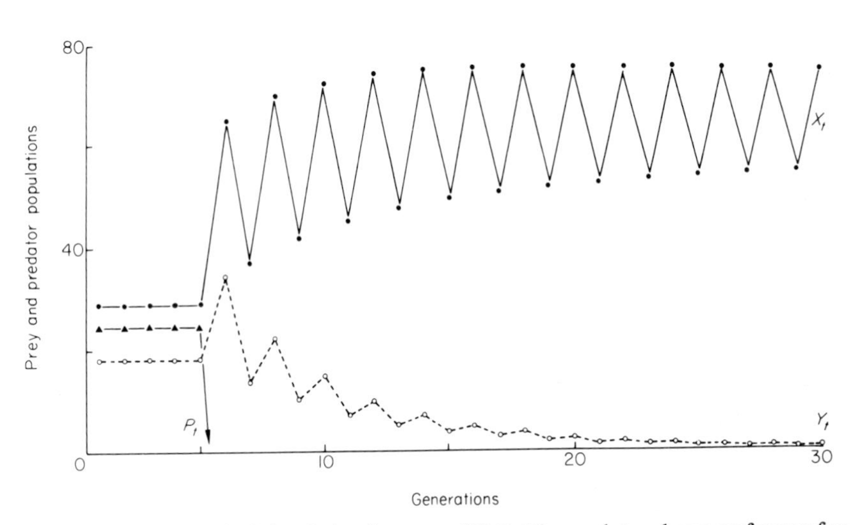


FIG. 13.7. A numerical simulation from eqn (13.5). The predator shows preference for species $X$ (the dominant competitor) and is thus preventing species $Y$ from becoming extinct. This is illustrated by removing the predator in generation 5. $a = 0{\cdot}033$, $a' = 0{\cdot}026$, $k = k' \to \infty$, $r = r' = 2{\cdot}0$, $K = 65$, $K' = 90$, $\alpha = 0{\cdot}54$ and $\beta = 1{\cdot}5$.

rates, $r \neq r'$. This arises because the carrying capacities $K$ and $K'$ are depressed by an amount $KaP^*/r$ and $K'a'P^*/r'$ respectively, as shown in Figure 13.6. We now have the interesting situation, recently pointed out by May (1977b), in which a potentially stable three-species equilibrium is possible with equal predation if the less able competitor has the greater population growth rate. Herein, therefore, is a mechanism which can encourage the more '$r$-selected' competitors to co-exist with their '$K$-selected' counterparts, and which may well provide the underlying explanation for situations where uniform grazing leads to enhanced species diversity in pastures, examples of which are discussed by Harper (1969).

### *Switching*

A further dimension to our discussion emerges when predators are allowed to 'switch' to whichever prey is the most abundant at a particular time. This switching implies that the proportion of a prey type attacked changes from less than would be expected on the basis of known $a$, $a'$, $k$ and $k'$ in eqn (13.5) to greater than expected as the proportion of that prey present increases. Clear examples of this are shown by Lawton, Beddington & Bonser (1974) and Murdoch, Avery & Smith (1975), and the subject is reviewed by Murdoch & Oaten (1975) and Murdoch (1977).

Switching is likely to assume its widest importance when the different prey types occur in somewhat disparate habitats, since it can now result simply from the allocation of a greater fraction of the searching time to whichever habitat is the more profitable. This is the mechanism suggested by Royama (1970) and observed by Murdoch, Avery & Smith (1975) for guppies feeding on a mixture of limbless, wingless *Drosophila* floating on the water surface and tubificid worms on the aquarium bottom. The guppies fed disproportionately on whichever prey was most abundant, spending increasing periods of time at the surface as the proportion of *Drosophila* increased.

Let us consider a version of eqn (13.5) with $f$ and $f'$ defined by

$$f(P_t) = \left[1 + \frac{(1 + E)aP_t}{k}\right]^{-k} \qquad f'(P_t) = \left[1 + \frac{(1 - E)a'P_t}{k'}\right]^{-k'} \tag{13.7}$$

where $E$ has the following dependence on the relative prey abundances

$$E = s(X_t - Y_t)/(X_t + Y_t) \tag{13.8}$$

and $s$ is a constant expressing the degree of switching as shown in Figure 13.8. The analytical treatment of this model for the special case of $k = k' \to \infty$ is again given in Comins & Hassell (1976).

The importance of switching in this model is that, like preference, it can create a potentially stable three-species equilibrium where none existed before

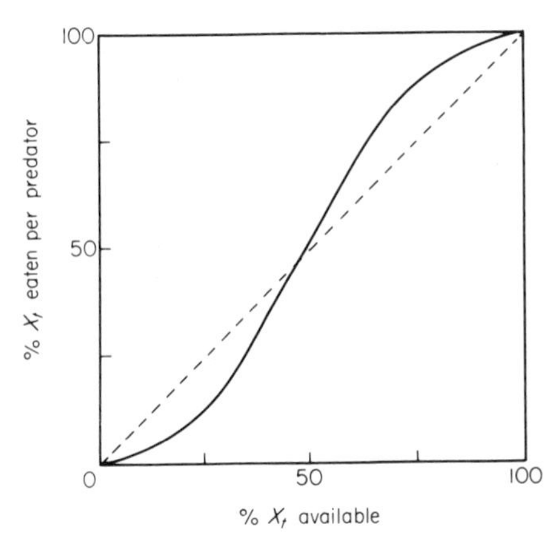


FIG. 13.8. An example of predator switching from eqns (13.7) and (13.8), where $a = a' = 0{\cdot}5$, $k = k' \to \infty$ and $s = 1{\cdot}0$. (From Comins & Hassell 1976.)

and furthermore, for the first time in our discussion, it can do so when the two prey alone show complete niche overlap ($\alpha\beta \geq 1$). This is achieved by switching tending to bend the prey isoclines which, if sufficient, can even convert the necessarily unstable configuration of Figure 13.2c to the potentially stable one in Figure 13.2a, as illustrated in Figure 13.9. Predation is therefore having much the same effect as illustrated by the simulation in Figure 13.7, with the very important difference that it becomes increasingly effective as prey niche overlap increases. The same conclusion applies to systems with more than two prey species. Roughgarden & Feldman (1975) found that switching could stabilize a differential equation system with one predator and three prey, and Comins & Gassell (1976) found the same effect for an $n$-prey extension of eqn (13.5).

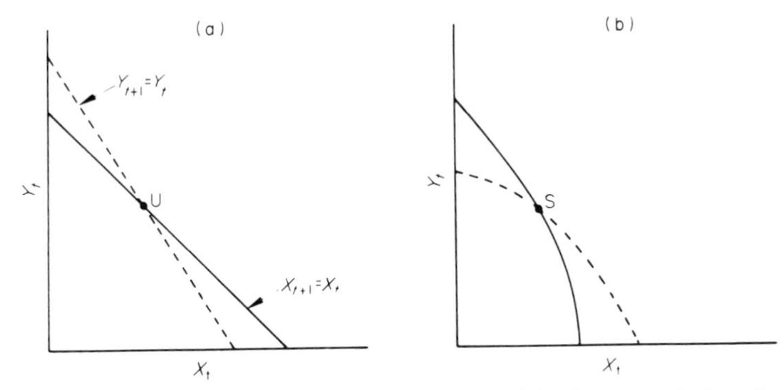


FIG. 13.9. Zero growth isoclines for species $X$ and $Y$. (a) without predation showing an unstable equilibrium ($U$). (b) with a switching predator, leading to curved isoclines and a potentially stable equilibrium ($S$). (After Comins & Hassell 1976.)

### *Conclusions*

The models of this section suggest three basic rules for the influence of polyphagous predators on prey co-existence.

1 'Equivalent predation' (with equal prey growth rates) can affect the local stability properties of an interaction only where the potential for a stable equilibrium already exists (as in Fig. 13.2a) and the prey on their own, therefore, exhibit either a locally stable equilibrium, limit cycles or higher order cyclic behaviour.

2 Predator preference can stabilize an interaction where no prey equilibrium would otherwise exist (as in Fig. 13.2b) as long as some niche separation occurs ($\alpha\beta < 1$). The same effect can occur with no predator preference if the less able competitors have greater population growth rates.

3 Predator switching has a similar effect to preference, but is also the sole mechanism that can stabilize an interaction in which the prey exhibit complete niche overlap (as in Fig. 13.2c).

In the light of these, it is of particular interest to revisit Paine's (1966) exclusion experiment, since Landenburger (1968) has shown *Pisaster* to have a marked preference for *Mytilus* species over a range of other molluscs tested, just the kind of behaviour required of a predator to maintain a larger prey community than could exist in its absence. Unfortunately, similar information on *Mitella*, the other dominant competitor in the system, and on whether *Pisaster* switches in the strict sense of the word, is unavailable.

## MULTI-PREDATOR SYSTEMS

In contrast to the multi-prey systems of the previous section, systems with more than one predator have so far received scant attention. We shall examine two of the simplest of such systems; those illustrated in Figure 13.1b,c. In doing this it is particularly appropriate that we should now think specifically of insect parasitoids rather than predators, for two reasons. Firstly, it is commonplace to find phytophagous insect species that are attacked by a complex of parasitoids, some of which are usually relatively specific (Zwölfer 1971). This would correspond to the system in Figure 13.1b, where two specific parasitoids attack the same host species. Secondly, the frequent occurrence of insect hyperparasitoids that attack primary parasitoid species makes them most appropriate to the three trophic level system of Figure 13.1c.

In both of these systems, a really interesting problem is again to define the conditions that permit the three species to co-exist in a stable interaction. Apart from its general ecological interest, the problem is relevant to biological control practices. In particular, it relates to the practice of introducing more than one parasitoid species into an area in an attempt to improve upon the existing biological control of a pest, and also to the practice of striving to prevent the accidental introduction of hyperparasitoids.

*Two parasitoids and one host*

Let us consider the general model discussed in Hassell (1978):

$$N_{t+1} = \lambda N_t f_1(P_t) f_2(Q_t)$$

$$P_{t+1} = N_t[1 - f_1(P_t)] \qquad (13.9)$$

$$Q_{t+1} = N_t f_1(P_t)[1 - f_2(Q_t)]$$

where $N$, $P$ and $Q$ denote the host and two parasitoid species in generations $t$ and $t + 1$, $\lambda$ is the finite host rate of increase and the functions $f_1$ and $f_2$ are the probabilities of a host not being found by $P_t$ or $Q_t$ parasitoids respectively. This model applies to two quite distinct types of interaction that are frequently to be found in real systems. It applies (1) to cases where $P$ acts first, to be followed by $Q$ acting only on the survivors. Such is the case where a host population with discrete generations is parasitized at different developmental stages. In addition, it applies (2) to cases where both $P$ and $Q$ act together on the same host stage, but the larvae of $P$ always out-compete those of $Q$ should multi-parasitism occur. Only where the outcome of multi-parasitism depends in part on the order of arrival within the host, as found by Anderson, Hoy & Weseloh (1977), is a rather different model structure required.

The first analysis of eqn (13.9) was due to Nicholson (1933) and Nicholson & Bailey (1935), who extended their single host–parasitoid model with random search (see eqn 13.2) to a system with two parasitoid species, each with a constant searching efficiency. They found that for given values of $\lambda$ and $a_1$, the searching efficiency of $P$, there is only a very narrow range of values for $a_2$, the searching efficiency of $Q$, that permits an equilibrium, albeit one that is always unstable.

Working with such an unstable model enables few conclusions to be drawn about the real world, and certainly prevents us considering the range of conditions that enable a locally *stable* three-species equilibrium to be achieved. Much more appropriate for this is a model in which parasitism is inherently stabilizing, such as that based on the negative binomial distribution, see eqns (13.1), (13.3), (13.6) and (13.7). The functions in eqn (13.9) now become

$$f_1(P_t) = \left[1 + \frac{a_1 P_t}{k_1}\right]^{-k_1} \qquad f_2(Q_t) = \left[1 + \frac{a_2 Q_t}{k_2}\right]^{-k_2} \qquad (13.10)$$

where $k_1$ and $k_2$, as before, capture the extent of non-random search by parasitoids in a patchy environment. Further details of this model are given in Hassell (1978) and May & Hassell (1980). We shall concentrate here on the emergent properties, and look in particular at three stability diagrams and some simulations.

The existence and local stability of a three-species equilibrium from eqns (13.9) and (13.10) depends in a complex way on the values of $\lambda$, $k_1$, $k_2$, $a_1$ and $a_2$. They may, however, be conveniently displayed by fixing the value of $\lambda$ (we shall let $\lambda = 2$ throughout) and then considering three special cases: (1) where both $P$ and $Q$ show the same degree of non-random search ($k_1 = k_2 = k$); (2) where $P$ can search non-randomly but $Q$ always searches randomly ($k_1 = k; k_2 \to \infty$); and (3) where $P$ searches randomly but $Q$ can search non-randomly ($k_1 \to \infty$; $k_2 = k$). For each case (1) to (3) it is now possible to draw the local stability conditions in terms of the ratio of the searching efficiencies $a_2/a_1$ and the value of $k$, as shown in Figures 13.10 and 13.11. The three-species equilibrium conditions lie between the upper and lower curves in each figure, but only in the hatched region is this equilibrium locally stable. Outside the hatched area, in regions A to E, several other possibilities arise.

(A) Species $P$ becomes extinct, leaving $Q$ in a two-species interaction that will only be stable if $k_2 < 1$ (see p. 285).

(B) As (A), but now it is $Q$ that becomes extinct.

(C) and (D) As (A) and (B) respectively, but the resulting two-species interaction is always unstable since $k > 1$.

(E) A three-species equilibrium exists, but one that is always locally unstable.

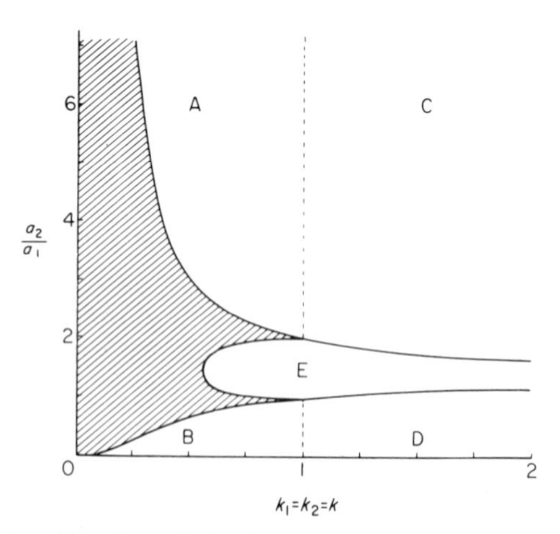


FIG. 13.10. Local stability boundaries for eqn (13.9), where $f_1$ and $f_2$ are defined in eqn (13.10) with $k_1 = k_2 = k$ and $\lambda = 2$. A three-species equilibrium may occur within the upper and lower boundaries but can only be locally stable within the hatched area. The stability properties within regions A to E are described in the text. (After Hassell 1978 and May & Hassell 1980, in which further details are given.)

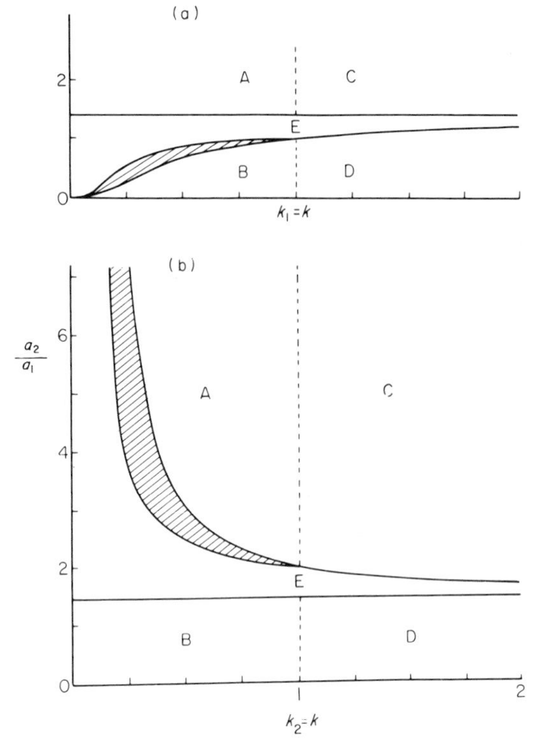


FIG. 13.11. Local stability boundaries as in Figure 13.10, but with (a) $k_1 = k$, $k_2 \to \infty$ and (b) $k_1 \to \infty$, $k_2 = k$. (After Hassell 1978 and May & Hassell 1980.)

An examination of these stability diagrams leads to several interesting conclusions.

1 A stable three-species equilibrium is most likely if both $P$ and $Q$ show marked non-random search ($k < 1$), and is impossible if they both tend towards random search ($k_1 = k_2 > 1$).

2 Should only one of the parasitoid species search randomly, a three-species equilibrium is still possible, given $k < 1$ for the other species, although only for a restricted range of $a_2/a_1$ values.

3 Figure 13.11 shows that it is not possible to have one species the cause of stability and the other to be the major cause of depression in the host equilibrium (i.e. to have the higher searching efficiency). This, if $P$ searches randomly a locally stable equilibrium, is only possible if $k_2 < 1$ *and* $a_2 > a_1$, and vice versa.

4 Given that both species contribute equally to stability ($k_1 = k_2 < 1$), the possibilities for stable co-existence are somewhat greater if $Q$ has the higher searching efficiency ($a_2 > a_1$). This is shown in Figure 13.12, in which $a_2$ is

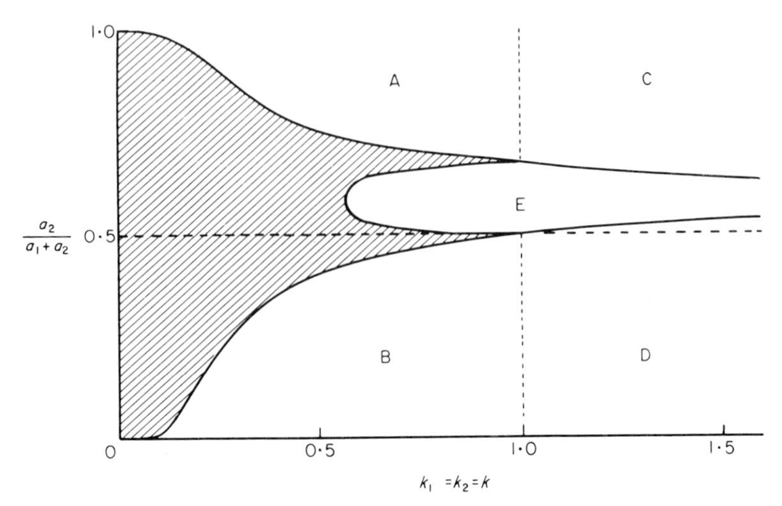


FIG. 13.12. Local stability boundaries as in Figure 13.10, but with the relative searching efficiencies of species *P* and *Q* normalized.

expressed as a proportion of the combined searching efficiencies. Only if the value for $k$ is very small (e.g. $k < 0{\cdot}3$) does the area of stable space where $a_1 > a_2$ approach that for $a_2 > a_1$.

Conclusion (4) hints at two things about the feasibility of multi-parasitoid systems. Where parasitoids attack a host species in temporal sequence, there are more possibilities for co-existence if the later acting species (*Q*) has the higher searching efficiency. If, on the other hand, the parasitoids attack the same host stage, then it is the species with the inferior larval competitor (again *Q*) which should have the higher searching efficiency. There is little relevant information to test these seemingly sensible conclusions, but what does exist is encouraging. Zwölfer (1971) gives several anecdotal examples where the 'superior' searchers have 'inferior' larvae, and vice versa. More recently, Miller (1977) has shown (Table 13.1) that for all combinations of multi-parasitism by three species of parasitoids attacking *Spodoptera praefica* the inferior larval competitor has the higher potential fecundity as an adult which, although by no means a reliable indicator of searching efficiency, may well scale in the same way in parasitoids with similar longevities and handling times.

One of the most interesting uses of this model is in relation to the biological control of insect pests and, in particular, to the contentious issue of whether single or multiple introductions of parasitoid species are preferable (Hassell 1978). Turnbull & Chant (1961) and Turnbull (1967) take the view that biological control is best served by only the single 'best' species being introduced, while van den Bosch (1968), Huffaker, Messenger & DeBach (1971) and others

TABLE 13.1. The relationship between larval competitive ability and the adult reproductive capacity in three species of parasitoids attacking *Spodoptera praefica*. The average eggs per adult female of each species are shown in brackets. (After Miller 1977.)

| Inferior larval competitor | | Superior larval competitor | |
|---|---|---|---|
| *Chelonus insularis* | (450) | *A. marginiventris* | (210) |
| *C. insularis* | (450) | *Hyposoter exiguae* | (130) |
| *Apanteles marginiventris* | (210) | *H. exiguae* | (130) |

believe that there is little to be lost and much to be gained by introducing a sequence of natural enemies if initially there is only partial control achieved. We commence with the situation shown in Figure 13.13, where a single parasitoid (species $P$) has been introduced to combat a pest and now exists in a stable interaction. This is the situation modelled by eqn (13.3) with $k < 1$. The host equilibrium, however, is not sufficiently depressed, and a further parasitoid (species $Q$) is to be introduced with the aim of lowering the host equilibrium. The three-species model now permits five possibilities, given in Table 13.2 (see also the simulations in Fig. 13.14).

The possible outcomes in Table 13.2 are also ranked in order of desirability. Only (1) and (2) lead to the objective of reducing the host's equilibrium, and are most likely to occur if species $Q$ is efficient at exploiting patches of high host density ($k_2 < 1$) and has a higher searching efficiency than species $P$ ($a_2 > a_1$).

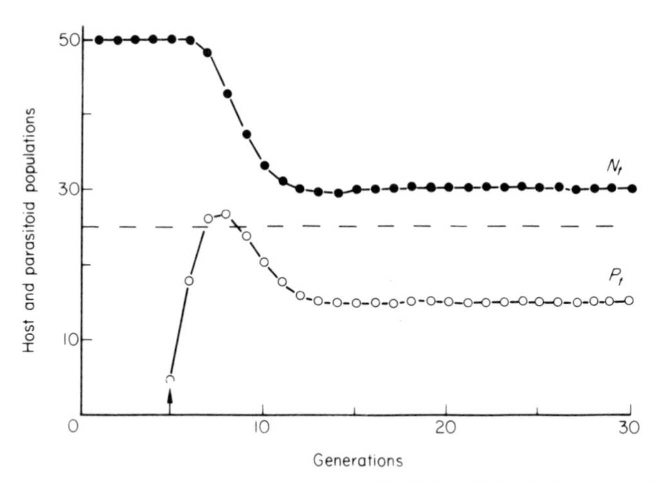


FIG. 13.13. A numerical simulation from eqn (13.3) in which the host population has a maximum size of 50 and $a = 0{\cdot}245$, $k = 0{\cdot}245$ and $\lambda = 2{\cdot}0$. The broken line indicates an arbitrary economic threshold.

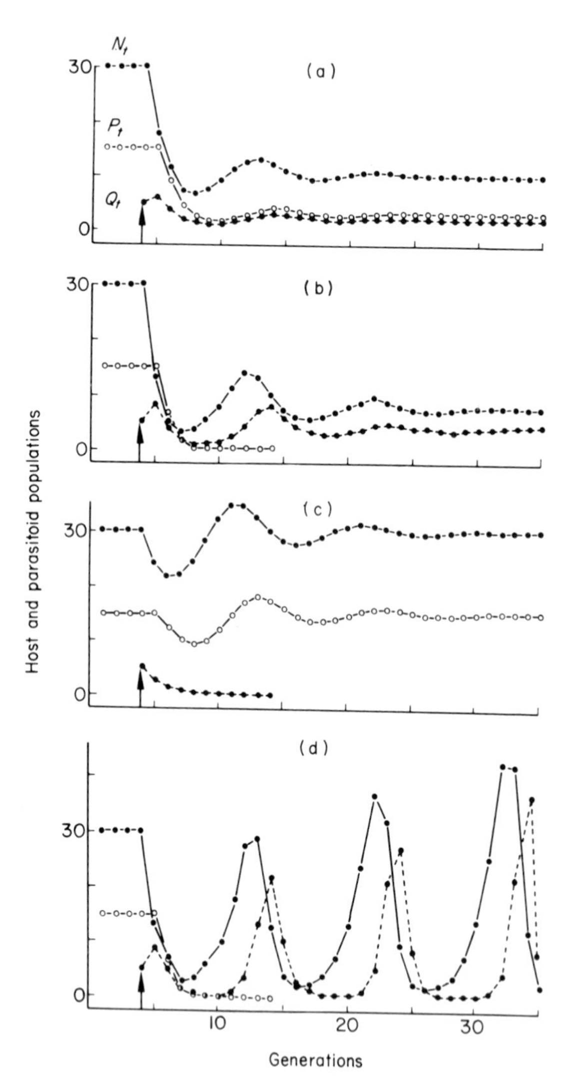


FIG. 13.14. Numerical simulations from eqns (13.9) and (13.10) illustrating four of the possible outcomes following the introduction of a further parasitoid species ($Q$) to the stable host-parasitoid interaction in Figure 13.13. $N$ and $P$ commence at their equilibrium values and $Q$ is introduced in generation 4. (a) comes from the hatched region of Figure 13.10, where $a_1 = 0{\cdot}25$, $a_2 = 0{\cdot}35$, $k_1 = k_2 = 0{\cdot}25$. (b) comes from region A of Figure 13.10, where $a_1 = 0{\cdot}1$, $a_2 = 0{\cdot}4$, $k_1 = k_2 = 0{\cdot}5$. (c) comes from region B of Figure 13.10, where $a_1 = 0{\cdot}1$, $a_2 = 0{\cdot}05$, $k_1 = k_2 = 0{\cdot}5$. (d) $a_1 = 0{\cdot}1$, $a_2 = 0{\cdot}25$, $k_1 = 0{\cdot}5$, $k_2 = 1{\cdot}1$. (After Hassell 1978.)

TABLE 13.2. Five possible outcomes following the introduction of a specific parasitoid ($Q$) where a stable single parasitoid ($P$)–host interaction already exists

| Outcome | Likely parameter combinations | Simulation |
|---|---|---|
| (1) $Q$ becomes established → locally stable 3-species equilibrium | $k_2 \ll 1; a_2 > a_1$ | Fig. 13.10a |
| (2) $Q$ replaces $P$ → stable 2-species equilibrium | $k_2 < 1; a_2 \gg a_1$ | Fig. 13.10b |
| (3) $Q$ fails to become established → former stable 2-species equilibrium | $k_2 <$ or $> 1; a_2 < a_1$ | Fig. 13.10c |
| (4) $Q$ forces $P$ to extinction → unstable 2-species interaction | $k_2 > 1; a_2 > a_1$ | Fig. 13.10d |
| (5) $Q$ and $P$ persist in a locally unstable 3-species interaction | $k_2 > 1; a_2 \gtrsim a_1$ | |

Examples of both outcomes exist in the biological control literature, but information on the searching parameters of the parasitoids involved is conspicuously lacking. One example, however, illustrated in Figure 13.15, provides some support for the condition that outcome (2) requires $a_2 > a_1$. The figure shows the percentage parasitism of fruit-flies (*Dacus dorsalis*) parasitized by three species of *Opius*, and the overall picture is one of successive replacement of the *Opius* spp., finally leaving *O. oophilus* as the dominant parasitoid. Of particular interest is that the maximum percentage parasitism achieved is greater for each successive species, which provides at least some support for the condition $a_2 > a_1$ applying in this system.

Of the remaining outcomes in Table 13.2, outcome (3) is a frequent result of biological control programmes, but the only loss is one of wasted effort. It is best avoided by ensuring that species $Q$ has a higher searching efficiency than $P$ (i.e. $a_2 > a_1$). The only highly undesirable outcomes are (4) and (5), in which the practical result will be the breakdown of what original control existed and the likely extinction of one or both parasitoid species. They can both be avoided by ensuring that $k_2 < 1$ and $a_2 > a_1$.

The message for biological control is therefore quite clear: seek natural enemies with high searching efficiencies and a marked ability to congregate in local areas of abundant hosts. Beddington, Free & Lawton (1978) and Hassell (1978) have recently argued that this is the ideal recipe for obtaining a stable single parasitoid–host interaction with low equilibrium populations, and now we see that it also maximizes the chances of establishing a further parasitoid species in a locally stable three-species interaction.

### *Hosts, parasitoids and hyperparasitoids*

An interesting feature of insect parasitoid–host interactions is the frequent occurrence of secondary or hyperparasitism, where a parasitoid seeks out the

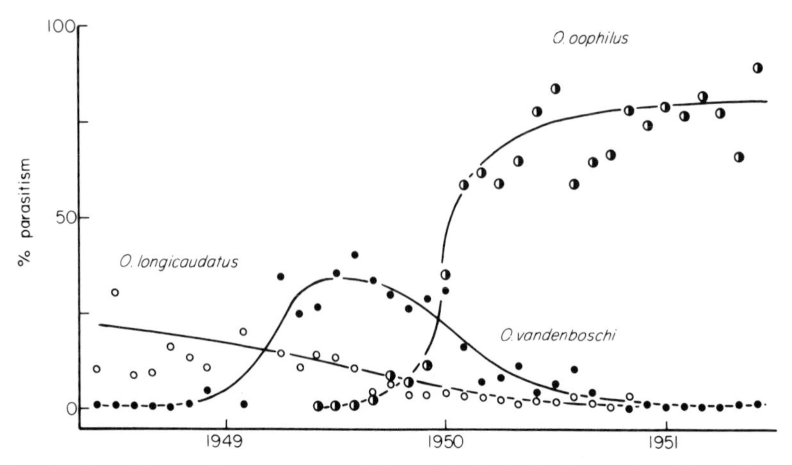


FIG. 13.15. Changes in per cent parasitism of the fruit fly, *Dacus dorsalis*, by three species of *Opius* parasitoids. Note that each successive parasitoid species causes a higher maximum level of parasitism. (Data from Bess, van den Bosch & Haramoto 1961; after Varley, Gradwell & Hassell 1973.)

immature stages of another parasitoid (the primary) as its host (Fig. 13.1c). An appropriate general model for such interactions is given by

$$\begin{aligned} N_{t+1} &= \lambda N_t f_1(P_t) \\ P_{t+1} &= N_t[1 - f_1(P_t)]f_2(Q_t) \\ Q_{t+1} &= N_t[1 - f_1(P_t)][1 - f_2(Q_t)] \end{aligned} \quad (13.11)$$

where $Q_t$ and $Q_{t+1}$ are now the numbers of hyperparasitoids in successive generations. The first analytical treatment of such a model is due to Beddington & Hammond (1977), who assumed that both $P$ and $Q$ search randomly (as in eqn (13.2)) and that the hosts exhibit a density-dependent rate of increase.

In this section, rather than rely on the host's density dependence for any stability, we follow in the mould of the previous models and define $f_1$ and $f_2$ from eqn (13.10). Once again, the detailed stability analysis is given in May & Hassell (1980) and here we merely display the local stability boundaries for the same particular case as in Figure 13.10, where $k_1 = k_2 = k$ (and $\lambda = 2$). The conditions for stability in this example are relatively straightforward, as shown in Figure 13.16. A locally stable, three-species equilibrium will always exist if $a_2 > a$ and $k < 1$, but co-existence in which the hyperparasitoid has the lower searching efficiency ($a_2 < a_1$) is only likely for very small values of $k$. Beddington & Hammond (1977) similarly concluded from their model that the likelihood of a three-species equilibrium is enhanced if the hyperparasitoid has the higher searching efficiency. Detailed studies on searching by hyperparasitoids

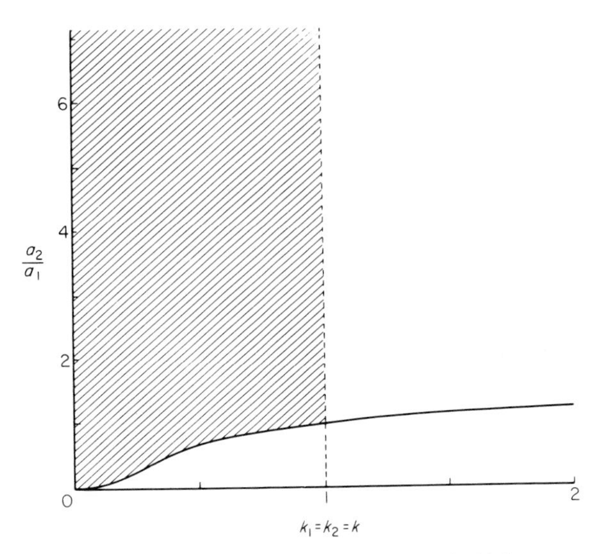


FIG. 13.16. Local stability boundaries for the host–parasitoid–hyperparasitoid system of eqn (13.11), where $f_1$ and $f_2$ are defined in eqn (13.10) and $k_1 = k_2 = k$. The hatched area shows where a locally stable three-species equilibrium occurs. (After Hassell 1978 and May & Hassell 1980, in which further details are given.)

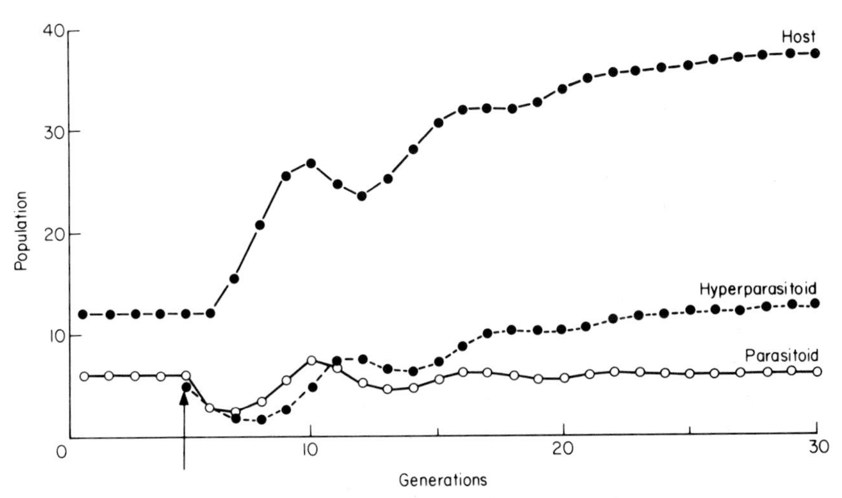


FIG. 13.17. A numerical simulation from eqn (13.11), where $f_1$ and $f_2$ are defined in eqn (13.10). $a_1 = 0{\cdot}25$, $a_2 = 0{\cdot}35$, $k_1 = k_2 = 0{\cdot}5$ and $\lambda = 2$. The host and parasitoid populations commence at their equilibrium values and the hyperparasitoid is introduced in generation 5.

have been all too few, but those that exist show the hyperparasitoids to have at least as high a searching efficiency as their primary parasitoid hosts.

The important conclusion to emerge for biological control practices is that a hyperparasitoid may prove even easier to establish in an existing host–parasitoid interaction than an additional primary species (cf. Figs 13.10 and 13.16). The consequences will always be in one direction: the host equilibrium will be raised, as seen from the simulation in Figure 13.17.

## CONCLUSION

Several of the properties of the multi-species models explored in this paper depend sensitively upon the assumptions on how predators search for their prey. Most important of all, these models show that, whenever possible, we should abandon the usual assumption of random search in favour of the more prudent strategy of tending to aggregate where prey are most abundant. Failure to do this can lead to quite incorrect generalizations on the dynamics of the system in question. Thus, we have seen that the outcome of a one-predator–two-prey model can be markedly influenced by predator switching, and the two-parasitoid–one-host models with non-random search are qualitatively very different to similar models with random search.

Most models with predator aggregation have been unwieldy in so far as prey and predator distributions have had to be specified. This has prevented their use in multi-species systems. In this paper, however, the problem has been bypassed by making use of May's (1978) detailed independent model describing the effects of predator aggregation. It should provide a very useful basis for the further exploration of these more complex predator–prey systems. Such developments will be all the more meaningful if they are accompanied by judicious experiments on the characteristics of predators and parasitoids in different types of interaction, against which the models may be moulded.

## ACKNOWLEDGMENTS

I am most grateful to R.M. Anderson and H.N. Comins for their perceptive comments on the manuscript.

## REFERENCES

**Anderson J.F., Hoy M.A. & Weseloh R.M. (1977)** Field cage assessment of the potential for establishment of *Rogas indiscretus* against the gipsy moth. *Environmental Entomology*, **6**, 375–380.

**Anderson R.M. & May R.M. (1978)** Regulation and stability of host-parasite population interactions. I. Regulatory processes. *Journal of Animal Ecology*, **47**, 219–247.

**Beddington J.R., Free C.A. & Lawton J.H. (1976)** Concepts of stability and resilience in predator–prey models. *Journal of Animal Ecology*, **45**, 791–816.

**Beddington J.R., Free C.A. & Lawton J.H. (1978)** Modelling biological control: on the characteristics of successful natural enemies. *Nature, London*, **273**, 513–519.

**Beddington J.R. & Hammond P.S. (1977)** On the dynamics of host-parasite-hyperparasite interactions. *Journal of Animal Ecology*, **46**, 811–821.

**Bess H.A., van den Bosch R. & Haramoto F.H. (1961)** Fruit fly parasites and their activities in Hawaii. *Proceedings of the Hawaiian Entomological Society*, **17**, 367–378.

**Bradley D.J. & May R.M. (1978)** Consequences of helminth aggregation for the dynamics of schistosomiasis. *Transactions of the Royal Society for Tropical Medicine and Hygiene*, **72**, 27–38.

**Comins H.N. & Hassell M.P. (1976)** Predation in multi-prey communities. *Journal of Theoretical Biology*, **62**, 93–114.

**Comins H.N. & Hassell M.P. (1979)** The dynamics of optimally foraging predators and parasitoids. *Journal of Animal Ecology*, **48** (in press).

**Connell J.H. (1975)** Some mechanisms producing structure in natural communities: a model and evidence from field experiments. *Ecology and Evolution of Communities* (Ed. by M.L. Cody & J.M. Diamond) pp. 460–490. Harvard University Press, Cambridge.

**Cramer N.F. & May R.M. (1972)** Interspecific competition, predation and species diversity: a comment. *Journal of Theoretical Biology*, **34**, 289–293.

**Crofton H.D. (1971a)** A quantitative approach to parasitism. *Parasitology*, **63**, 179–193.

**Crofton H.D. (1971b)** A model of host–parasite relationships. *Parasitology*, **63**, 343–364.

**Fujii K. (1977)** Complexity–stability relationship of two-prey–one-predator species system model: local and global stability. *Journal of Theoretical Biology*, **69**, 613–623.

**Harper J.L. (1969)** The role of predation in vegetational diversity. *Diversity and Stability in Ecological Systems*, Brookhaven Symposia in Biology, **22**, 48–62.

**Hassell M.P. (1978)** *The Dynamics of Arthropod Predator–Prey Systems.* Princeton University Press, Princeton.

**Hassell M.P. & Comins H.N. (1976)** Discrete time models for two-species competition. *Theoretical Population Biology*, **9**, 202–221.

**Hassell M.P. & Comins H.N. (1978)** Sigmoid functional responses and population stability. *Theoretical Population Biology*, **14**, 62–67.

**Hassell M.P. & May R.M. (1973)** Stability in insect host-parasite models. *Journal of Animal Ecology*, **42**, 693–726.

**Hassell M.P. & May R.M. (1974)** Aggregation in predators and insect parasites and its effect on stability. *Journal of Animal Ecology*, **43**, 567–594.

**Huffaker C.B., Luck R.F. & Messenger P.S. (1977)** The ecological basis for biological control. *Proceedings XV International Congress of Ecology* (*Washington, 1976*) pp. 560–586.

**Huffaker C.B., Messenger P.S. & DeBach P. (1971)** The natural enemy component in natural control and the theory of biological control. *Biological Control* (Ed. by C.B. Huffaker) pp. 16–67. Academic Press, New York.

**Landenburger D.E. (1968)** Studies on selective feeding in the Pacific starfish *Pisaster* in Southern California. *Ecology*, **49**, 1062–1075.

**Lawton J.H., Beddington J.R. & Bonser R. (1974)** Switching in invertebrate predators, *Ecological Stability* (Ed. by M.B. Usher) pp. 141–158. Chapman & Hall, London.

**May R.M. (1974)** Biological populations with non-overlapping generations: stable points. stable cycles, and chaos. *Science*, **186**, 645–647.

**May R.M. (1975)** Biological populations obeying difference equations: stable points, stable cycles, and chaos. *Journal of Theoretical Biology*, **49**, 511–524.

**May R.M. (1977a)** Togetherness among schistosomes: its effects on the dynamics of the infection. *Mathematical Biosciences*, **35**, 301–343.

**May R.M. (1977b)** Predators that switch. *Nature*, **269**, 103–104.

**May R.M. (1978)** Host-parasitoid systems in patchy environments: a phenomenological model. *Journal of Animal Ecology*, **47**, 833–843.

**May R.M. & Hassell M.P. (1980)** The dynamics of multiparasitoid-host interactions. *American Naturalist*, **48** (in press).

**Miller J.C. (1977)** Ecological relationships among parasites of *Spodoptera praefica*. *Environmental Entomology*, **6**, 856–859.

**Murdoch W.W. (1977)** Stabilizing effects of spatial heterogeneity in predator–prey systems. *Theoretical Population Biology*, **11**, 252–273.

**Murdoch W.W., Avery S. & Smith M.E.B. (1975)** Switching in predatory fish. *Ecology*, **56**, 1094–1105.

**Murdoch W.W. & Oaten A. (1975)** Predation and population stability. *Advances in Ecological Research*, **9**, 2–131.

**Nicholson A.J. (1933)** The balance of animal populations. *Journal of Animal Ecology*, **2**, 132–178.

**Nicholson A.J. & Bailey V.A. (1935)** The balance of animal populations. Part I. *Proceedings of the Zoological Society of London*, **1935**, 551–598.

**Paine R.T. (1966)** Food web complexity and species diversity. *American Naturalist*, **100**, 65–75.

**Paine R.T. (1974)** Intertidal community structure. *Oecologia*, **15**, 93–120.

**Parrish J.D. & Saila S.B. (1970)** Interspecific competition, predation and species diversity. *Journal of Theoretical Biology*, **27**, 207–220.

**Roughgarden J. & Feldman M. (1975)** Species packing and predation pressure. *Ecology*, **56**, 489–492.

**Royama T. (1970)** Factors governing the hunting behaviour and selection of food by the great tit (*Parus major* L.). *Journal of Animal Ecology*, **39**, 619–668.

**Steele J.H. (1974)** *The Structure of Marine Ecosystems*. Harvard University Press, Cambridge, Massachusetts.

**Turnbull A.L. (1967)** Population dynamics of exotic insects. *Bulletin of the Entomological Society of America*, **13**, 333–337.

**Turnbull A.L. & Chant D.A. (1961)** The practice and theory of biological control of insects in Canada. *Canadian Journal of Zoology*, **39**, 697–753.

**van den Bosch R. (1968)** Comments population dynamics of exotic insects. *Bulletin of the Entomological Society of America*, **14**, 112–115.

**van Valen L. (1974)** Predation and species diversity. *Journal of Theoretical Biology*, **44**, 19–21.

**Varley G.C., Gradwell G.R. & Hassell M.P. (1973)** *Insect Population Ecology*. Blackwell Scientific Publications, Oxford.

**Zwölfer H. (1971)** The structure and effect of parasite complexes attacking phytophagous host insects. *Proceedings of the Advanced Study Institute on 'Dynamics of Numbers in Populations' (Oosterbeek, 1970)* pp. 405–418.

# 14. HARVESTING AND POPULATION DYNAMICS

J. R. BEDDINGTON

*Department of Biology, University of York,*
*Heslington, York YO1 5DD*

## INTRODUCTION

In this paper I will consider how ecological theory provides insights into the problems that occur in harvesting a species for sustainable yield. By treating yield simply in biological terms either as biomass or as numbers of individuals I am excluding important economic considerations. Similarly, in treating harvesting as a predator–prey process in which the numerical response of the predator is determined exogenously I am capriciously ignoring the important work of Clark (1976) and others in analysing the economic forces that determine the level of fishing effort. The paper is divided into two major sections: in the first, I explain the ways in which increased knowledge of the underlying dynamics of a population and its ecological role facilitate its exploitation for sustainable yield; in the second part, this essentially static picture is considered in the light of the dynamics of the system and the problems of multiple equilibria, stability and environmental variability are specifically addressed.

In what follows I will largely confine the discussion to the harvesting of single species of marine organisms, but much of the analysis can be extended to hunted, but not ranched, populations of terrestrial animals.

## HARVESTING FOR SUSTAINABLE YIELD

### *Homogenous models*

The most widely analysed and discussed models of harvested populations are simple first-order differential equations which specify the rate of change of a population according to the equation

$$\frac{dN}{dt} = f(N) \tag{14.1}$$

The sustainable yield is then determined by the value of $f(N)$ at any population level $N$. Thus, for the logistic equation $f(N) \equiv rN(1 - N/K)$ analysed by

Graham (1935), Schaefer (1954) and many others, the sustainable yield as a function of the population level is a parabolic curve with a maximum value $rK/4$ reached at a population level of $K/2$. The maximum sustainable yield is clearly important, if only for the reason that continued harvesting above that level will result in depletion of the population. However, Holt (1977), reviewing a catalogue of such models derived from the fisheries literature, makes the following point. Such models may be classified according to whether they have a maximum sustainable yield at a higher or lower population level than the logistic. If the attempt is made to harvest at maximum sustainable yield following an incorrect choice of functional form for the model, then severe depletion of the population can occur. This problem is likely to occur in its most acute form when harvesting a species whose true MSY level is at a considerably higher population level than that assumed. It is part of the lore of the subject that long-lived species high in the trophic web have growth curves skewed to the right compared with the logistic. Hence, harvesting such species on the assumptions of the logistic equation may lead to depletion. Holt makes the further point that fisheries data are usually not sufficiently good to be able to discriminate between functional forms. This point is revisited in the light of the effects of environmental fluctuation later in this paper. Discrete time analogues of such models are treated in a similar way, e.g. Ricker (1958).

### *Models involving age structure*

It has been known by farmers for a long time, and by ecologists for a somewhat shorter time, that judicious manipulation of the age structure of a population can substantially increase the growth rate and hence sustainable yield. Thus, for a simple Leslie matrix model of the form

$$V_{t+1} = MV_t \tag{14.2}$$

in the stable age structure the population can sustain a harvest rate of $(\lambda - 1)/\lambda$, where $\lambda$, the finite growth rate, is the largest eigenvalue of the matrix (Leslie 1945, 1948; Usher 1972). However, if that harvest is aimed differentially at different age groups, an increased yield may be obtained, with the necessary conditions for maximum sustained yield being that a maximum of two age groups are harvested, one being completely removed (Beddington & Taylor 1973). Recently, Reed (unpublished) has indicated that these results, applicable to linear models, in fact hold for a wide class of density-dependent Leslie models. However, the utility of such results for a fishery, where it is impossible to guarantee complete removal of one age class or indeed to discriminate much between the mortality rates imposed on different age classes, is problematic. Of more use are models which, following Beverton & Holt (1957), permit the effect on yield of altering the age at first fishing for given fishing mortality rates to be explored. Figure 14.1 illustrates the pattern of yield obtained as a function of the age of first fishing and fishing mortality for one such model of the general form

$$N_{t+1} = S(1 - f)N_t + S^k(1 - f)^j N_{t-k}F(N_{t-k}) \qquad (14.3)$$

which describes the dynamics of an adult population $N_t$, with constant natural survival rate $S$ subject to a proportional fishing mortality $f$ operating on the previous $j$ age groups. Fecundity is determined by the density-dependent function $F(N_{t-k})$, $k$ being the age of sexual maturity. This model may be derived from a simple extension of a non-linear Leslie matrix and has application in the analysis of some baleen whale populations (Beddington 1978).

For a given age of first fishing the figure shows an increase in yield up to a peak, followed by a decrease as fishing mortality increases. If the fishing mortality axis were to be scaled as a function of the population size, the figures would then be comparable to the relationship in the homogeneous models, the effect of manipulating the age structure being to produce a set of such curves from which the one giving the largest yield may be chosen.

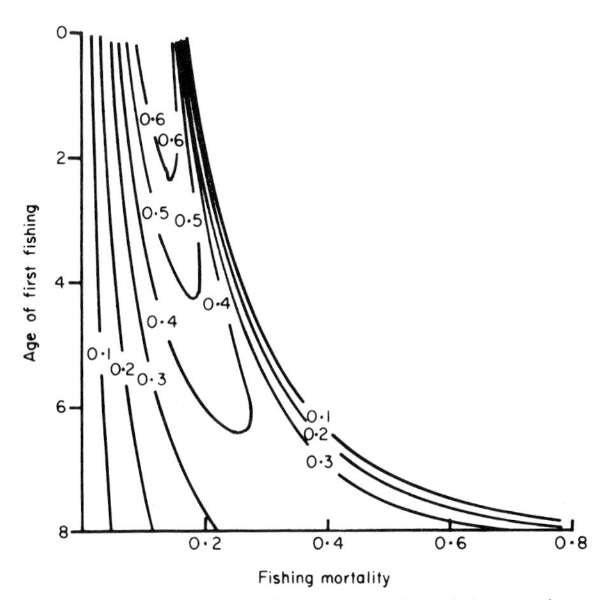


FIG. 14.1. The yield contours expressed as a proportion of the carrying capacity are shown as functions of the fishing mortality and the age of first fishing. The specific functional form used is $F(N_{t-k}) \equiv \exp[r(1 - aN_{t-k})]$, with $k = 8$, $S = 0{\cdot}9$, $r = 1, a = 1$.

*Models involving sexual dimorphism*

In the models considered previously it has been implicitly assumed that the sexes are not exploited differentially, and that the female population produces recruits independent of the number and nature of the male population. For a polygamous species with pronounced sexual dimorphism there is a clear possibility for manipulating the sex composition of the population to produce increased sustainable yields. Models of such a process demand that the effect of the male population on the fecundity of the females be explicitly incorporated. This can be done simply by demanding as a constraint that the ratio of males to females should not fall to a level where fecundity is adversely affected (Beddington 1974). A typical model of this form is used in the management of the sperm whale (*Physeter catodon*) fishery and can be expressed as the equation system

$$\begin{aligned} N_{1t+1} &= N_{1t}S_1 + \tfrac{1}{2}S_1^k N_{1t-k}F(N_{1t-k}, N_{2t-k}) \\ N_{2t+1} &= N_{2t}S_2 + \tfrac{1}{2}S_2^j N_{1t-j}F(N_{1t-j}, N_{2t-j}) \end{aligned} \qquad (14.4)$$

where $N_1$ and $N_2$ are the female and male populations with ages of sexual maturity of $k$ and $j$ and survival rates $S_1$ and $S_2$ respectively. The function $F(N_{1t}, N_{2t})$ expresses the fecundity rate as a function of both male and female density (Beddington & Kirkwood 1979). In general, the function $F(N_{1t}, N_{2t})$ is likely to be determined not only by the relative density of males and females but also by their absolute density, and one such form involving the searching process of males for females has been proposed by Beddington & May (1978). It is then possible to manipulate these models to investigate the effect on yield of different levels of male and female fishing mortality. Once again the ability to distinguish between individuals and their ecological roles permits an increase in sustainable yield to be obtained.

All the analyses considered above involve estimating the potential output of a population at various levels of density and demographic structure. The second stage is to consider the way in which the fishery operates as a predation process in removing individuals from the population; this, the functional response of the fishery, is considered in the next section.

*The functional response of a fishery*

In this section, for simplicity, I will consider the functional response of a fishery in terms of the continuous time models of population growth. Such models involve the growth function $f(N)$ which has associated with it the death rate imposed by fishing $h(N)$, giving the equation

$$\frac{dN}{dt} = f(N) - h(N) \qquad (14.5)$$

The regulation of a fishery essentially involves setting some form for $h(N)$ and two basic strategies exist; one sets $h(N)$ as constant so that a constant yield is taken, the other sets the effort of the fishery constant and implicitly assumes that $h(N) = qEN$, where $E$ is the measure of effort, usually in terms of the number of boats or types of fishing operation per unit time, and $q$ is termed the catchability coefficient (Rothschild 1977). Viewed as a predator–prey problem, one strategy alters the number of predators and the duration of the predation process to ensure a constant death rate, the other sets the number of predators to be a constant. Usually both such analyses assume that there is a type 1 functional response (Holling 1965). It is clear that if regulation operates in terms of predator numbers in this way, then necessarily a finite handling time $h$ will produce a form for the death rate such that

$$h(N) \equiv \frac{qEN}{1 + qEhN} \tag{14.6}$$

Indeed, it is worth remarking that once it is recognized that a fishery operates like a set of predators we may pose the question: what may be expected for the functional response? For example, by analogy with natural predators where alternative prey species are available, a sigmoid or type 3 functional response may occur (Holling 1965; Hassell, Lawton & Beddington 1977). Indeed, the 'switching' phenomena noted in many natural predators (Murdoch 1969; Lawton, Beddington & Bonser 1974) may be observed in multi-species fisheries. An example of this concerns the baleen whale fishery in the southern ocean, where depletion of stocks of larger whales resulted in a concentration of the fishing upon the smaller species; Laws (1962) specifically documents data of this type for the movement of the fishery from blue to fin whales. The time scale of recovery of the baleen whales is too long to permit observation of the reverse behaviour, but it is plausible that such a phenomenon exists in some mixed fisheries (and operates on a much shorter time scale). Similarly, where a harvested species is patchily distributed and fishermen aggregate in areas of high concentration, an encounter rate determined by some aggregated distribution such as the negative binomial may be more appropriate than the normal Poisson assumption. These questions are important not only for estimation of the correct effort levels to produce a specific yield (Rothschild 1977) but also for estimation of population abundance. In many fisheries the catch per unit of effort is used as an index of population abundance. This implicitly assumes a type 1 functional response and, if there is a significant handling time, can result in the overestimation of population abundance (Beddington 1979). Figure 14.2 illustrates this point.

In addition to these problems, the form of the fisheries' functional response can be crucial in determining the dynamic behaviour of the exploited population. This is considered in the next section.

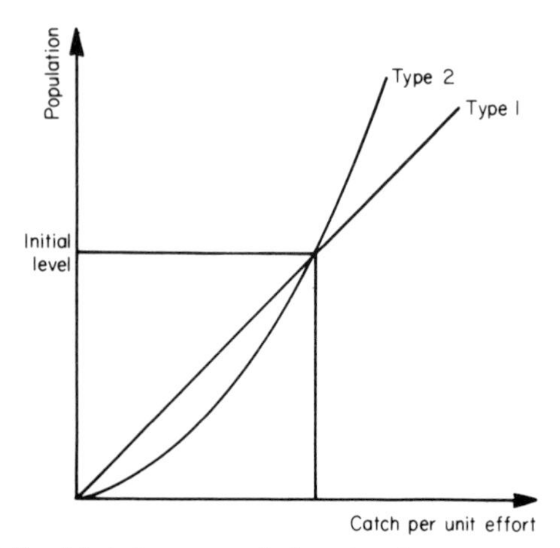


FIG. 14.2. The relationship between population abundance and catch per unit of effort is illustrated for a type 1 and a type 2 functional response. The assumption that there is no significant handling time leads to underestimation of the degree of depletion of a population.

## THE DYNAMICS OF EXPLOITED POPULATIONS

### *Multiple equilibrium*

Species of commercial importance to man are no different from other species, being embedded in an ecosystem with a large number of possible trophic links. Accordingly, it is reasonable to inquire whether the single-species models so far discussed are at all applicable. Some theoretical considerations reviewed by May (1979) and Goh (1978) point to the likelihood that complex ecosystems consist of a set of self-regulating subsystems loosely coupled to each other. This, together with the observation that in a surprising number of cases single-species studies have been successful in explaining the observed behaviour of populations, indicates that this problem may not be so acute. However, even for models of small numbers of species, a bewilderingly large number of possible equilibrium states can occur. For a discussion of some of the mechanisms that produce such states see May (1977) and Peterman, Clark & Holling (1979). In the context of a single-species model it is therefore necessary to recognize that an appropriate model may contain a number of equilibrium states generated by the subsumed or ignored interspecies interactions. In populations with complex age and sex structure the likelihood of such a multiplicity of equilibria is increased (Gulland 1975).

For a harvested population the situation may be yet more complex. Figure 14.3 illustrates the possible equilibria for a model of the form of eqn (14.5),

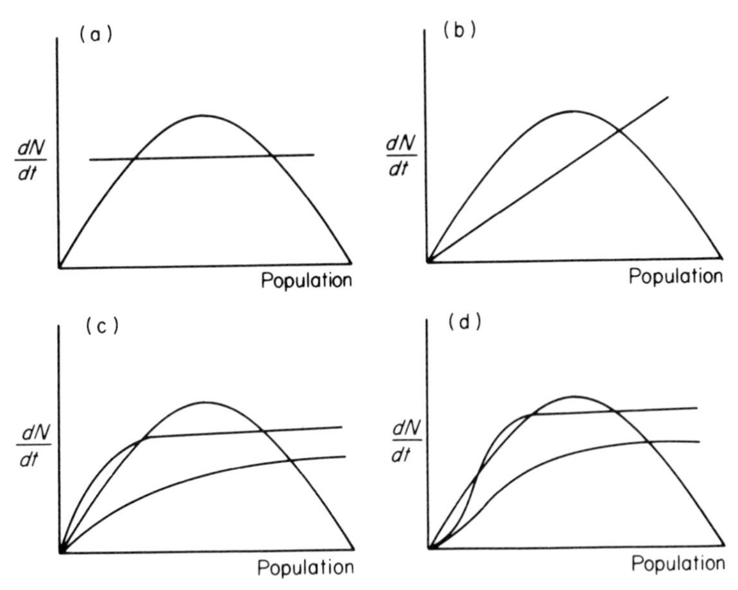


FIG. 14.3. The relationship between the population growth curve (assumed here to be a logistic) and the death rate imposed by a fishery is shown. The intersection of the growth curve and the various functional responses are the possible equilibria.
(a) A constant quota strategy in which, irrespective of fishing effort or population size, the same yield is taken. This produces two equilibrium points, the upper one stable and the lower unstable.
(b) A constant effort strategy on the assumption that the functional response is type 1. There is a single stable equilibrium.
(c) A constant effort strategy where there is a significant handling time effect and the overall functional response is type 2. For low levels of effort the behaviour is similar to that of 14.3(b) with a single stable equilibrium. For higher effort a second unstable equilibrium is generated, and if the population falls below this level it will continue to be depleted as long as the fishery operates.
(d) A constant effort strategy where a sigmoid or type 3 functional response is operating. For low levels of effort the behaviour is again analogous to that of 14.3(b), but when effort increases complications occur. Two new equilibria may be generated, the lower stable and the upper unstable. This latter equilibrium then acts as a threshold between the two stable equilibria and reduction of the population below this level will result in it moving to the lower relatively unproductive equilibrium.

the intersections of the population growth curve and the functional response of the fishery producing the equilibrium points. All the illustrations involve attempts to harvest at some yield below the maximum; here the equilibrium point nearest to the carrying capacity is always stable, but its domain of attraction is reduced as the yield is increased. For example, increasing the quota strategy in Figure 14.3a decreases the domain of attraction of this upper equilibrium point until at the maximum sustainable yield this domain vanishes and the equilibrium is unstable. Even assuming that the harvesting operates in the most favourable way, producing only one stable equilibrium point (e.g. in Fig. 14.3b), it is still necessary to recognize that the underlying population growth curve may possess other equilibria and that harvesting for a sustainable yield should avoid the population moving to these lower unproductive states. This is essentially a problem of the system's ability to withstand random environmental perturbations.

### *Environmental stochasticity*

All the analyses so far assume a deterministic world, but ecosystems and their component populations are continually subject to successive perturbations caused by a fluctuating environment. Hence, the simple deterministic models with fixed parameters and equilibrium states need to be replaced with randomly varying parameters and equilibrium probability distributions. When dealing with complicated models, analytical treatment in this way becomes intractable and the attempt has been made to gain a qualitative insight into a system's ability to absorb perturbations by considering the characteristic return time of the deterministic model. This idea, common in ecological theory (e.g. May 1974; May *et al.* 1974; Beddington, Free & Lawton 1976; Pimm & Lawton 1979), has been extended to the fisheries' literature by Doubleday (1976), Beddington & May (1977) and Sissenwine (1977), who consider the effect of harvesting on the characteristic return time of a population. The essential idea is that long return times indicate a population or system unresistant to environmental perturbations, a short return time, one capable of absorbing them. For simple growth models of the form (14.5) this analysis can be complemented by a full analytical treatment of the stochastic differential equations. Such a treatment enables the mean and variance of population and yield to be calculated for different levels of harvesting. The analysis of Beddington & May (1977) for the logistic equation has been supplemented by that of May *et al.* (1978) for a wide class of population growth functions with the following conclusions. The characteristic return time of the system will usually increase as harvesting takes the system to and beyond its maximum sustainable yield level. For growth functions skewed to the left there may be little change in the characteristic return time as the MSY level is approached. Indeed, in some cases a slight decrease may occur; however, once the MSY level is substantially exceeded, the return time increases markedly. For growth functions skewed to the right the return time increases with harvesting rate. If environmental fluctuations enter the system in a density-independent way (for example, on the linear term in $r$ in the logistic equation), then the characteristic return time will reflect the variability of the stochastic system, long return times being associated with high variability in the population and yield. However, if environmental noise affects the system in a density-dependent way (for example, as a randomly varying carrying capacity in the logistic equation), then the system will respond in a different way, becoming less responsive to environmental variability as return times lengthen. These points, made in this context by Shepherd (1977), have been made in the more general context by, among others, Roughgarden (1975) and Turelli (1977). Discussion on whether populations are likely to be affected by environmental noise in a density-dependent or -independent way is made more difficult by the considerable oversimplification of the continuous time models used. It is, therefore, reasonable to ask whether these general results apply to more realistic models of population dynamics involving age structure and time delays. Figure 14.4 illustrates the characteristic return time contours of the model of eqn (14.3) corresponding to the same parameters as the yield contours illustrated in Figure 14.1. The characteristic return times for the model are obtained by linearizing about the equilibrium in the usual way and obtaining the characteristic equation (Beddington 1978). The return times are then defined as $1/|\log \lambda|$, where $\lambda$ is the root of largest modulus of the characteristic equation. It should be noted that it is possible for certain parameter values for eqn (14.3) to be unstable in the unexploited state. In this situation intermediate levels of harvesting can produce a stabilizing effect. Indeed, the typical pattern of return times with fishing effort is a decrease in return time followed by an increase. It is now possible to choose for a given

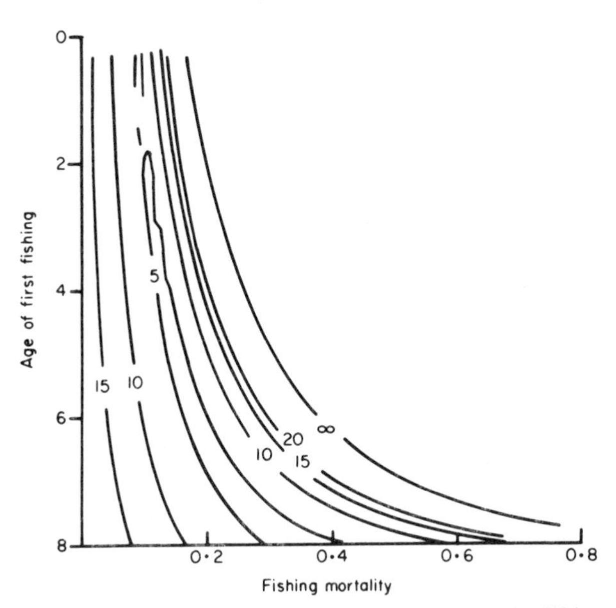


FIG. 14.4. The relationship between the characteristic return time, fishing mortality and age of first fishing is illustrated as contours of equal return time. The model form used and the parameter values are the same as for Figure 14.1.

level of yield from a set of return times by moving along the yield contour. Hence, by simulating the dynamics of the equation, it is possible to investigate the way in which random fluctuations in the parameters of the model affect the overall variability. Figure 14.4 illustrates some typical results of this analysis which corroborate the results of the continuous time models. Variation in the density-independent parameter produces levels of fluctuation inversely correlated with the return time, variation in the density-dependent parameters the reverse (Fig. 14.5).

In May *et al.* (1978) the relationship between yield and effort for a variety of continuous time models with density-independent variation is presented. Typically, these figures show that an increased amount of variability in yield occurs as harvesting effort increases. Extrapolation of this to real world situations is made problematic by the actual applicability of the models. It is, therefore, gratifying to note that these qualitative relationships are still holding in

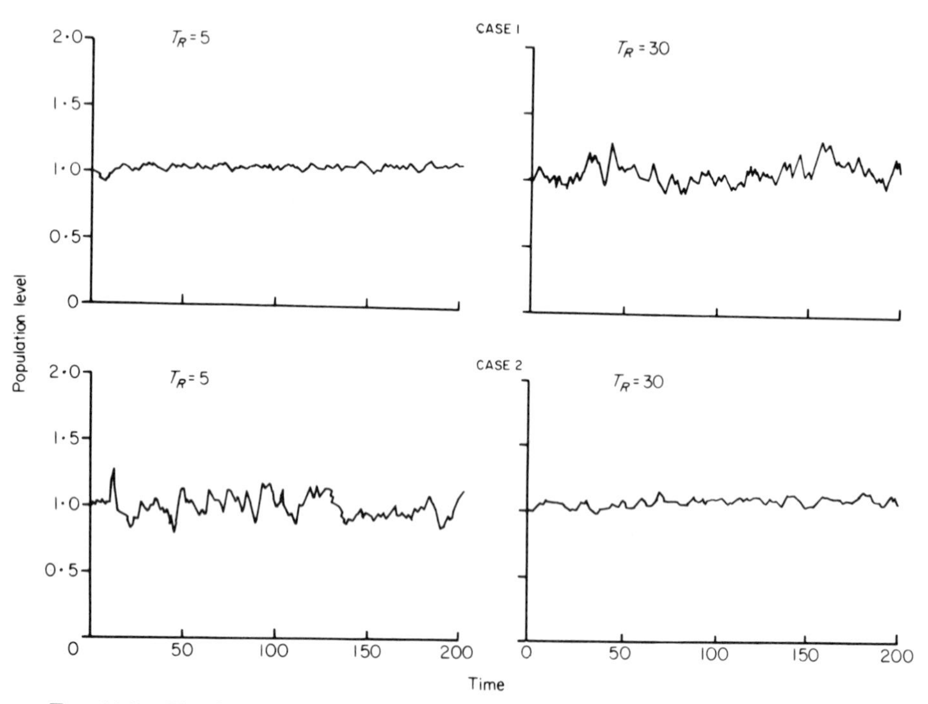


FIG. 14.5. The time trajectory of a population subject to harvesting for the same sustainable yield is shown for two different return times, 5 and 30. In case 1, subject to density-independent noise on the parameter $r$ which is assumed to be normally distributed with mean unity and standard deviation 0·2. In case 2, subject to density-dependent noise on the parameter $a$ which is assumed to be normally distributed with mean unity and standard deviation 0·2.

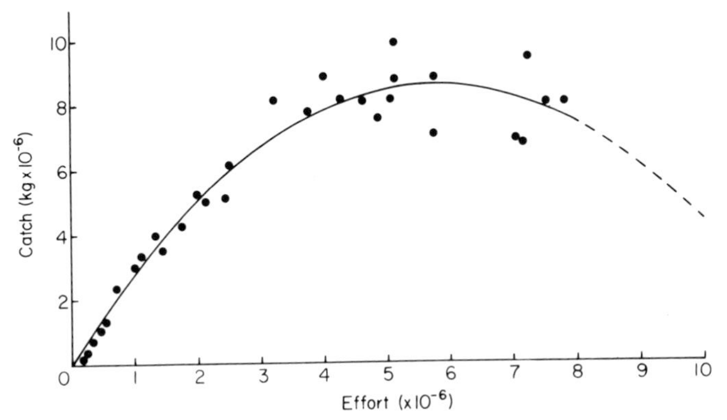


FIG. 14.6. The observed relationship between effort and yield from the fishery for the western rock lobster *Panulirus cygnus* George redrawn from Hancock (1978). The smooth curve assumes a logistic relationship, but the increasing variability as the MSY level is approached makes it difficult to assess the correct functional form.

more complicated and realistic models. This indicates that there are reasons to believe that many fish populations may be subject to density-independent noise as data on yield effort relationships mostly show this type of pattern. Figure 14.6 illustrates a typical example.

Given that such data are often needed to estimate both the sustainable yield and the underlying model of the fisheries' dynamics, the situation, seen in the light of Holt's analysis, is disturbing. It implies not only that there is uncertainty but that such uncertainty is likely to last. Accordingly, the policy goal of a substantial sustainable yield may require either continued adjustment of the level of harvest to produce the necessary feedback mechanisms to stabilize the population in the range where it produces substantial yields; or, where there are strong reasons for requiring a time-invariant harvesting strategy, a functional response determined by management that is in itself stabilizing. Such questions have been addressed from somewhat different viewpoints by Tautz, Larkin & Ricker (1969), Allen (1973), Walters (1975), Peterman, Walters & Hilborn (1978) and Harwood (1978).

## *The choice of trophic level*

In this concluding section it is worth noting that most of the models considered so far assume either explicitly or implicitly that, when harvested, a species is capable of producing a density-dependent response caused by the reduction in competitive pressure on some limiting resource. Such models are most applicable to organisms high in the trophic web which are themselves not subject to significant levels of predation. For example, by considering a simple prey–predator model, MacArthur (1969) was able to derive an approximation to the logistic model for the predator dynamics, on the assumption that the prey rate of change was small. However, there is currently considerable interest in exploiting unconventional fishery resources and moving down the trophic web to gain the substantial increase in ecological efficiency. The most obvious example under discussion is the Antarctic krill *Euphausia superba*, which forms the major food organism of a substantial multi-species complex in the southern ocean; for a review see Everson (1977). Whether such a prey species is capable of showing an immediate density-dependent response depends on whether it is being held above or below its own maximum sustainable yield level by predation. If it is being held at or below this level, harvesting will be additive to the overall level of predation and a decline in abundance will occur, which will be halted, only after a time delay, as predator density decreases. Such problems are likely to be extremely important where there is a significant difference in the generation time of the prey and predator species and are likely to result in any equilibrium levels being overshot by both population trajectories. Analysis of such problems necessitates a multi-, or at least two-, species approach. In concluding, it is interesting to note that the levels of depression at which a species is held by predation, considered in an entirely different context by Lawton & McNeill (1979), is likely also to be an important question for multi-species fisheries.

## ACKNOWLEDGMENTS

I would like to acknowledge helpful discussion with S.J. Holt, R.M. May and J.H. Lawton and the assistance of B. Grenfell with computation.

## REFERENCES

**Allen K.R. (1973)** The influence of random fluctuations in the stock recruitment relation on the economic return from salmon fisheries. *Conseil Permanent International pour l'Exploration de la Mer. Rapports et Procès Verbaux de Réunions*, **164**, 351–359.

**Beddington J.R. (1974)** Age structure, sex ratio and population density in the harvesting of natural animal populations. *Journal of Applied Ecology*, **11**, 915–924.

**Beddington J.R. (1978a)** On the dynamics of Sei whales under exploitation. *Report of the International Whaling Commission*, **28**, 169–172.

**Beddington J.R. (1978b)** On the risks associated with different harvesting strategies. *Report of the International Whaling Commission*, **28**, 165–167.

**Beddington J.R. (1979)** On some problems of estimating population abundance from catch data. *Bulletin of the International Whaling Commission*, **29** (in press).

**Beddington J.R., Free C.A. & Lawton J.H. (1976)** Concepts of stability and resilience in predator–prey models. *Journal of Animal Ecology*, **45**, 791–816.

**Beddington J.R. & Kirkwood G.P. (1979)** On the structure of possible sperm whale models. *Bulletin of the International Whaling Commission*, **29** (in press).

**Beddington J.R. & May R.M.C. (1977)** Harvesting natural populations in a randomly fluctuating environment. *Science*, **197**, 463–465.

**Beddington J.R. & May R.M. (1978)** A possible model for the effect of sex ratio on sperm whale fertility. *Bulletin of the International Whaling Commission*, **29** (in press).

**Beddington J.R. & Taylor D.B. (1973)** Optimal age specific harvesting of a population. *Biometrics*, **29**, 801–809.

**Beverton R.J.H. & Holt S.J. (1957)** On the dynamics of exploited fish populations. *Fisheries Investigation Series* II, **19**.

**Clark C.W. (1976)** *Mathematical Bioeconomics*. Wiley, London.

**Doubleday W.G. (1976)** Environmental fluctuations and fisheries management. *International Commission for the Northwest Atlantic Fisheries Special Publication*, **1**, 141–150.

**Everson I. (1977)** The living resources of the southern ocean. *Food and Agriculture Organisation of the United Nations, Rome.*

**Goh B.S. (1978)** Robust stability concepts for ecosystem models. *Theoretical Systems Ecology* (Ed. by E. Halton). Academic Press, New York.

**Graham M. (1935)** Modern theory of exploiting a fishery and application to North Sea trawling. *Journal du Conceil. Conseil permanent international pour l'exploration de la mer*, **10**, 264–274.

**Gulland J.A. (1975)** The stability of fish stocks. *Journal du Conseil. Conseil permanent international pour l'exploration de la mer*, **37**, 199–204.

**Hancock D.A. (1979)** Population dynamics and management of shellfish stocks. *I.C.E.S. Special Meeting on Population Assessment of Shellfish* (in press).

**Harwood J. (1978)** The effect of management policies on the stability and resilience of British grey seal populations. *Journal of Applied Ecology*, **15**, 413–421.

**Hassell M.P., Lawton J.H. & Beddington J.R. (1977)** Sigmoid functional responses by invertebrate predators and parasitoids. *Journal of Animal Ecology*, **46**, 249–262.

**Holling C.S. (1965)** The functional response of predators to prey density and its role in mimicry and population regulation. *Memoirs of the Entomological Society of Canada*, **45**, 1–60.

**Holt S.J. (1977)** Aspects of determining the stock level for maximum sustainable yield. *ACMRR/MM/SC/29 Food and Agriculture Organisation of the United Nations, Rome.*

**Laws R.M. (1962)** The effects of whaling on the southern stocks of Baleen whales. *The Exploitation of Natural Animal Populations* (Ed. by E.D. LeCren & M.W. Holdgate) pp. 137–158. Blackwell, Oxford.

**Lawton J.H., Beddington J.R. & Bonser R. (1974)** Switching in invertebrate predators. *Ecological Stability* (Ed. by M.B. Usher & M.H. Williamson) pp. 141–158. Chapman & Hall, London.

**Lawton J.H. & McNeill S. (1979)** Between the devil and the deep blue sea: the problem of being a herbivore. *Population Dynamics* (Ed. by R.M. Anderson, B.D. Turner & L.R. Taylor) pp. 223–244. Blackwell Scientific Publications, Oxford.

**Leslie P.H. (1945)** On the use of matrices in certain population mathematics. *Biometrika*, **33**, 183–212.

**Leslie P.H. (1948)** Some further notes on the use of matrices in population mathematics. *Biometrika*, **35**, 213–245.

**MacArthur R.H. (1969)** Species packing and what interspecies competition minimises. *Proceedings of the National Academy of Science*, **64**, 1369–1371.

**May R.M. (1974)** *Stability and Complexity in Model Ecosystems*, 2nd edition. Princeton University Press, Princeton.

**May R.M. (1977)** Thresholds and break points in ecosystems with a multiplicity of stable states. *Nature, London*, **269**, 471–477.

**May R.M. (1979)** The structure and dynamics of ecological communities. *Population Dynamics* (Ed. by R.M. Anderson, B.D. Turner & L.R. Taylor) pp. 385–407. Blackwell Scientific Publications, Oxford.

**May R.M., Beddington J.R., Horwood J.R. & Shepherd J. (1978)** Exploiting natural populations in an uncertain world. *Mathematical Biosciences*, **42**, 219–252.

**May R.M., Conway G.R., Hassell M.P. & Southwood T.R.E. (1974)** Time delays, density dependence and single species oscillations. *Journal of Animal Ecology*, **43**, 747–770.

**Murdoch W.W. (1969)** Switching in general predators: experiments on predator specificity and stability of prey populations. *Ecological Monographs*, **39**, 335–354.

**Peterman R.M., Clark W.C. & Holling C.S. (1979)** The dynamics of resilience: shifting stability domains in fish and insect systems. *Population Dynamics* (Ed. by R.M. Anderson, B.D. Turner & L.R. Taylor) pp. 321–341. Blackwell Scientific Publications, Oxford.

**Peterman R.M., Walters C.J. & Hilborn R. (1978)** Systems analysis of Pacific salmon management problems. *Adaptive Environmental Assessment and Management* (Ed. by C.S. Holling). Wiley, London.

**Pimm S.L. & Lawton J.H. (1979)** Number of trophic levels in ecological communities. *Nature*, **268**, 329–331.

**Ricker W.E. (1958)** Handbook of computations for biological statistics of fish populations. *Bulletin of the Fisheries Research Board of Canada*, **119**.

**Rothschild B.J. (1977)** Fishing effort. *Fish Population Dynamics* (Ed. by J.A. Gulland) pp. 96–115. Wiley, New York.

**Roughgarden J. (1975)** A simple model for population dynamics in stochastic environments. *American Naturalist*, **109**, 713–736.

**Schaefer M.B. (1954)** Some aspects of the dynamics of populations important to the management of commercial fish populations. *Inter-American Tropical Tuna Commission Bulletin*, **1**, 26–56.

**Shepherd J.G. (1977)** The sensitivity of exploited populations to environmental 'noise' and the implications for management. *I.C.E.S. CM 1977/F:27.*

**Sissenwine M.P. (1977)** The effect of random fluctuations on a hypothetical fishery. *International Commission for the Northwest Atlantic Fisheries Special Publication*, **29**, 137–144.

**Tautz A., Larkin P.A. & Ricker W.E. (1969)** Some effects of simulated long-term environmental fluctuations on maximum sustained yield. *Journal of the Fisheries Research Board of Canada*, **26**, 2715–2726.

**Turelli M. (1977)** Random environments and stochastic calculus. *Theoretical Population Biology*, **12**, 140–178.

**Usher M.B. (1972)** Developments in the Leslie matrix model. *Mathematical Models in Ecology* (Ed. by J.N.R. Jeffers) pp. 29–60. Blackwell Scientific Publications, Oxford.

**Walters C.J. (1975)** Optimal harvest strategies for salmon in relation to environmental variability and uncertain production parameters. *Journal of the Fisheries Research Board of Canada*, **32**, 1777–1784.

# 15. THE DYNAMICS OF RESILIENCE: SHIFTING STABILITY DOMAINS IN FISH AND INSECT SYSTEMS

RANDALL M. PETERMAN,* WILLIAM C. CLARK
AND C. S. HOLLING

*Institute of Animal Resource Ecology,*
*University of British Columbia,*
*Vancouver, B.C., Canada V6T 1W5*

## INTRODUCTION

Views of stability characteristics of ecological systems differ among ecologists. Some believe that ecosystems are globally stable, others think they are highly unstable and fragile, while others presume that there are bounds to a system's response capabilities. In the last case perturbations can be absorbed only up to some limit, and past that limit the system changes its characteristics. A fourth and more recently emerging point of view is that the boundaries or limits to a system's response will move in relation to both management perturbations and internal changes in the system (Holling 1978).

Our understanding of ecosystem dynamics is strongly coloured by the stability concept we hold. History demonstrates that our understanding and management schemes often fail, in part because we are viewing the system with the wrong conceptual framework. For example, what we once thought were globally stable fishery systems turned out to have multiple equilibria when critical harvest rates were exceeded.

Let us briefly expand on the first three views of ecosystem stability. The concept of global stability is well represented by the work of Patten (1975). His ecosystem models, which assume linear interactions between the components, have single equilibria and the system can be perturbed in many ways and still remain in the same attractor region.

At the opposite end of the spectrum is the view that ecosystems are highly fragile and unstable and that small perturbations result in collapse of the present system. Tropical rain forests are often held to fit this description (Gomez-Pompa, Vazquez-Yanes & Guevara 1972).

The compromise between these two extremes says that there may be more

* Also of Canadian Department of Fisheries and Environment.

than one equilibrium or attractor, and that each pair of attractors is separated by a boundary. As long as the system does not cross a boundary, it will tend to move back towards the same equilibrium point. If a boundary is crossed, the system tends towards the equilibrium of the new resident attractor region. Several theoretical models have been developed that demonstrate the possible existence of multiple domains of attraction when several non-linear system components are linked (Takahashi 1964; Paulik 1973; Austin & Cook 1974; Bazykin 1974; Sutherland 1974; Levin 1978). Some of these models predict the qualitative patterns of behaviour observed in real ecosystems (e.g. Holling 1973; Noy-Meir 1975). However, these theoretical models can exhibit almost any type of behaviour depending on the particular parameter values (Bazykin 1974). More relevant are studies that correctly predict observed multiple domain behaviour by combining actual measured system components (Southwood & Comins 1976; McLeod 1978).

All of the above stability concepts are actually myths, simplified views of the world that are not necessarily true but which bring order to the variety of our experience. No single myth applies to all ecological systems, yet each embodies a perspective that has proved rewarding under certain circumstances in the past.

Our own work has forced the realization that these three views of stability are inadequate for understanding many ecological systems, some as different as forest insects and fish. Therefore we have developed an additional myth focused on the movement of boundaries and the consequent change in shapes of stability domains and in system behaviour. The present paper documents this dynamic aspect of the concept of multiple domains of attraction. Boundary movement occurs not only as model parameter values are changed (Noy-Meir 1975; Southwood & Comins 1976), but also in response to the dynamics of internal state variables and exogenous driving variables. We show that process studies can be synthesized into predictive models (as in Gilbert *et al.* 1976; Steele & Frost 1977; Holling 1978) which can quantify boundary movements, suggest critical experiments to test our understanding and create a more solid framework for making management decisions. Two examples will illustrate the approach: the spruce budworm–forest system in eastern Canada and Pacific salmon along with its natural and human predators.

## THE SPRUCE BUDWORM ECOSYSTEM

The spruce budworm (*Choristoneura fumiferana*) is a defoliating lepidopteran insect of the North American boreal forest. Throughout most of its range it is usually rare, presumably held in check by a combination of natural enemies, inadequate resource levels and weather (Morris 1963). These controls periodically break down, however, and the system experiences budworm outbreaks that inflict extensive mortality on the preferred host trees, balsam fir (*Abies balsamea*) and white spruce (*Picea glauca*) (Davidson & Prentice 1967). Outbreaks spread rapidly in space since budworm moths may disperse tens or even hundreds of kilometres in a single generation. Detailed studies on the system have been carried out since the late 1940's, especially in eastern Canada (summarized in Morris 1963 and Miller & Ketella 1975).

Our approach to understanding the budworm–forest system was to combine the data from these detailed process studies into a complex simulation model representing a 'dynamic life table' (Gilbert *et al.* 1976) for the budworm and its local interaction with trees, weather, natural enemies and human management. The seven million hectares of New Brunswick were treated as a mosaic of 393 patches governed by local interactions, and individual patches were linked via insect dispersal. The detailed results of this work are reported elsewhere (Baskerville 1976; Holling, Jones & Clark 1976; Jones 1977; Clark, Jones & Holling 1978, 1979) and an extensive monograph on the work is in preparation. The resulting model adequately predicts various qualitative aspects of observed behaviour patterns for several different regions of North America.

The insight into budworm stability characteristics comes from the dynamics of system components in a representative spatial patch. Equations governing local interactions of the budworm, its food host and its natural enemies are summarized in Jones (1977), but these are too complex to be solved analytically. Furthermore, it is almost impossible to comprehend the effect of various system components on the total system's behaviour merely by performing numerous long-term simulations; realistic ecosystem models have too wide a range of behaviours. Instead we have used the simulation model to generate succinct summaries of interacting components. A starting insect population density, $N_t$, is chosen as well as a given forest condition, $F$. The model is run one generation, and net changes are calculated. This is done iteratively for several $N_t$ and curves of net recruitment rate, $R$ $(= N_{t+1}/N_t)$, are plotted. Recruitment rate will hereafter be referred to simply as recruitment, but one should be cautioned that this is slightly different usage from Ricker (1954). Figure 15.1a shows the recruitment curve generated when only the effect of food deprivation is included in the simulation model. This process creates a high-density stable equilibrium for budworm at a larval density of $P_1$, where population growth rate is at $R = 1$. Mortality due to parasites modifies this recruitment rate mainly at low densities (Fig. 15.1b). Avian predators that show type III functional responses (Holling 1959) modify the recruitment curve further but only at low insect densities (Fig. 15.1c).

This predation component, augmented by parasitism, adds two new potential equilibria (crossovers at $R = 1$). $P_3$ is another stable point, or attractor, while $P_2$ is unstable, and it forms the boundary between the two domains of attraction, $D_1$ and $D_2$. $P_2$ is an unstable point because slightly larger densities lead to continued increase (since $R > 1$) and slightly smaller densities to continued decrease ($R < 1$). Thus, the population moves away

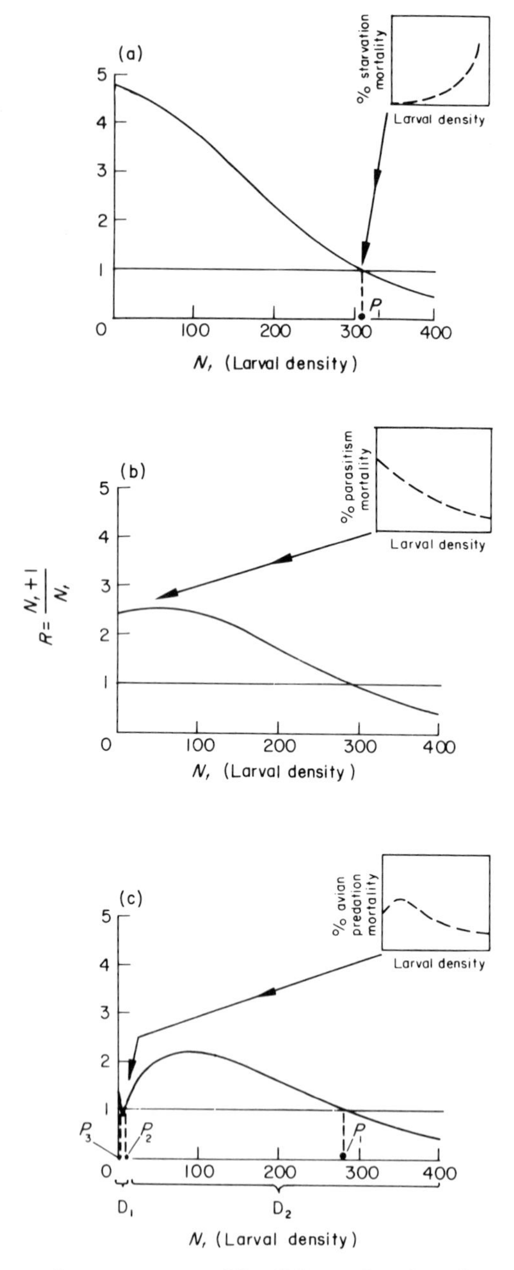


FIG. 15.1. The recruitment rate curve ($N_{t+1}/N_t$) as a function of $N_t$ for spruce budworm (units of large larvae per square metre). (a) shows the effect of starvation mortality, (b) the additional effect of larval parasites and (c) incorporates the additional process of avian predation on larvae.

from $P_2$. The low- and high-density equilibria, $P_1$ and $P_3$, are stable points or attractors by analogous arguments. Densities slightly larger than either will tend to a decreased population in the next generation since $R < 1$, and slightly smaller densities will increase it because $R > 1$. Thus, each stable point creates its own domain of attraction and the two domains are separated by a boundary population density, $P_2$. The position of this boundary is of great importance to the behaviour of the system.

However, the host forest is not static, and as branch density changes and the forest matures budworm recruitment rate increases for all insect densities. A family of recruitment curves is shown in Figure 15.2a; each reflects the effect of a different level of forest maturity or branch density. Note that the sizes of the domains of attraction change as do the locations of the boundary between them.

A more succinct way to represent these changing boundaries and domains is to plot the equilibrium points for each forest condition, thus creating a manifold (Fig. 15.2b) (see Jones 1975; Holling 1978 for further details). These equilibria are never precisely attained, but their location dictates the direction and rate of change of the system. Just as an animal's skeleton determines much of the body's appearance, the structure of the equilibrium states organizes the system's dynamic behaviour. The manifold of Figure 15.2b is replotted in Figure 15.3 and it illustrates how the dynamics of the budworm system can be interpreted in terms of the equilibria. An immature forest (low $F$) can support only an endemic insect population since only the lower attractor domain exists for forest conditions less than $F_1$. As the forest matures, the insect population density tends to stay low since it is still in the lower attraction domain. Forest maturity greater than $F_2$ leads to a rapid population explosion since past that point only the upper attractor exists (as in the top two curves of Fig. 15.2b).

The upper equilibrium is stable for the insect but not for the forest. After a few years of outbreak densities, the forest deteriorates and its effective maturity, $F$, decreases. The arrows at the top of the graph show that insect populations decrease as $F$ decreases, until once again only the endemic equilibrium exists.

Note that incremental changes in forest maturity have no appreciable effect on insect numbers until one final increment causes maturity to become greater than $F_2$, and an outbreak results. The manifold also 'explains' a frequent observation. If an outbreak population at $C$ is reduced to an endemic density at $S$, it explodes again when insecticide spraying is stopped because only the upper equilibrium exists under that forest condition. What used to be a stable, low endemic population density is no longer stable because the boundary between the domains is eliminated at that forest maturity.

Finally, Figure 15.3 shows that immigration of insects can initiate an outbreak, even in only moderately mature forests, by boosting insect density above the boundary density (dashed line) that separates the two attractor

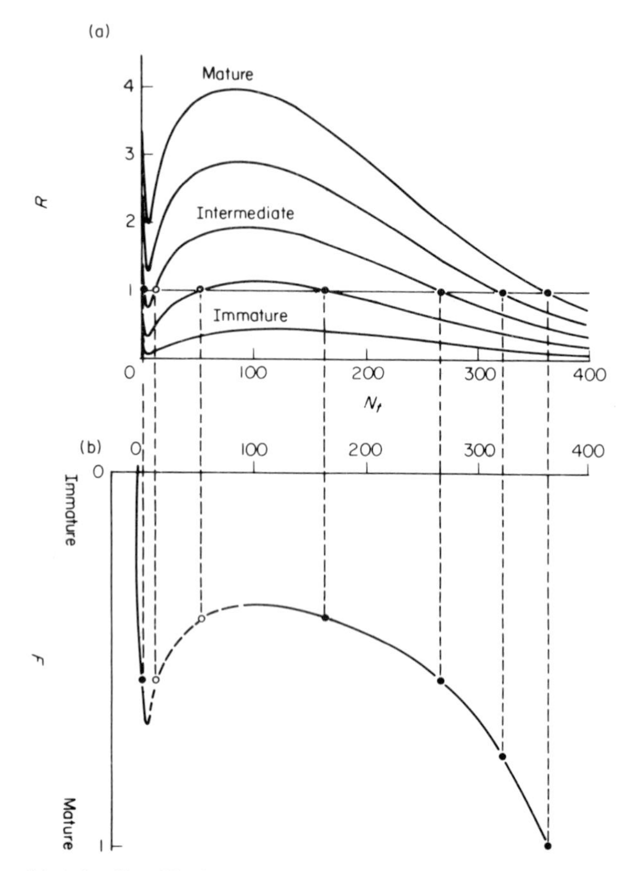


FIG. 15.2. (a) A family of budworm recruitment curves for different levels of forest maturity as measured by branch density. The stable (solid circle) and unstable (open circle) points or equilibrium crossovers of each of these curves are projected onto the axes in (b) to generate a manifold of equilibrium points. (b) should be viewed sideways, with *F*, or forest maturity, as the *X*-axis and $N_t$, or the potential equilibrium insect population density, as *Y*. The dashed line on the underside of the manifold represents the locations of the boundary population size which separates the upper and lower domains.

domains (see arrow to point I). Note how the required pulse of immigrants diminishes as the forest ages.

However, boundaries can move for reasons besides changing forest conditions. A family of manifolds is shown for varying intensities of avian predation in Figure 15.4. Note that the location of the boundary and depth of the 'predation pit' change. Such effects could be a direct consequence of management if insecticides, as they do, affect bird populations and behaviour. Thus, the concern about insecticidal effects on birds should be not only for the sake of

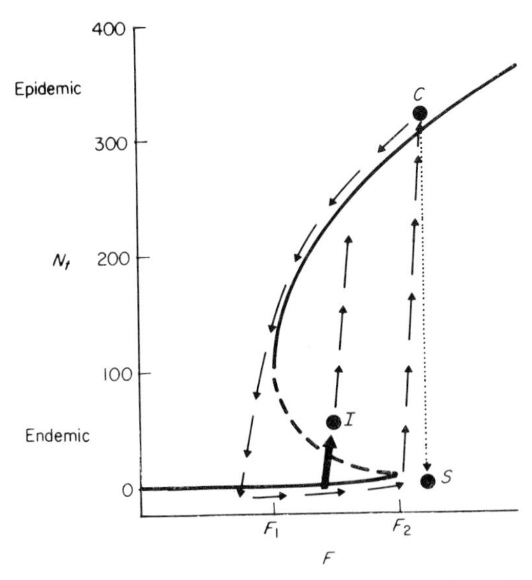


FIG. 15.3. The manifold of Figure 15.2b turned sideways and showing the normal oscillatory pattern of the spruce budworm system (outside loop of dashed arrows). As the forest changes in maturity, *F*, the insect population tends to be drawn towards the stable equilibrium (solid manifold lines) of the resident attractor domain. See text. If an outbreak population at point C is reduced by insecticide to point S, the population will be drawn back towards the upper equilibrium when spraying is stopped. Immigration of insects can move the system across the boundary (dashed line) to point I which is in the upper attractor domain, and an outbreak results.

the birds but also for the importance of birds in determining budworm dynamics.

### *Improvement of our understanding*

Several benefits have resulted from focusing on the boundary and its associated domains of attraction. We can now explain how forest maturation can inevitably trigger an outbreak, and why control measures applied to outbreak populations are locked into a Sysiphus-like fate. The role of insect dispersal is also clarified.

The effects of normal weather variation can also be represented by a family of recruitment curves, though over a much more limited range than in Figure 15.2a. These only result in minor variants on the manifold already shown, and the net effect of weather is to delay outbreaks or initiate them slightly sooner than expected.

The vital role of avian predators emerges from this study, since they are largely responsible for the lower stable attractor. The effect of new control

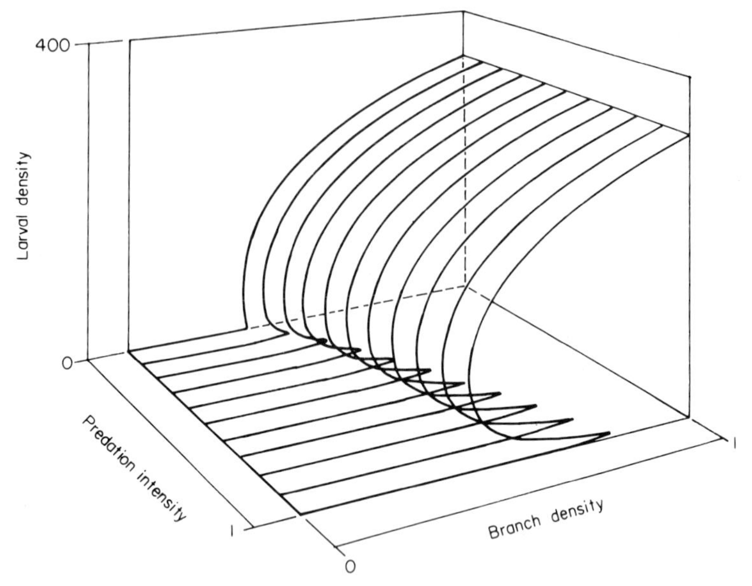


FIG. 15.4. Modifying the intensity of avian predation produces a family of manifolds with different shapes. Predation intensity of 1 is the level observed in nature; zero is no predation.

measures which provide a modest increase in mortality at *low* budworm density can be quantified (Watt 1964; Winkler 1975). Such measures deepen the 'predator pit' and move the boundary upward. Explorations of such an agent (virus) in the simulation model show dramatic improvement in various management indicators (Clark, Jones & Holling 1979).

Another management scheme is suggested by our understanding of the manifold. Particular kinds of logging strategies can reduce the effective forest maturity which helps maintain the lower stability domain and the potential for natural control of budworm populations (Fig. 15.3). Recognition of this effect of harvesting strategies has resulted in design of new management policies which explicitly incorporate forest manipulation into budworm control methods (Winkler 1975; Peterman 1977b; Walters & Hilborn 1978; Clark, Jones & Holling 1979).

## PACIFIC SALMON

The population dynamics of Pacific salmon (*Oncorhynchus* spp.) have been studied intensively for decades, largely because of the economic importance of these fish. As yet no single population has been studied sufficiently to provide as deep an insight into system components as we have for the spruce budworm. However, the component processes such as freshwater growth and survival, predation, marine survival and fishing mortality have been studied for different systems and they permit a synthesis into an overall picture for a representative salmon stock. A fuller discussion of some of these component processes is given in Peterman (1977a).

A recruitment rate curve is produced from a simulation model of this salmon population in exactly the same way as for the budworm. Figure 15.5a shows that an upper equilibrium is established at $P_1$ due to competition for limited resources among the offspring. Predation by larger fish is very intense at low prey abundances and in some cases appears to be due to a type III functional response (Peterman & Gatto 1978). Adding this predation process creates a system with two stable equilibria and their associated attraction domains, as well as the boundary separating them (Fig. 15.5b). This qualitative effect of predation was suspected by Neave (1953) and Ricker (1954). If we for the moment assume that harvesting by man can be set at a constant rate, independent of fish density, then we generate a family of recruitment rate curves (Fig. 15.5c). The Skeena River sockeye salmon system from 1920 to 1950 demonstrated such a constant harvest rate (Larkin & MacDonald 1968). Just as with the budworm, the crossover points or equilibria from Figure 15.5c are projected into a manifold which describes the location of all equilibria as a function of exploitation rate (Fig. 15.6). Note again that domain sizes and the boundary location change with exploitation rate.

Salmon with different fecundities generate slightly differently-shaped manifolds and at high fecundities begin to reduce the severe effect of predation (Fig. 15.7). This figure also shows that classically-defined maximum-sustainable-yield (MSY) harvest rates are dangerously close to exploitation rates that cause population collapse.

Let us now remove the assumption of fixed exploitation and instead assume that the fishing fleet is dynamic. Recent analyses of commercial trollers on British Columbia chinook salmon show functional and numerical (within-season aggregation) responses remarkably similar to those of natural predators (Fig. 15.8). Management regulations clearly put bounds on the influence of such responses, but we can at least conclude, on the basis of past observations, that the effect of fishing can be to deepen and broaden the 'pit' caused by predation, which in turn enlarges the lower stability domain. There are examples of this mortality effect in both commercial and native Indian food fisheries (Larkin & MacDonald 1968; Ward & Larkin 1964).

Some of our colleagues are at present investigating longer term processes that are equivalent to predator–reproductive numerical responses that result in changes in salmon fishing efficiency, fleet size and composition, and other fleet characteristics. While we cannot yet quantitatively evaluate the influence of these longer term processes, it is fairly clear that they can result in moving

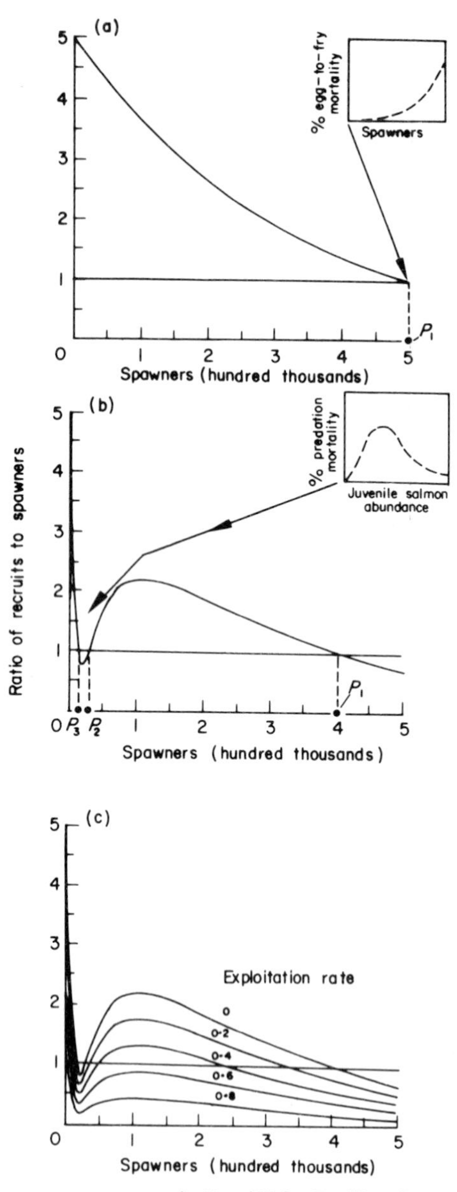


FIG. 15.5. Recruits per spawner equals $N_{t+1}/N_t$ for Pacific salmon. Recruitment rate curves are shown as each process is added to the model. (a) Resource limitation on egg–fry life stages creates an upper stable equilibrium, $P_1$. (b) Adding Type III predation mortality creates a lower domain of attraction. (c) A constant exploitation rate on adults generates a family of recruitment curves for different harvest rates.

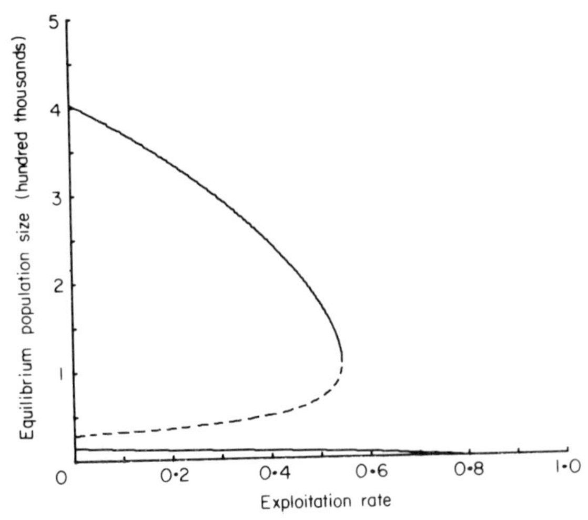


FIG. 15.6. By plotting the equilibrium crossovers of the family of curves shown in Figure 15.5(c) we obtain, as in Figure 15.2, an equilibrium manifold which shows the potential equilibrium population sizes for given *sustainable* exploitation rates.

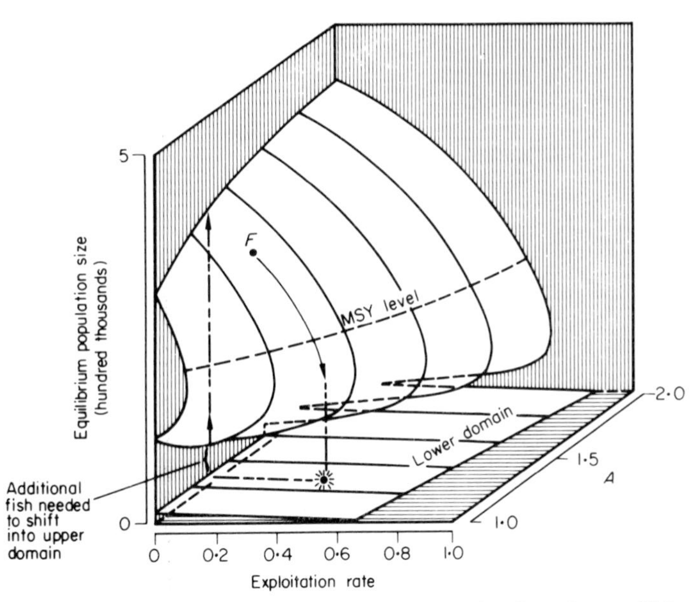


FIG. 15.7. Salmon stocks with different fecundities ('*A* values' on the graph) have distinct manifolds and depth of 'predation pits'. A system beginning at point F can be overharvested if exploitation rate becomes too large, and the system drops into the lower domain of attraction. The population cannot move back into the upper domain even if the harvest rate is reduced to zero. To do this, enhancement must supply the number of fish indicated in the lower left portion of the figure. (Reproduced, with permission, from Peterman 1977a.)

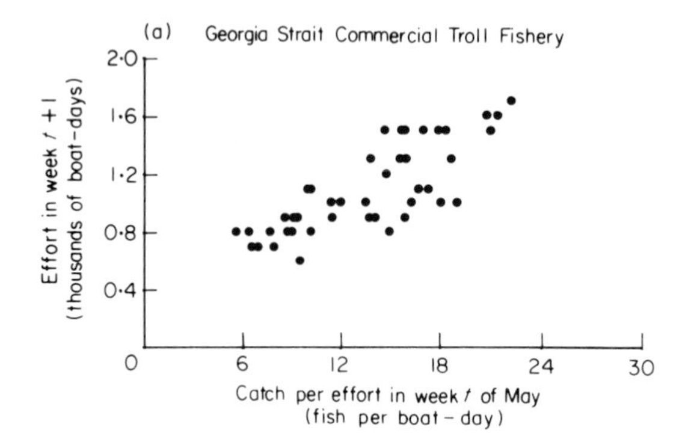


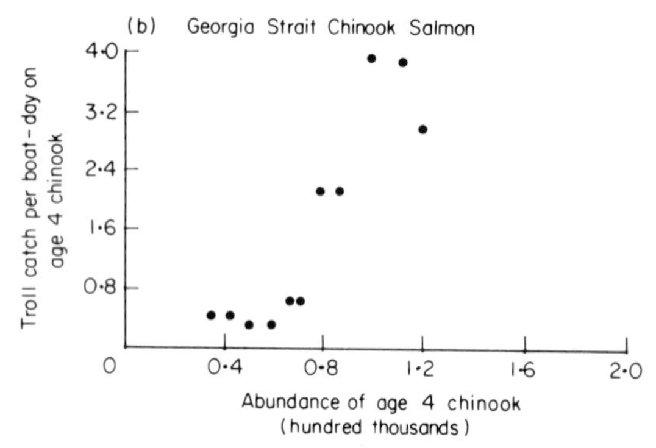


FIG. 15.8. Salmon fishing fleets show aggregation (a) and functional responses (b) similar to natural predators. In (a) high catch per boat-day results in higher effort in the following week. (a) covers ten years of data and (b) one year. (Data courtesy of A.W. Argue, Fisheries and Marine Service, Vancouver.)

boundaries and shifting attractors (Jones & Walters 1976). A fish population that could sustain exploitation by a fishing fleet of a given size and efficiency may no longer be able to sustain the higher exploitation rate of a much more efficient fleet of the same size.

### *Improvements in understanding*

The preceding view of salmon populations explains why some salmon stocks, after collapse, have not recovered although exploitation has been relaxed (Holling 1973; Peterman 1977a). The population is stuck in the lower domain of attraction and cannot increase until it crosses above the boundary population size. Such models can quantify just how many fish would need to be added to push the population into the upper domain (see Fig. 15.7). In another situation, where neither hatcheries nor reduction of exploitation rate is practical, the manifold analysis can quantify how much predation pressure would need to be eliminated in order to result in the same population jump.

Our focus on boundaries furthermore confirms a previous interpretation of the failures of several past enhancement attempts to recover fish populations that were trapped at small sizes. Perhaps they simply were not boosted high enough to overcome the effect of predators (Loftus 1976).

Finally, the dramatic and rapid collapse of many fish stocks can be interpreted as the system crossing past the point where there is an upper equilibrium and attractor region (path from point $F$ in Fig. 15.7).

## SIMILARITIES AND SHIFTING BOUNDARIES

### *Process views*

The budworm and salmon systems have a striking similarity in their stability and resilience characteristics as reflected by the shapes of their recruitment curves and manifolds. Yet these were built up in each case from measurements of component processes. Both systems have two domains of attraction separated by a boundary, and both cases demonstrate the significance of certain biological processes in creating those domains. In budworm the strength and location of the low-density stability domain is determined largely by avian predation on older larval instars. In salmon the same effect is caused by predation on juveniles. Similarly, intraspecific competition in early life stages shapes the high-density stability domain for both systems. Unlike budworm, however, salmon do not 'destroy' their habitat at high densities.

Further comparisons between budworm and salmon focus on the *dynamics* of the basic equilibrium manifold structure. In both cases factors governing appearance, disappearance and shifting of equilibria are critical determinants of the system's natural and managed behaviours. In budworm the parameters of avian predation intensity modify the manifold, as does the internal system state variable of forest defoliation. The exogenous driving variable of weather also moves the boundary location. In salmon we have the same three classes of causes of changing topology: stock productivity (a parameter), fishing fleet size or efficiency (a state variable) and environmental noise in marine survival rate (an exogenous variable).

### *Sensitivity analysis and experimental design*

Manifolds are a means to compress complex ecosystem models into a comprehensible form. Because our analysis concentrates on the structure of the manifold *per se*, a much smaller number of parameters needs to be explored than in classical approaches to sensitivity analysis. Also, examination of changes in manifold structure provides a more concise measure of the effect of parameter changes than temporal or phase space behaviour.

This approach has proved to be a more effective way to achieve two particular aims of sensitivity analysis: to create realistic management schemes and to design more carefully experiments that test our understanding as it is encapsulated in the model (Peterman 1977a; Holling 1978; Clark, Jones & Holling 1979). We have already discussed several management issues in previous sections. Exploration of manifolds, which are based on detailed understanding of processes, results in specific statements of conditions for diagnostic experiments. For example, a critical test of the importance of avian predators at low budworm density is to do a bird shoot-out under particular forest and insect population conditions. This would have the effect of removing the 'predator pit' and causing a local outbreak. Just such an experiment has been done, with predicted results, for testing the importance of birds and small mammals in controlling gypsy moth populations (Campbell & Sloan 1977). A similar experiment could be done on salmon by removing a specified proportion of the freshwater predators of a stock which is trapped in the lower domain. A rapid population growth should result (Larkin 1974; Loftus 1976).

## ADAPTIVE MANAGEMENT

It is no coincidence that most existing examples of ecological systems that show multiple domains of stability come from entomology, fisheries, grazing systems and epidemiology (Noy-Meir 1975; May 1977). Large oscillations in some insect and disease populations occur naturally and attract wide attention and study because of the damage they do. Also, many fisheries have experienced a wide range of population conditions because of intensive harvesting. In both the natural and managed cases population sizes have varied over a wide enough range to give us some understanding of the shapes of relationships between system components, which leads to a comprehension of stability domain characteristics as shown here.

It may be time, therefore, to start viewing management manipulations seriously as experimental perturbations (Larkin 1974) that can be done in prescribed ways to provide information on existence and locations of boundaries and shapes of manifolds. For example, a deliberate overharvesting of small stocks may lead to a 'mapping out' of the manifold for a larger system because this permits us to learn where boundaries are (see Jones & Walters 1976; Peterman 1977a). This fits under the general title of 'adaptive management procedures' that aim to provide information as well as yield (or whatever is being obtained from the system). Adaptive management concepts are more fully explored in the two papers cited above, and as well in Walters & Hilborn (1976) and Holling (1978).

## SPECULATIONS ON THE MYTHOLOGY

There is now enough accumulated evidence to suggest that many ecological systems have multiple equilibrium states (Holling 1973; Southwood & Comins 1976; May 1977). However much that point may have been argued in the past, it certainly has been implicitly recognized in applied environmental management. The efforts to set environmental and health standards, for example, are based on the desire to keep variables well away from some boundary that separates a desired state from one that is qualitatively different and undesirable. The efforts to keep species from even local extinction are equally in harmony with the view that boundaries separate extinction domains from other stability domains. These are efforts to keep the unexpected at bay.

However, we see from the two case studies described earlier that the environmental manager's job is made more difficult because boundaries that define stability regions move in relation to the values of certain state variables. When the forest is young, there is a defined lower stability region for budworm, but as the forest grows this becomes smaller and smaller until it vanishes altogether. Then populations move to the upper attractor.

Yet another feature of ecological systems is natural movement of variables, such as populations, from one domain to another. Depleted fish populations can recover because of favourable environmental conditions, even after many generations in a lower attractor domain. Colonizing species of the '*r*-selected' type may even frequently journey into extinction regions, only to be rescued by events such as dispersal (Southwood & Comins 1976; Southwood 1977).

Thus, the variables of natural systems are *not* typically frozen into one stability region, and yet many management approaches such as selective breeding, insect pest control and setting health standards attempt to achieve just that. This raises the question: what do these variability-reducing approaches do to the biological system?

That question becomes all the more important because the stability notion of moving boundaries described above has yet another dimension of change: evolution. Certainly the position of boundaries can move or even disappear as variables change, but the positions of those boundaries are set by the parameter values (e.g. intensity of fish predation). Parameter values, fixed only for convenience in our models, are in fact the result of genetic selection and perhaps, as well, of other non-genetic mechanisms that retain a memory of the past. In either case the parameters have the values they do in nature because of a balance achieved as a consequence of the past history of experience and of variability (Gilbert *et al.* 1976). A management activity that modifies the selection pressures can therefore result in parameter changes which in turn produce boundary movement. The danger is that, because of this link between natural selection, parameter values and boundary locations, the system may evolve in response to management regimes so that boundaries move to meet the present system state and not the reverse. All management efforts to avoid the boundary might then be subverted.

Hence our concept of dynamic attractors emphasizes the following features. First, there can be multiple stability regions. Second, as both variables and parameters change, these stability regions change their size and shape. Third, apparently fixed parameter values are the result of natural selection. Fourth, those selection pressures can shift if the system is moved to a new part of state space (new range of population densities, new mortality agents, etc.) and boundaries can move.

In our view of stability, resilience is defined not simply as an ability to absorb unexpected events and variability but, in addition, to *benefit* from such changes.

How extensive is the evidence for these four properties? In searching the literature for indications of the importance or not of our concept of shifting domains, we need answers to four questions.

1 Is there evidence for more than one equilibrium; or, better, is there evidence for the existence of a boundary separating two qualitatively different states (such as point $P_2$ in the recruitment curve of Figure 15.5b)?

2 Is there evidence that identifies the processes responsible for the boundary and the features of the functional relations that produce the boundary (e.g. depensatory mortality)?

3 Is there evidence that changes in the form or magnitude of those relations (e.g. the parameters of the function) will move the position of the boundaries?

4 Is there evidence that moving or restricting a system to a different region of state space (say by limiting variability of population density) leads to evolutionary change in these parameters and, as a consequence, in movements of the boundary?

We have been unsuccessful in finding even one case that provides convincing answers to all four questions. That means either that this concept of shifting domains is a figment of our imaginations or that ecologists have not had this concept in mind and so have not posed the questions. We are naturally inclined to the latter possibility. Relevant examples are available, however, that independently confirm the premise in each of the four questions separately. To assess the first two questions, for example, there are several cases where insecticide application destroyed the parasites or predators of a formerly rare insect which then became a pest species. One such group of 'secondary' pests is the *Lecanium* scale insects on English walnut in California (Michelbacher 1962). One paper (Messenger, Biliotti & van der Bosch 1976) even states that the controlling effect of the parasites may have gone unnoticed had it not been for the insecticide application, which provides an excellent example of 'unplanned' adaptive management. In another case the introduction of a predator of the cottony cushion scale was quite successful, but when DDT came into use the predator was more severely affected and the scale again became a serious pest (Metcalf 1975). Apparently, in these cases removal of the parasites or predators also removed the lower attractor domain for the host species, leaving only the upper attractor (Question 1). Moreover, the form of their response is known to have the features required to create a lower equilibrium (Question 2).

The third and fourth questions concerning the effects of parameters on stability boundaries and of changes in selection pressures on those parameters find some support from a few quantified studies of co-evolution. For example, the parasite *Mesoleius tenthredinis* was introduced into Canada to control the larch sawfly and was initially reported to be highly successful in reducing outbreak populations (Muldrew 1953). However, within 20 to 30 years it was found that the host sawfly had developed the ability to encapsulate the parasite eggs (Muldrew 1953) and the host population again reverted to pre-parasite levels (Turnock 1972). This appears to be an example of moving boundaries because the changing degree of presence and effectiveness of a natural enemy changed the qualitative state of the pest population. Another example of moving boundaries comes from the myxomatosis–rabbit system in Australia. Rabbit abundance and rate of virus infection changed as the virus and its host underwent a series of evolutionary adjustments (Fenner & Ratcliffe 1965). Finally, Pimentel's genetic feedback work provides yet another example of evolutionary adjustment that apparently changed the size, shape and location of the attractor domains. Experiments with the housefly and a parasite showed that, after a number of generations, the host became more resistant to the parasite, the parasite became less virulent and the population fluctuations were much less violent (Pimentel, Nagel & Madden 1963; Pimentel & Stone 1968).

Our concept of shifting stability characteristics is of more than just theoretical interest; it has potential consequences for the management of renewable resources profound enough that the four questions deserve intensive exploration. This concept suggests, for example, that environmental standards should not be fixed and immutable, but that variables should occasionally be allowed to exceed limits so long as recovery mechanisms are encouraged (Fiering & Holling 1974; Burton, Kates & White 1977; Clark 1978). It suggests that fisheries' enhancement should not concentrate on enhancing all less fecund stocks to the same level of productivity. In that event, if overharvesting occurs, all stocks might be involved. Rather, some less productive local stocks could be left to their own devices and monitored to provide 'early-warning' signals of more extensive problems. Again, investment in procedures to recover such stocks would be necessary for the maintenance of an adaptive response (Peterman 1977a). Finally, the concept suggests that, rather than attempting to suppress forest insect pests uniformly, managers should use such insects as

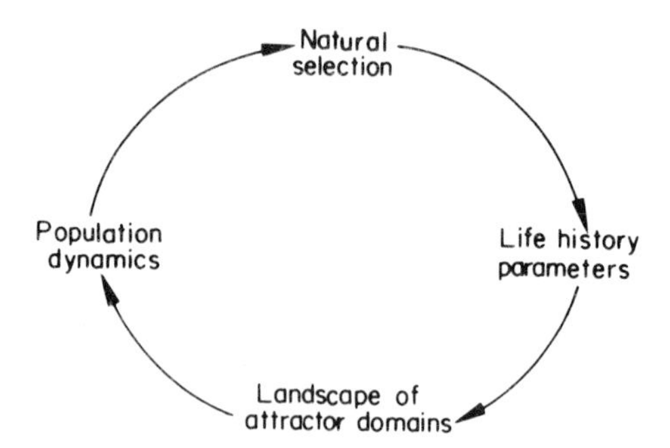


FIG. 15.9. Population changes can modify selection pressures, which can change life-history parameters, which in turn determine shapes of attractor domains. Shifts in the landscape of attractors modify the dynamics of the population, which again affect selection pressures.

part of the forest renewal mechanism (Clark, Jones & Holling 1979; Peterman 1978).

This dynamic resilience concept turns variability and the unexpected from a threat to a potential benefit. This compares with the static view of a fixed stability landscape that leads to traditional approaches aimed at tight control. Quantification of boundary movements and of shifting topology as discussed here can add another dimension to the methods and experience of researchers and managers.

We have discussed how various management manipulations can alter the landscape of attractor domains through the causal links between selection pressures, life-history parameters and stability characteristics. From this foundation it is easy to imagine that similar changes could occur in the absence of any management intervention. In different parts of their journeys across the stability landscape, populations experience different selection pressures, and the resultant changes in competitive ability, fecundity and other life-history parameters can cause a change in that landscape (Fig. 15.9). This reciprocal effect between populations and shapes of their stability domains should provide a focus for future complementary research in population genetics and population dynamics.

## ACKNOWLEDGMENTS

We sincerely thank our colleagues at the Institute of Animal Resource Ecology for helpful discussions and four readers, P.A. Larkin, J. Myers, M. Jones and J. Anderson, who provided valuable suggestions on the manuscript. We particularly thank Joan Anderson and Wendy Courtice for their efficient typing of the drafts of the manuscript and Ulrike Hilborn for drawing the original figures.

## REFERENCES


**Austin M.P. & Cook B.G. (1974)** Ecosystem stability: a result from an abstract simulation. *Journal o) Theoretical Biology*, **45**, 435–458.

**Baskerville G.L. (Ed.) (1976)** *Report of the Task Force for Evaluation of Budworm Control Alternatives*. Department of Natural Resources, Federicton, New Brunswick, Canada.

**Bazykin A. (1974)** Volterra's system and the Michaelis–Menten equation (in Russian). *Problems in Mathematical Genetics* pp. 103–143. U.S.S.R. Academy of Sciences. Novosibirsk. (Available in English as Structural and dynamic stability of model predator–prey systems. 1976, IIASA RM-76-8.)

**Burton I., Kates R.W. & White G.F. (1977)** *The Environment as Hazard*. Oxford University Press, New York.

**Campbell R.W. & Sloan R.J. (1977)** Natural regulation of innocuous gypsy moth populations. *Environmental Entomology*, **6** (2), 315–322.

**Clark W.C. (1978)** Managing the unknown. *Managing Technological Hazard: Research Needs and Opportunities* (Ed. by R.W. Kates) pp. 109–142. Institute of Behavioral Science, Boulder, Colorado.

**Clark W.C., Jones D.D. & Holling C.S. (1978)** Patches, movements, and population dynamics in ecological systems: a terrestrial perspective. *Spatial Pattern in Plankton Communities* (Ed. by J.H. Steele) pp. 385–432. Plenum Press, New York & London (in press).

**Clark W.C., Jones D.D. & Holling C.S. (1979)** Lessons for ecological policy design: a case study of ecosystem management. *Ecological Modelling* (in press).

**Davidson A.G. & Prentice R.M. (Eds) (1967)** *Important forest insects and diseases of mutual concern to Canada, the United States and Mexico*. Department of Forestry and Rural Development, Canada.

**Fenner F. & Ratcliffe F.N. (1965)** *Myxomatosis*. Cambridge University Press, Cambridge.

**Fiering M.B. & Holling C.S. (1974)** Management and standards for perturbed ecosystems. *Agro-Ecosystems*, **1**, 301–321.

**Gilbert N., Gutierrez B.D., Fraser B.D. & Jones R.E. (1976)** *Ecological Relationships*. W.H. Freeman, Reading.

**Gomez-Pompa A., Vazquez-Yanes C. & Guevara S. (1972)** The tropical rain forest: a non-renewable resource. *Science*, **177**, 762–765.

**Holling C.S. (1959)** The components of predation as revealed by a study of small mammal predation of the European pine sawfly. *Canadian Entomologist*, **91** (5), 293–320.

**Holling C.S. (1973)** Resilience and stability of ecological systems. *Annual Review of Ecology and Systematics*, **4**, 1–23.

**Holling C.S. (Ed.) (1978)** *Adaptive Environmental Assessment and Management*. John Wiley & Sons, London.

**Holling C.S., Jones D.D. & Clark W.C. (1976)** Ecological policy design: a case study of forest and pest management. *International Institute of Applied Systems Analysis Conference*, **1**, 139–158.

**Jones D.D. (1975)** The application of catastrophe theory to ecological systems. *New Directions in the Analysis of Ecological Systems, Part 2* (Ed. by G.S. Innes) pp. 133–148. *Simulation Councils, Inc., La Jolla*. [Reprinted in *Simulation*, **29** (1), 1–15, 1977.]

**Jones D.D. (1977)** The budworm site model. *Proceedings of the Conference on Pest Management* (Ed. by G.A. Norton & C.S. Holling) pp. 91–155. International Institute of Applied Systems Analysis, Laxenburg, Austria.

**Jones D.D. & Walters C.J. (1976)** Catastrophe theory and fisheries regulation. *Journal of the Fisheries Research Board of Canada*, **33** (12), 2829–2833.

**Larkin P.A. (1974)** Play it again Sam—an essay on salmon enhancement. *Journal of the Fisheries Research Board of Canada*, **31**, 1433–1459.

**Larkin P.A. & MacDonald J.G. (1968)** Factors in the population biology of the sockeye salmon of the Skeena River. *Journal of Animal Ecology*, **37**, 229–258.

**Levin S.A. (1978)** Pattern formation in ecological communities. *Spatial Pattern in Plankton Communities* (Ed. by J.H. Steele) pp. 433–465. Plenum Press, New York & London.

**Loftus K.H. (1976)** Science for Canada's fisheries rehabilitation needs. *Journal of the Fisheries Research Board of Canada*, **33**, 1822–1857.

**McLeod J. (1978)** Discontinuous stability in a sawfly life system and its relevance to pest management strategies. *Selected Papers in Forest Entomology from XV International Entomological Congress, Washington, D.C.* (Ed. by D.L. Wood) pp. 71–84. U.S. Forest Service.

**May R.M. (1977)** Thresholds and breakpoints in ecosystems with a multiplicity of stable states. *Nature, London*, **269** (5628), 471–477.

**Messenger P.S., Biliotti E. & van der Bosch R. (1976)** The importance of natural enemies in integrated control. *Theory and Practice of Biological Control* (Ed. by C.B. Huffaker & P.S. Messenger) pp. 543–563. Academic Press, London.

**Metcalf R.L. (1975)** Insecticides in pest management. *Introduction to Insect Pest Management* (Ed. by R.L. Metcalf & W.H. Luckmann) 242 pp. John Wiley & Sons, New York.

**Michelbacher A.E. (1962)** Influence of natural factors on insect spider mite populations. *Proceedings of the 11th International Entomological Congress, Vienna*, **2**, 694.

**Miller C.A. & Ketella E.G. (1975)** Aerial control operations against the spruce budworm in New Brunswick, 1952–1973. *Aerial Control of Forest Insects in Canada* (Ed. by M.L. Prebble) pp. 94–112. Department of the Environment, Ottawa.

**Morris R.F. (Ed.) (1963)** The dynamics of epidemic spruce budworm populations. *Memoirs of the Entomological Society of Canada No. 31.*

**Muldrew J.A. (1953)** The natural immunity of the larch sawfly (*Pristiphora erichsonii*) to the introduced parasite *Mesoleius tenthredinis* in Manitoba and Saskatchewan. *Canadian Journal of Zoology*, **31**, 313–332.

**Neave F. (1953)** Principles affecting the size of pink and chum salmon populations in British Columbia. *Journal of the Fisheries Research Board of Canada*, **9**, 450–491.

**Noy-Meir I. (1975)** Stability of grazing systems: an application of predator–prey graphs. *Journal of Ecology*, **63**, 459–481.

**Patten B.C. (1975)** Ecosystem linearization: an evolutionary design problem. *American Naturalist*, **109** (969), 529–539.

**Paulik G. (1973)** Studies of the possible form of the stock and recruitment curve. *Fish Stocks and Recruitment* (Ed. by B.B. Parrish) pp. 302–315. *Rapports et Proces-Verbaux des Reunions, Conseil International pour l'Exploration de la Mer*, **164**.

**Peterman R.M. (1977a)** A simple mechanism that causes collapsing stability regions in exploited salmonid populations. *Journal of the Fisheries Research Board of Canada*, **34** (8), 1130–1142.

**Peterman R.M. (1977b)** Graphical evaluation of environmental management options: examples from a forest-insect pest system. *Ecological Modelling*, **3**, 133–148.

**Peterman R.M. (1978)** The ecological role of mountain pine beetle in lodgepole pine forests. *Proceedings of the Symposium on Mountain Pine Beetle Management in Lodgepole Pine Forests* (Ed. by R. Stark, A.A. Berryman & G.D. Amman) Washington State University Coop. Exten. Service (in press).

**Peterman R.M. & Gatto M. (1978)** Estimation of functional responses of predators on juvenile salmon. *Journal of the Fisheries Research Board of Canada*, **35**, 797–808.

**Pimentel D., Nagel W.P. & Madden J.L. (1963)** Space-time structure of the environment and the survival of parasite-host systems. *American Naturalist*, **97**, 141–167.

**Pimentel D. & Stone F.A. (1968)** Evolution and population ecology of parasite–host systems. *Canadian Entomologist*, **100**, 655–662.

**Ricker W.E. (1954)** Stock and recruitment. *Journal of the Fisheries Research Board of Canada*, **11**, 559–623.

**Southwood T.R.E. (1977)** The relevance of population dynamic theory to pest status. *Origins of Pest, Parasite, Disease and Weed Problems* (Ed. by J.M. Cherrett & G.R. Sagar) pp. 35–54. 18th Symposium of the British Ecological Society. Blackwell Scientific Publications, Oxford.

**Southwood T.R.E. & Comins H.N. (1976)** A synoptic population model. *Journal of Animal Ecology*, **45**, 949–965.

**Steele J.H. & Frost B.W. (1977)** The structure of plankton communities. *Philosophical Transactions of the Royal Society of London, Series B*, **280**, 485–534.

**Sutherland J.P. (1974)** Multiple stable points in natural communities. *American Naturalist*, **108**, 859–873.

**Takahashi F. (1964)** Reproduction curve with two equilibrium points: a consideration on the fluctuation of insect population. *Researches on Population Ecology*, **6**, 28–36.

**Turnock W.J. (1972)** Geographical and historical variability in population patterns and life systems of the larch sawfly (*Hymenoptera: Tenthredinidae*). *Canadian Entomologist*, **104**, 1883–1900.

**Walters C.J. & Hilborn R. (1976)** Adaptive control of fishing systems. *Journal of the Fisheries Research Board of Canada*, **33**, 145–159.

**Walters C.J. & Hilborn R. (1978)** Ecological optimization and adaptive management. *Annual Review of Ecology and Systematics*, **9**, 157–189.

**Ward F. & Larkin P.A. (1964)** Cyclic dominance in Adams River sockeye salmon. *Progress Report of the Int. Pacific Salmon Fish. Comm.*, **11**, 1–116.

**Watt K.E.F. (1964)** The use of mathematics and computers to determine the optimal strategy and tactics for a given pest control problem. *Canadian Entomologist*, **96**, 202–220.

**Winkler C. (1975)** *An optimization technique for the budworm forest-pest model*. International Institute of Applied Systems Analysis, Laxenburg, Austria, RM-75-11.

# 16. INTERACTIONS IN MARINE ECOSYSTEMS

JOHN H. STEELE

*Woods Hole Oceanographic Institute,*
*Woods Hole, Massachusetts 02543, U.S.A.*

## INTRODUCTION

The study of interactions within an ecosystem depends on knowledge of the external factors affecting or driving the system. In any system the structure and variability of the physical environment is likely to be crucial. This is certainly true for marine systems.

In the oceans, because of technical and theoretical limitations, earlier work portrayed a relatively smooth environment or tended to study relatively large-scale features culminating in theories of the westward intensification of currents such as the Gulf Stream (Stommel 1948). To quantify these theories required knowledge of how energy derived from wind stress at the sea surface was transferred at the largest scales. As a result of developments in instrumentation, it became apparent that most of the kinetic energy in the oceans occurred at intermediate scales in eddy-like motions with sizes of a few hundred kilometres. Thus, attention has shifted from the apparent uniformity of large-scale currents to the highly energetic but highly variable motions on smaller scales and, especially, to the problems of exchange of energy between different scales of motion. These problems are similar to those in meteorology (Somerville 1977), where the relatively rapid transfer of perturbations from the very small to intermediate scales set limits to weather prediction in space and time, even though the knowledge of these processes permits some prediction of the statistical properties of meteorological systems. This growth of small-scale 'errors' is a fundamental non-linear property of turbulent flows (Lorenz 1969). Until these smaller-scale processes are included in larger-scale models, 'we can have little confidence that these models (for oceans and atmosphere) will be useful for prediction purposes, for climate studies or for biological processes' (Holland 1977).

These aspects of physical oceanography are directly relevant to the study of marine ecosystems. They also provide, by analogy, an indication of the problems facing the study of any whole ecosystem where the non-linearities of interactions in space and time are likely to be comparable with, and include as a component, those in the physical environment. Probably it is for this reason that, following the direction of physical studies, there has been a shift from large-scale biome programmes to smaller-scale research studies. There has also been a corresponding shift in emphasis from concepts which consider average populations and their rates of change to ideas derived from the spatial and temporal variability (Platt & Denman 1975).

The earlier studies were based on general ideas of energy flow through food webs (Lindemann 1942). This paradigm tended to ignore details of species composition, spatial variations and, often, seasonal or shorter-term changes. In the marine field it has been used in attempts to predict, for example, potential fish yields in the world oceans (Ryther 1969) or in parts of the sea (Steele 1965; Gulland 1971). It has spawned several minor industries such as energy transfer measurement. Energy flow diagrams can be regarded as descriptions of certain features of ecosystems rather than explanations of the underlying dynamics. The use of the energy flow concept, however, usually required the assumption of a world relatively uniform in space and time. These requirements of energy flow ideas might be considered to have led to the emphasis on the rather ill-fated and semantically ambiguous concept of 'stability'. Although much of the criticism of this latter concept has been in terms of its theoretical naivety, the main objection is that the underlying idea of a 'steady state' has little relevance to much of the observation of the real world. This has become apparent, especially in the oceans, once large sets of data were available through increased sampling or improved technology.

Thus, at present, a study of interactions in marine ecosystems involves a consideration of the factors, without as well as within, which may produce or ameliorate the observed variability.

## OBSERVATIONS

The general variability in the North Sea has been discussed elsewhere (Steele & Henderson 1977; Steele 1978) and is illustrated here by plankton data (Fig. 16.1). The smoothing obtained by averaging these data demonstrates the inadequacies in considering averaged processes as representations of population changes with time. Similar conclusions can be derived by comparing the very great variations in numbers of recruits of commercial species such as haddock (Steele 1978).

There are also the directly related problems of variability in space, and the fine horizontal structure can be illustrated by chlorophyll data from the northern North Sea (Fig. 16.2). Spectral analysis of such data (Platt 1972; Denman & Platt 1976; Steele & Henderson 1977) shows that the spatial variation is partly, but not wholly, explicable in terms of physical variability. Also, on occasion, certain more regular features are observable which can persist over periods of days (Fig. 16.2). Small-scale variability in the herbivores is more difficult to

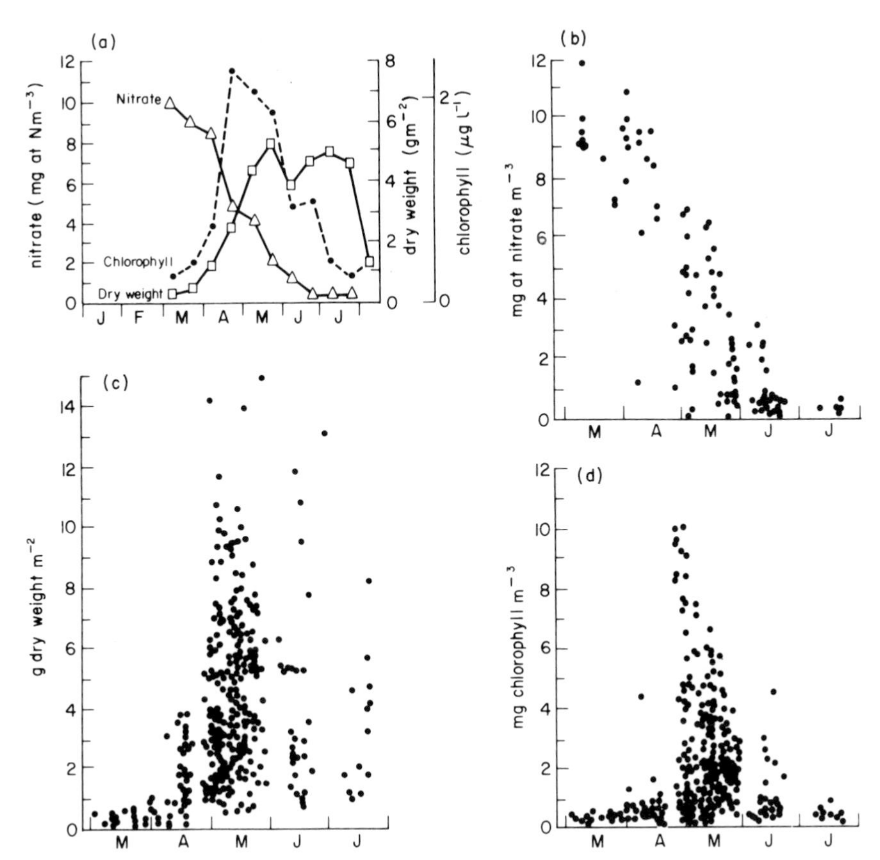


FIG. 16.1. (a) Average cycle of nitrate, chlorophyll and zooplankton dry weight in the centre of the northern North Sea, 1961–70. (b) Raw data for nitrate. (c) Raw data for chlorophyll. (d) Raw data for zooplankton (Adams & Martin unpublished).

observe but counts of copepods near the surface showed that it is at least as great (Mackas 1976).

Chlorophyll data are not always so variable. Lorenzen (1971) found that chlorophyll was uniformly low in large areas of the open subtropical Pacific. These were probably areas with little physical variability in space or time and low biological productivity. Compared with the North Sea, such regions may differ in having a fairly uniform environment, and areas of the central North Pacific have been compared to a chemostat (Eppley *et al.* 1973). Such environments may satisfy the conditions of predictability proposed by Slobodkin & Sanders (1969) with the species present being biologically accommodated.

Areas nearer shore generally have greater primary productivity and this usually arises from meteorological events inducing vertical mixing which will

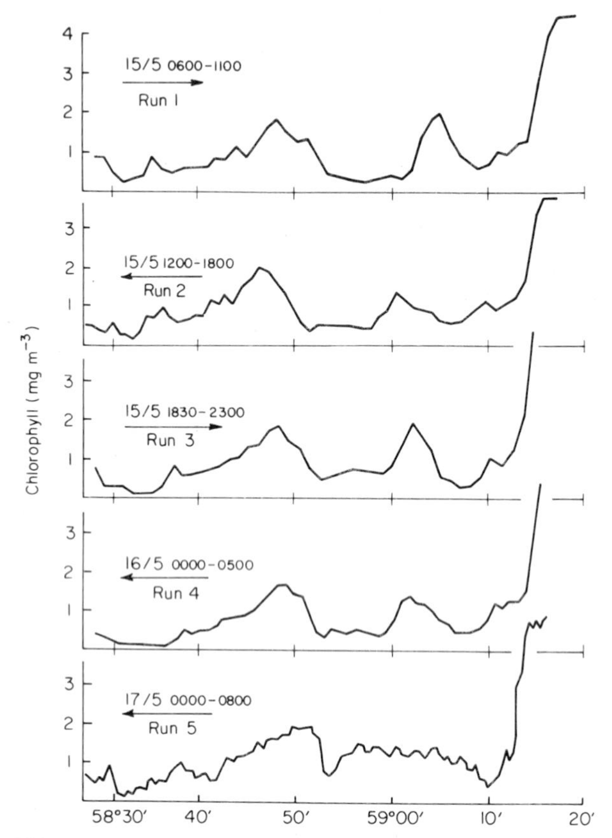


FIG. 16.2. Chlorophyll data collected in the North Sea by fluorometer at 3-m depth from a moving ship. Repeated runs were made which illustrate the spatial variability and also the short-term persistence in some of the main features.

have the variability and unpredictability associated with such phenomena as wind strength and solar radiation. These inputs of energy will produce temporal variations in the vertical structure of the water column.

Wind stress also induces horizontal motion of the upper layers and this, combined with tidal currents, will move and disperse organisms laterally. Such movements are affected by bottom topography, including the presence of coastal boundaries, and by freshwater input. These factors complicate the horizontal movements and can produce horizontal variation in the vertical structure at a wide range of scales.

The effect of the physically-induced variability of phytoplankton on the herbivores and then on higher trophic levels may be exemplified by the consequences of variation in timing of the spring outburst on herbivores and on predatory fish larvae (Fig. 16.3). Climatic variation could produce changes in timing of phytoplankton growth (Fig. 16.3a). Herbivorous copepods such as *Calanus* overwinter as small populations in the pre-adult stage. It is likely that, at the first appearance of enhanced food concentrations, they metamorphose to adults and use the food to produce eggs which develop in a few days to feeding nauplii. Thus, within limits, the reproductive potential of the overwintering stock is independent of timing (Fig. 16.3b). Other factors such as food particle size may determine the relative success of different species *A* and *B*. Later in the year succeeding cohorts may reverse the levels of *A* and *B* because of changes in food particle size (Steele & Frost 1977).

Thus, in terms of reproductive potential, such species can cope with year-to-year variations in timing of the spring outburst. The same reasoning would apply to spatial variations in timing within any year. In terms of field surveys of an area at any time after the start of the outburst, this effect would *enhance* the observed variability since population variations would then be out of phase in both plants and herbivores. Further, the variability would be seen as occurring in terms of differences in numbers of the same dominant species. In addition, the variable timing in any one year should be imprinted on later herbivore populations in terms of their age structure. For the northern North Sea *Calanus finmarchicus* is the dominant species in the spring. The changes in age structure can be seen, partly on finer scales but especially on the larger scales corresponding to major differences in timing of the phytoplankton bloom which occurs earlier on the eastern side of the North Sea (Fig. 16.4). This difference is not apparent in the biomass but is still present in the age structure in early summer. Such variability is evidence of mechanisms which will help to maintain the populations rather than of inadequate response to their environment.

For the main gadoid species in the North Sea spawning occurs in the spring before the outburst starts and so the larvae will reach the end of the yolk sac stage, when feeding must begin (Rosenthal & Hempel 1970), at a time independent of the progress of the spring outburst. This match or mismatch (Cushing 1976) will affect the survival of these larvae (Jones & Hall 1973) so that, very schematically, the suitable food (Fig. 16.3c) arising from an early bloom (1) will affect the survival of species I (Fig. 16.3d) in a manner represented in Figure 16.3e. In this way the effect of variation in timing is amplified rather than damped out as it is with the herbivores. There can be a similar match or mismatch in spatial distribution of food and larvae.

Observed variability at this stage in the life cycle of any one species would appear to indicate a factor which, on its own, could lead to extinction. The possible nature of other necessary factors is discussed elsewhere (Steele 1978).

The consequence for any one species of fish is the possibility of a large year-to-year variation in survival due to food availability at the larval stage. On the other hand, the environmental variability will enhance the longer-term survival for a range of gadoid (and other) species with different spatial and

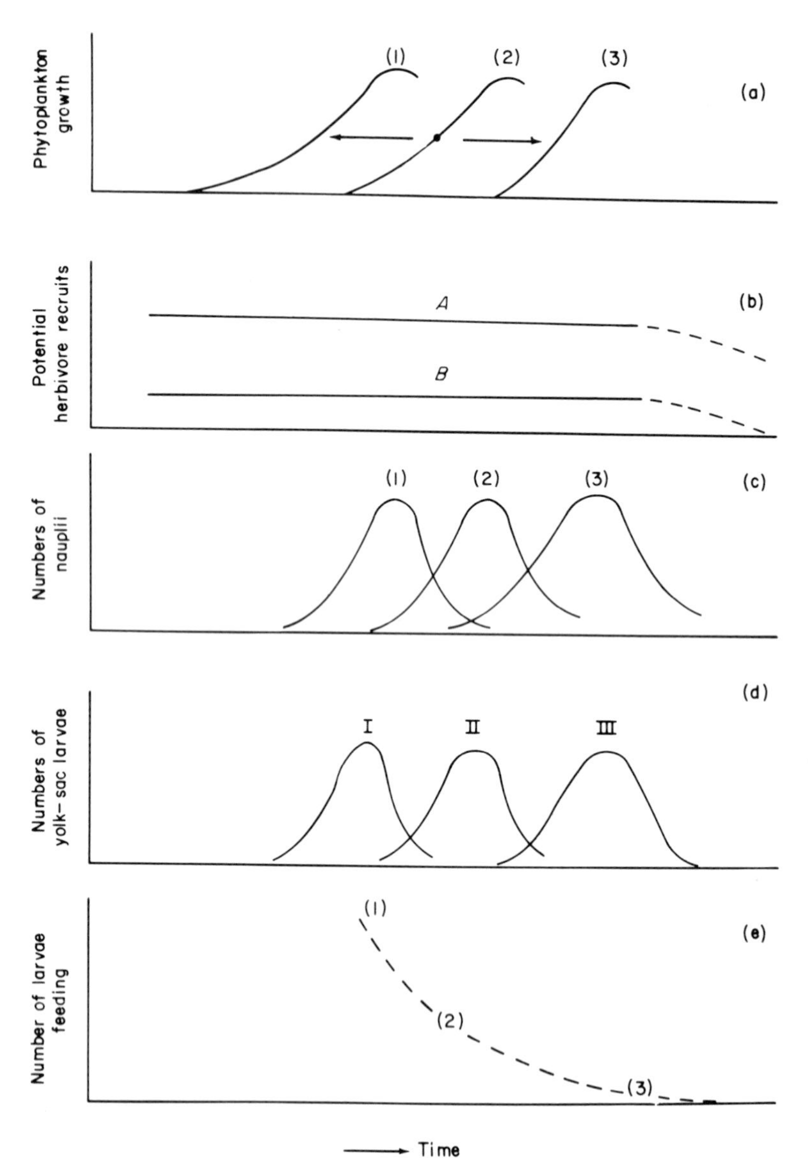


FIG. 16.3. Schematic representations of the consequences of variable climatic conditions in the spring. (a) Variations in timing of phytoplankton growth. (b) Numbers of recruits of two species, *A* and *B*. (c) Numbers of nauplii of one species of herbivore arising from different timing of the spring outburst. (d) Numbers of yolk sac larvae of three different species of demersal fish. (e) Numbers of larvae of fish species I feeding on nauplii from the differently timed broods.

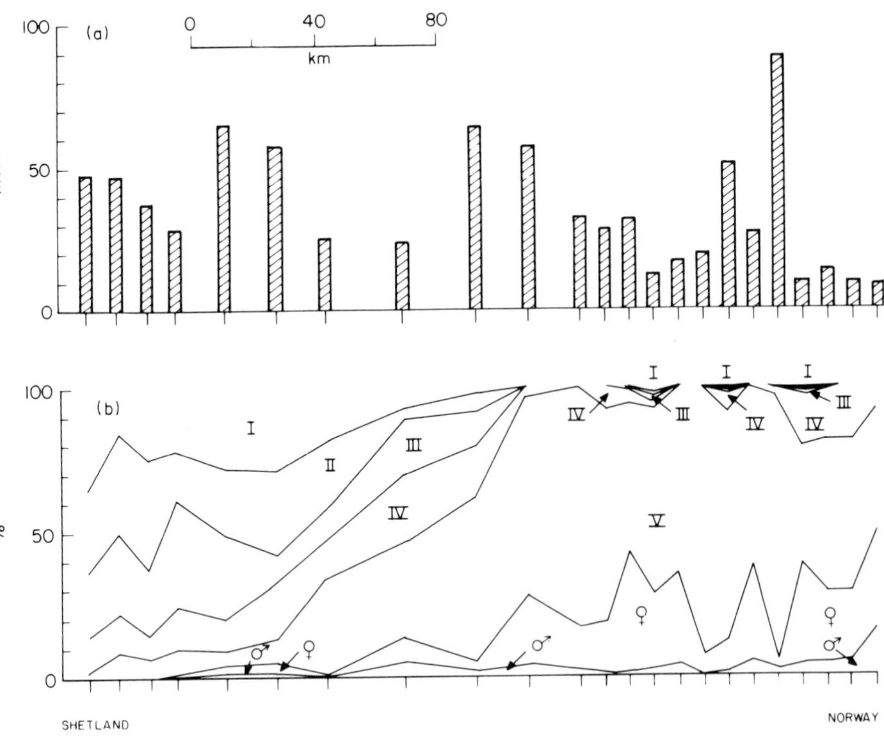


FIG. 16.4. (a) Plankton volumes from a line of stations in the North Sea between Shetland and Norway. (b) Distribution of *Calanus finmarchicus* among the copepodite stages. (From Report 1/76 of the Norwegian Coastal Current Project, Geophysical Institute, University of Bergen.)

temporal distributions of larvae in the spring. In this way large structural changes in communities can be generated within the same general energy flow pattern. At least in theory (Steele & Frost 1977) more than one herbivorous species, and a size range of phytoplankton, was a necessary condition for population persistence after the spring outburst. In this context the combination of the concepts of energy flow between trophic levels and of competition—for nutrients, phytoplankton or herbivores—gives considerable scope for variability in species composition and in age structure of particular herbivore species.

These simple illustrations are intended to demonstrate, first, that observed variability may be an indication of a basically unstable relationship for one species. However, it can also be the consequence of compensatory mechanisms. Secondly, the details of the biological responses are important and can derive from a particular part of the life cycle. Thirdly, the consequences may be somewhat different when considered in terms of a multi-species group such as demersal gadoid fish.

## EXPERIMENTS

Even if part of the observed variability in herbivorous zooplankton derives from adaptations to variable timing of increase in food, there is still the question whether, as the herbivores grow up, interactions between plants and herbivores incorporate further compensatory mechanisms or whether the continued mixing and dispersion of different populations plays an important role in preventing major imbalance in both trophic levels. Earlier laboratory work on zooplankton feeding is inconclusive in terms of threshold or 'S'-shaped functional responses of copepods to variations in phytoplankton concentration. There was some indication that a threshold existed in experiments with copepods fed with natural populations (Parsons *et al.* 1969; Adams & Steele 1966) but not with single-species cultures of phytoplankton (Frost 1972). Possible mechanisms which produce 'S'-shaped curves such as maximization of net energy intake (Lam & Frost 1976) have been suggested. Variations in vertical migration by copepods might also produce the same effect. It is of interest that a review of 'S'-shaped responses in terrestrial organisms has shown their existence in a wide range of species (Hassell, Lawton & Beddington 1977), suggesting that this may be the rule rather than an exceptional form of predatory behaviour.

A recent larger-scale experimental technique involves the use of long plastic columns suspended vertically in the sea, in which three trophic levels—phytoplankton, herbivorous copepods and invertebrate carnivores—can be isolated for 30–60 days (see Davies, Gamble & Steele 1975; Gamble, Davies & Steele 1977; Menzel 1977). Such ecosystems can be maintained for the 30–60-day periods with some decrease in diversity but with the main features of such communities remaining intact. The general conclusion is that horizontal mixing and exchange of populations is not necessary to damp out potentially large fluctuations arising from phytoplankton–herbivore interactions. Thus, some components of the interactions must have a damping effect on population variations. These results do *not* imply that horizontal variations cannot arise from biological factors such as variable vertical migration combined with shear flow of horizontal layers (Evans 1977). Nor do they imply that longer-term effects of the predator level on herbivores cannot be significant and, in fact, there are suggestions that differences between initially similar enclosures may be due to differing development of predator populations.

Similar work in freshwater enclosures (Hall, Cooper & Werner 1970) suggested that varying nutrient enrichments altered the productivity but not the species composition, whereas changes in the predators produced marked changes in species composition of the herbivores. The latter conclusion agrees in principle with the ideas of Brooks & Dodson (1965) on the effects of predatory fish in lakes. The same tentative conclusions have been obtained from the work in marine enclosures (Gamble, Davies & Steele 1977). These ideas stress the importance of the predators not merely as a necessary upper level to 'close' the planktonic ecosystem by removing herbivores but as a component which will affect the internal structure of the lower trophic levels both through the quantity of predation and also through size selection of prey. In turn, this emphasizes the need to look in more detail at the species structure of each trophic component.

## THEORY

In meteorology and physical oceanography, although the basic Navier–Stokes equations are known, the problem of including motion on all scales introduces conceptual and technical difficulties which, at present, do not permit an adequate simulation of the physical structure of, say, a whole ocean basin. Thus, theoretical studies and simulation models have concentrated on certain aspects such as vertical structure in response to wind stress and heating (Niiler & Kraus 1977) or the energy transfer through meso-scale eddies (Rhines 1977). If both these aspects can be understood, then the consequences may be parameterized for use in larger three-dimensional models. Again there is an analogy in, and a direct impact of these developments on, theoretical studies of marine ecosystems since these physical motions have considerable effects on growth and distribution of plankton.

For the ecological models there is the further complication of species diversity. This can be eliminated by considering only the biomass of particular trophic levels such as 'herbivores' or 'primary carnivores'. Species can be introduced as a major component with each species given a separate biomass parameter. Or, as in models of fish dynamics, the growth and reproduction of a particular species can be considered. Obviously, a detailed study of the dynamics of one or more species can introduce considerable complexity. Thus, ecological theory and modelling have tended to concentrate on one of these three aspects, vertical distribution, horizontal dispersion or species structure.

There are theoretical models of vertical distribution of phytoplankton (Radach & Maier-Remier 1975) and zooplankton (Steele & Mullin 1977) which can be used to demonstrate the possible effect of vertical variations on population response. Steele and Mullin introduce the age structure of one species of copepod and follow the development of six cohorts. Age structure is completely deterministic in the sense that each cohort has a single 'age' and all animals within it reproduce on the same day to define a new cohort. Given this determinism, earlier work with a single cohort (Steele 1974) showed that limit cycles could be generated in the plant–herbivore interactions. The period of these cycles was the lifetime of each cohort. The recent studies demonstrate that, with more complex cohort structure, complicated interrelations emerge which can produce long-term variability with periods much greater than the average cohort lifetime.

Instead of using a cohort structure for the herbivores, the animals can be made to 'grow' through a number of weight categories. In this simplified representation growth at each time step from one weight to the next depends only on the food taken in and is independent of how recently animals entered that weight class. It has been pointed out (Evans personal communication) that this induces effective 'upward diffusion' of animals through the age structure. This is a smoothing process comparable, numerically, to that used to simulate horizontal physical dispersion of organisms. In consequence, a model using this technique has a time-smoothed herbivore biomass (Steele & Frost 1977).

A more accurate picture of herbivore growth would be intermediate between these extreme representations. This comparison demonstrates how the details of formulating growth and age structure affect the resultant population variability in the model.

Studies of the effects of horizontal dispersion have usually represented the physical processes as diffusion. With the simplest prey–predator relations (Lotka–Volterra) Murray (1975) has shown that initial spatial variations in prey–predator interactions will be damped out by diffusion. However, by making specific assumptions such as different diffusion rates for prey and predator, and special non-linear biological interactions, long-term patchiness may be induced (Okubo 1974). Other, possibly more realistic, physical models combine shear between two or more layers in the sea with vertical mixing between these layers to induce dispersion (Kullenberg 1972; Evans 1977). Such physical models can be combined with nutrient–phytoplankton–herbivore dynamics (including vertical migration of the herbivores) to study possible patch formation (Evans, Steele & Kullenberg 1977; Evans 1978). This simulation has shown evidence of selection for particular patch sizes and long-term generation of patchiness. However, in this last model, for technical reasons, the representation of the herbivore age structure used the weight class, rather than the cohort, approach. For this reason an excessive 'biological' smoothing may have been introduced. Again this illustrates how problems of detailed representation interact with more general ecological problems.

The representation of species composition in the context of population dynamics requires prior assumptions about details of interactions both within and between trophic levels. For marine systems a simplifying assumption (Parsons 1969) is that size (or weight) of organisms can be used as a parameter to define 'species' both for phytoplankton (as cell diameter) and for herbivores (as initial and final, adult, weights). Further, the food selection can also be defined in terms of relative size of the herbivore and its food. The basis for these assumptions and a resulting simulation model are described fully in Steele & Frost (1977). Some of the results support the observations that structure at lower trophic levels, including phytoplankton, can be dependent on the magnitude and selectivity of predation. Further, the persistence of the herbivore species is dependent on the existence of suitably sized food for the reproducing adults or for the first naupliar feeding stages.

## DISCUSSION

I have tried to show how particular factors in the system may contribute to the observed variability. Variability can be seen in terms of changes in total biomass or as changes in species composition within a trophic level or as changes in age structure of a particular species. It can occur at a range of space and time scales. The examples I have given have been from an area and from that part of the marine ecosystem which displays the greatest variations, the coastal pelagic environment.

*A priori*, it is reasonable to suppose that such variability is driven by the unpredictability of the physical environment, and the evidence supports this. The observations also indicate (as do general concepts) that some features of the biological system are involved in moderating, or possibly amplifying, this variability.

For terrestrial systems Southwood (1977) has shown the range of population variation which can occur and has pointed out the importance of habitat variability in determining population response. For single populations, and using simple deterministic models of population dynamics, May (1976) has shown how changes in population growth parameters can produce a wide range in the types of response. Although these include limit cycles and apparently random fluctuations, observations on populations whose generations do not overlap (Hassell, Lawton & May 1976) indicate that monotonic damping is the expected response to an initial perturbation. For this and other reasons it appears necessary to think of imposed variability as a recurring process. Theoretically, this could be simulated by stochastic perturbation on the general population level or on one of the population growth parameters. The evidence would suggest that the unpredictability of the environment affects a particular part of the life cycle and the response may be quite different for different components of the ecosystem. The period of recruitment is an obvious candidate and the response may be to absorb the unpredictability, as suggested for copepods, or to enhance it, as proposed for larval fish. In the latter case it is necessary to consider the response of several related species and to involve later stages in life when the variability is moderated (Steele 1978). For certain birds the opposite pattern may hold (Krebs 1970) with enhanced variability at an intermediate stage in the life cycle.

Within any trophic structure there will be short- and long-lived species. These will tend to have smaller and larger spatial scales. By using techniques of spectral analysis it may be possible to represent the variance of all these populations in space and time. There are, however, obvious limitations in thinking about, or modelling, interactions with temporal scales from one day to several years, and appropriate spatial scales from hundreds of metres to thousands of kilometres (Steele 1978). It is, in principle, the same problem which faces the meteorologists where either the smallest- or the largest-scale interactions must be simplified so that they can be introduced as inputs or as non-interacting parameters.

As Holling (1973) has pointed out, resilient populations which can absorb relatively large environmental fluctuations may be more variable than relatively constant populations in comparatively unvarying environments. However, the examples of copepods and larval fish would imply that there is no straightforward relation of resilience and observed variability, particularly for multi-species association.

Because of the conceptual and technical limitations on our studies, it is necessary also to bound the system by ignoring or parameterizing certain ecological interactions. Thus, the 'environment' of the system under study includes not only physical factors but ecological boundaries. For many theoretical studies it is necessary to close the system at some prey–predator interaction by converting the predator into a functional response dependent only on the number or biomass of the prey. The exact functional form chosen may play a major role in determining the internal response of the system to perturbation; for example, by removing the need for damping mechanisms within the system. This may be done intentionally or unintentionally (Steele 1976) and have a profound effect on the apparent conclusions in relation to particular natural ecosystems.

It is also necessary to consider the role of ecologically adjacent systems; for example, faecal material from the pelagic herbivores fuel the underlying benthic communities. These, in turn, release to the water nutrients which provide some of the requirements for the phytoplankton. Such loosely coupled cycles may be important as control mechanisms for both systems which may be altered by decoupling them (Steele 1976).

Thus, in ecology, not only do we have the physical problems of the effect of smaller-scale temporal and spatial events at larger scales, but within the biological system we need to include details of fine structure involving variable animal behaviour or species composition, without excluding the larger-scale interactions which can also have a profound effect. This can be seen sometimes in the differing approach of the theoreticians and those making detailed observations in the field. In ecology very large simulation models do not seem to be a solution to this problem. On occasion it may be possible to encapsulate the small-scale events in simple functional forms as Ludwig, Jones & Holling (1977) have done for the spruce budworm problem. For the pelagic marine ecosystems an understanding of the small-scale variations seems necessary for the interpretation of large-scale changes such as those observed in commercial fisheries (Steele 1978). The methodology for intensive sampling on these scales is available and theoretical problems can be posed in terms of spectral rather than spatial distributions, providing a basis for quantitative comparison with observations. Comparable studies exist in the field of human geography (Cliff & Ord 1976; Haggett 1972). As Southwood's (1977) review indicates, there is the need for a similar approach in terrestrial ecology to marry detailed observations of spatial patterns with general theory.

## REFERENCES

**Adams J.A. & Steele J.H. (1966)** Shipboard experiments on the feeding of *Calanus finmarchicus* (Gunnerus). *Some Contemporary Studies in Marine Science* (Ed. by H. Barnes) pp. 19–35. Allen & Unwin, London.

**Brooks J.L. & Dodson S.I. (1965)** Predation, body size and composition of plankton. *Science*, **150**, 28–35.

**Cliff A.D. & Ord J.K. (1976)** Model building and the analysis of spatial pattern in human geography. *Journal of the Royal Statistical Society*, **37**, 297–348.

**Cushing D.H. (1976)** Biology of fishes in the pelagic community. *The Ecology of the Seas* (Ed. by D.H. Cushing & J.J. Walsh) pp. 317–340. Blackwell Scientific Publications, Oxford, London, Edinburgh & Melbourne.

**Davies J.M., Gamble J.C. & Steele J.H. (1975)** Preliminary studies with a large plastic enclosure. *Estuarine Research* (Ed. by L.E. Cronin) pp. 251–264. Academic Press, New York.

**Denman K.L. & Platt T. (1976)** The variance spectrum of phytoplankton in a turbulent ocean. *Journal of Marine Research*, **34**, 593–601.

**Eppley R.W., Renger E.H., Venrick E.L. & Mullin M.M. (1973)** A study of plankton dynamics and nutrient cycling in the central gyre of the North Pacific Ocean. *Limnology & Oceanography*, **18**, 534–551.

**Evans G.T. (1978)** Biological effects of vertical–horizontal interactions. *Spatial Pattern in Plankton Communities* (Ed. by J.H. Steele). Plenum Press, New York, pp. 157–179.

**Evans G.T. (1979)** A two-layer shear diffusion model. *Deep-Sea Research* (in press).

**Evans G.T., Steele J.H. & Kullenberg G. (1977)** A preliminary model of shear diffusion and plankton populations. *Scottish Fisheries Research Report L09*.

**Frost B.W. (1972)** Effects of size and concentration of food particles on the feeding behaviour of the marine plankton copepod *Calanus pacificus*. *Limnology & Oceanography*, **17**, 805–815.

**Gamble J.C., Davies J.M. & Steele J.H. (1977)** Loch Ewe bag experiment, 1974. *Bulletin of Marine Science*, **27**, 146–175.

**Gulland J.A. (1971)** *The Fish Resources of the Ocean*. Fishing News (Books) Ltd., West Byfleet, Surrey.

**Haggett P. (1972)** *Geography: A Modern Synthesis*. Harper & Row, London.

**Hall D.J., Cooper W.E. & Werner E.E. (1970)** An experimental approach to the production dynamics and structure of freshwater animal communities. *Limnology & Oceanography*, **15**, 839–928.

**Hassell M.P., Lawton J.H. & May R.M. (1976)** Patterns of dynamical behaviour in single-species populations. *Journal of Animal Ecology*, **45**, 471–486.

**Hassell M.P., Lawton J.H. & Beddington J.R. (1977)** Sigmoid functional responses by invertebrate predators and parasitoids. *Journal of Animal Ecology*, **46**, 249–262.

**Holland W.R. (1977)** The role of the upper ocean as a boundary layer in models of the oceanic general circulation. *Modelling and Prediction of the Upper Layers of the Ocean* (Ed. by E.B. Kraus) pp. 7–30. Pergamon Press, Oxford.

**Holling C.S. (1973)** Resilience and stability of ecological systems. *Annual Review of Ecology and Systematics*, **4**, 1–23.

**Jones R. & Hall W.B. (1973)** A simulation model for studying the population dynamics of some fish species. *The Mathematical Theory of the Dynamics of Biological Populations* (Ed. by M.S. Bartlett & R.W. Hiorns) pp. 347. Academic Press, London & New York.

**Krebs J.R. (1970)** Regulation of numbers in the great tit (Aves: Passeriformes). *Journal of Zoology, London*, **162**, 317–333.

**Kullenberg G. (1972)** Apparent horizontal diffusion in stratified vertical shear flow. *Tellus*, **24**, 17–28.

**Lam R.K. & Frost B.W. (1976)** Model of copepod filtering response to changes in size and concentration of food. *Limnology & Oceanography*, **21**, 490–500.

**Lindemann R.L. (1942)** The trophic-dynamic aspect of ecology. *Ecology*, **23**, 399–408.

**Lorenz E.N. (1969)** The predictability of a flow which possesses many scales of motion. *Tellus*, **21**, 289–307.

**Lorenzen C.J. (1971)** Continuity in the distribution of surface chlorophyll. *Journal du Conseil. Conseil permanent international pour l'exploration de la mer*, **34**, 18–23.

**Ludwig D., Jones D.D. & Holling C.S. (1977)** Qualitative analysis of insect outbreak systems: the spruce budworm and forest. *Institute of Resource Ecology, University of British Columbia, Working Paper* **W-17**, 45 pp.

**Mackas D. (1976)** Horizontal spatial heterogeneity of zooplankton on the Fladen Ground. *I.C.E.S. CM 1976/L:20*.

**May R.M. (1976)** Models for single populations. *Theoretical Ecology: Principles and Applications* (Ed. by R.M. May) pp. 4–25. Blackwell Scientific Publications, Oxford.

**Menzel D.W. (1977)** Summary of experimental results: controlled ecosystem pollution experiment. *Bulletin of Marine Science*, **27**, 142–145.

**Murray J.D. (1975)** Non-existence of wave solutions for the class of reaction-diffusion equations given by the Volterra interacting-population equations with diffusion. *Journal of Theoretical Biology*, **52**, 459–469.

**Niiler P.P. & Kraus E.B. (1977)** One-dimensional models of the upper ocean. *Modelling and Prediction of the Upper Layers of the Ocean* (Ed. by E.B. Kraus) pp. 143–172. Pergamon Press, Oxford.

**Okubo A. (1974)** Diffusion-induced instability in model ecosystems: another possible explanation of patchiness. *Technical Report, 86 Chesapeake Bay Institute*. Johns Hopkins University, Baltimore.

**Parsons T.R. (1969)** The use of particle size spectra in determining structure of a plankton community. *Journal of the Oceanographic Society of Japan*, **25**, 172–181.

**Parsons T.R., LeBrasseur R.J., Fulton J.D. & Kennedy O.D. (1969)** Production studies in the Strait of Georgia. Part II. Secondary production under the Fraser River plume, February–May 1967. *Journal of Experimental Marine Biology and Ecology*, **3**, 39–50.

**Platt T. (1972)** Local phytoplankton abundance and turbulence. *Deep-Sea Research*, **19**, 183–188.

**Platt T. & Denman K.L. (1975)** Spectral analysis in ecology. *Annual Review of Ecology and Systematics*, **6**, 189–210.

**Radach G. & Maier-Reimer E. (1975)** The vertical structure of phytoplankton growth dynamics, a mathematical model. *Mémoires Société Royale des Sciences de Liège* (Ed. by J. Nihoul) pp. 113–146. 6$^{e}$ Serie, **7**.

**Rhines P.B. (1977)** The dynamics of unsteady currents. *The Sea* (Ed. by Edward D. Goldberg, I.N. McCave, J.J. O'Brien & J.H. Steele) pp. 1048. Wiley-Interscience, Vol. 6.

**Rosenthal H. & Hempel G. (1970)** Experimental studies in feeding and food requirements of herring larvae (*Clupea harengus* L.). *Marine Food Chains* (Ed. by J.H. Steele) pp. 344–364. Oliver & Boyd, Edinburgh.

**Ryther J.H. (1969)** Photosynthesis and fish production in the sea. *Science*, **166**, 72–77.

**Slobodkin L.B. & Sanders H.L. (1969)** On the contribution of environmental predictability to species diversity. *Diversity and Stability in Ecological Systems*. Report of Brookhaven Symposia in Biology No. 22.

**Somerville R.C.J. (1977)** The role of the upper ocean in large-scale numerical prediction of the atmosphere. *Modelling and Prediction of the Upper Layers of the Ocean* (Ed. by E.B. Kraus) pp. 143–172. Pergamon Press, Oxford.

**Southwood T.R.E. (1977)** Habitat, the templet for ecological strategies? Presidential address to the British Ecological Society, January 5, 1977. *Journal of Animal Ecology*, **46**, 337–365.

**Steele J.H. (1965)** Some problems in the study of marine resources. *Special Publications International Commission for the Northwest Atlantic Fisheries*, **6**, 463–474.

**Steele J.H. (1974)** *The Structure of Marine Ecosystems*. Harvard University Press, Cambridge, Massachusetts.

**Steele J.H. (1976)** The role of predation in ecosystem models. *Marine Biology*, **35**, 9–11.

**Steele J.H. (1978)** Some problems in the management of marine resources. *Applied Biology* (Ed. by T.H. Coaker), Vol. II (in press).

**Steele J.H. & Henderson E.W. (1977)** Plankton patchiness in the northern North Sea. *Fisheries Mathematics* (Ed. by J.H. Steele) pp. 1–19. Academic Press, London.

**Steele J.H. & Frost B.W. (1977)** The structure of plankton communities. *Philosophical Transactions of the Royal Society of London*, **280**, 485–534.

**Steele J.H. & Mullin M.M. (1977)** Zooplankton dynamics. *The Sea: Ideas and Observations in Progress in the Study of the Seas* (Ed. by E.D. Goldberg, I.N. McCave, J.J. O'Brien & J.H. Steele) pp. 857–890. Wiley-Interscience, Vol. 6.

**Stommel H. (1948)** The westward intensification of wind-driven ocean currents. *Transactions of the American Geophysics Union*, **29**, 202–206.

# 17. A PROFUSION OF SPECIES? APPROACHES TOWARDS UNDERSTANDING THE DYNAMICS OF THE POPULATIONS OF THE MICRO-ARTHROPODS IN DECOMPOSER COMMUNITIES

M. B. USHER, P. R. DAVIS,
J. R. W. HARRIS* AND B. C. LONGSTAFF†

*Department of Biology, University of York,*
*Heslington, York YO1 5DD*

## INTRODUCTION

The diversity of communities of soil micro-arthropods need not be reviewed: it has long been the subject of speculation at Congresses of Soil Zoology (e.g. Anderson 1975; Ghilarov 1977). One can, however, ask if the communities are unusually diverse, or if the diversity is an artefact of the way in which soil biologists view the communities. Several studies of the Collembola in the British Isles are shown in Figure 17.1, where it can be seen that no community, defined in terms of the dominant plant species, contained more than 25 species. Woodland communities tend to be more species-rich than grassland communities, which are richer than moorland communities. None of the data in Figure 17.1, which are restricted to insects in a single order, show a community that is particularly diverse. Indeed, Wood (1967) recorded a total of only 128 species of springtails and mites in a limestone grassland soil in Yorkshire, whereas Morris (1969, 1971) recorded more than 28 species of Heteroptera and 45 species of Auchenorhyncha in a chalk grassland habitat in Bedfordshire (these 73 species account for only part of the order Hemiptera). Comparisons, order by order, between the micro-arthropods of the soil and litter communities and the arthropods of the above-ground communities will probably indicate that the 'decomposer' communities are not as diverse as the 'grazer' communities. The habits of

* Present address: Centre for Overseas Pest Research, College House, Wrights Lane, London W8 5SJ.
† Present address: CSIRO Division of Entomology, P.O. Box 1700, Canberra City, ACT 2601, Australia.

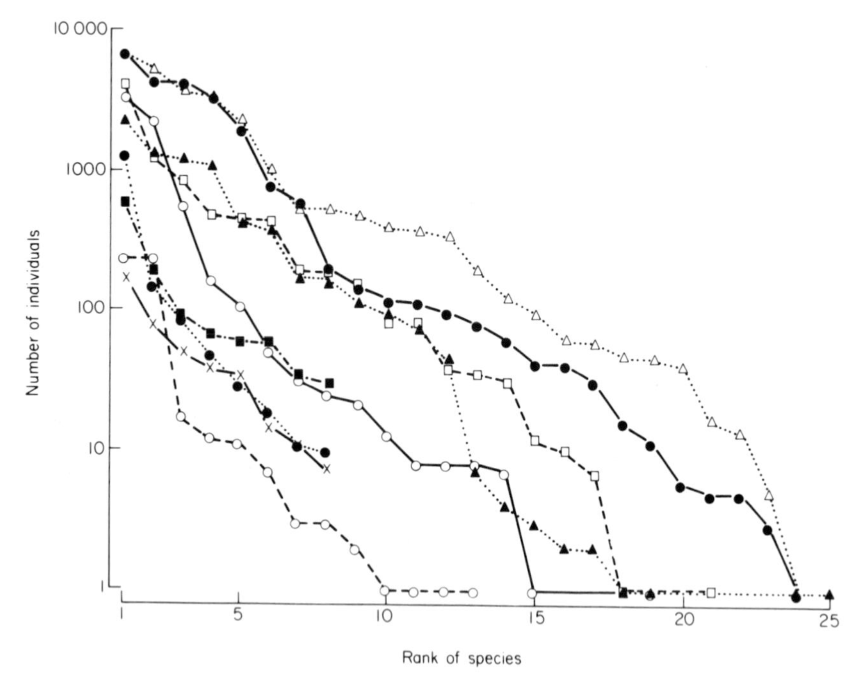


FIG. 17.1. Species abundance curves for Collembola in nine British habitats. From right to left, approximately, the lines represent: a mixed coniferous forest (Poole 1961—dotted line, open triangles); Scots pine forest (Usher 1970—continuous line, filled circles); various moorland habitats (Hale 1966)—alluvial grassland (dashed line, open squares), limestone grassland (dotted line, filled triangles), *Calluna* (continuous line, open circles) and *Juncus* (dashed line, open circles); and lowland habitats (Macfadyen 1952) in which only the eight commonest species were recorded—*Molinia* (dot and dash line, filled squares), *Deschampsia* (dotted line, filled circles) and *Juncus* (continuous line, crosses).

ecologists looking at only one order and of soil biologists looking at all orders have almost certainly led to an erroneous belief in an unusually diverse decomposer community. Soil and litter communities usually contain less than 150 micro-arthropod species within any habitat defined in terms of a reasonably homogeneous area of vegetation.

Generally, two fundamentally different approaches have been used to investigate the structure and functioning of communities. To borrow terms from the field of numerical classification, these could be called 'divisive' and 'agglomerative'. The divisive approach consists of sampling a community, and analysing the results in order to pick out temporal, spatial and other patterns and correlations. In other words, the unit of study is the community and the aim is the analysis of patterns of individual species within it. The agglomerative approach consists of taking individual species of a community and studying their biology, and their paired and higher-order interactions. The unit of study is thus the species, and the aim is to build predictive models of successively more and more complex artificial communities. Both approaches are reviewed in the following sections, and results are given from some of our studies, either in the Black Wood of Rannoch, Perthshire (divisive studies), or in the grasslands on the Yorkshire Wolds (agglomerative studies).

## THE DIVISIVE APPROACH

The aim of the divisive approach is to find patterns of species within samples drawn from the whole community. These patterns range from traditional ecological investigations, such as the seasonal, vertical and horizontal distribution of individual species, to multivariate studies which relate species distributions to the environmental influences. Following the ideas of Hutchinson (1957) and the work of MacArthur & Levins (1967), it seems possible that multivariate analyses might also yield information on the niches occupied by the individual species, and on the extent of niche overlap.

### *Seasonal distribution*

The seasonal distribution, or phenology, of the species of soil micro-arthropods has been described by many authors: Hale (1967) and Wallwork (1967) have reviewed the literature relating to Collembola and mites respectively. Many of the studies of seasonal distribution can be criticized for their short duration. Most studies span merely a single year (such as those of Usher 1970, 1971b, 1975b or Niijima 1971 on Collembola) or a period only slightly in excess of a year (such as those of Moritz 1963 and Lions 1973 on Cryptostigmata populations). Hale's (1966) study of moorland Collembola spanned two years, but only studies like Tamura's (1976) of *Folsomia octoculata* for three-and-a-half years and van der Drift's (1959) of various beetles for six years provide data which can be analysed to estimate both the true seasonal distribution and the year-to-year variation. Descriptive studies of individual species yield little information that can be interpreted in terms of community structure, and it appears that a study of the phenology of a whole soil arthropod community has not been attempted.

The data used by Usher (1970, 1971b, 1975b) have been reworked and are shown in Figure 17.2. The illustration shows the total number of Collembola, Mesostigmata and Cryptostigmata (and the five most abundant species in each group) in 1 536 $cm^3$ of Scots pine litter sampled at twelve times during one year. Some of the statistics for comparison of these graphs are listed in Table 17.1.

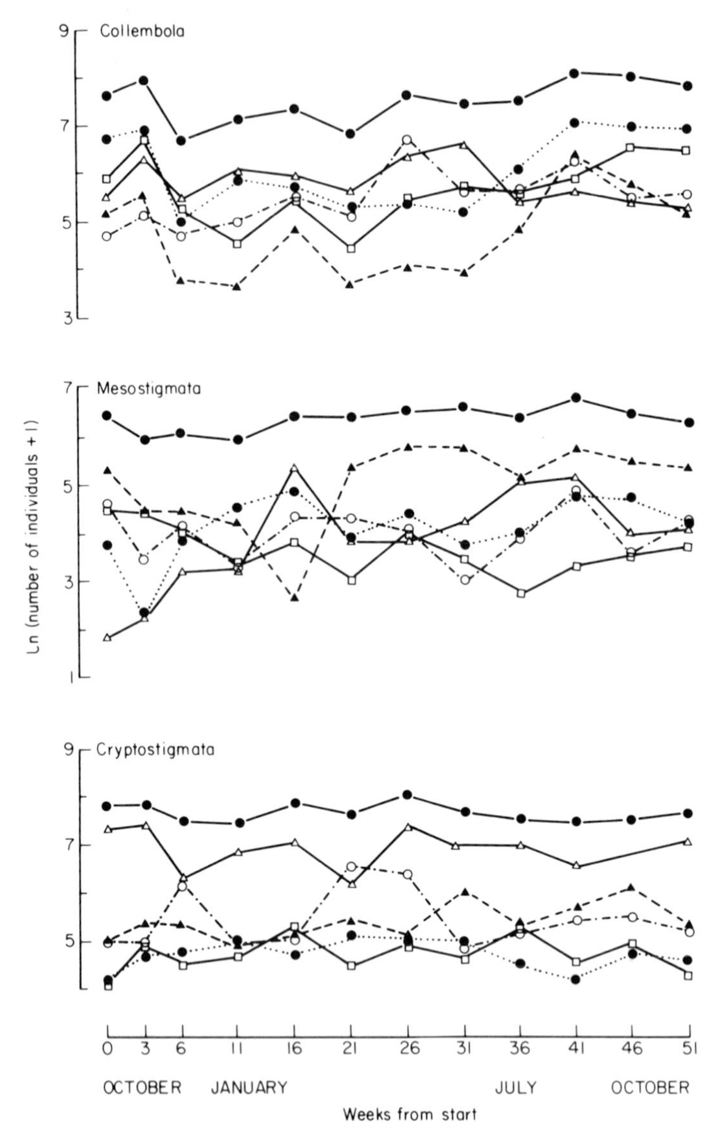


FIG. 17.2. Seasonal variability in populations and communities of soil arthropods (data from Usher 1970, 1971b, 1975b). The upper set of lines represent the Collembola, the centre the Mesostigmata and the lower the Cryptostigmata. In each case the total count for that group (continuous thick line joining filled circles) is shown together with the five most abundant species in the group. The species codes (shown in Table 17.1) respectively for the three groups are: dotted line, filled circles (Oa, Vn, Ap); continuous line, open squares (Fm, Pl, Cb); continuous line, open triangles (Is, Pr, Tv); dashed line, filled triangles (Tc, Om, Sm); and dot and dash line, open circles (Fq, Tp, Pp).

TABLE 17.1. The mean ($m$), the coefficient of variation ($CV$) and code (used in Figures 17.2, 17.4 and 17.5) of the total of all Collembola species, all Mesostigmata and all Cryptostigmata, as well as the five most frequent species in each group, in twelve samples of 1 536 $cm^3$ of Scots pine forest litter and humus. All data have been transformed by $\ln(x + 1)$ before analysis, and no back-transformation has been carried out. The 'maximum $CV$' for the three totals represents what the $CV$ would have been if all of the species in the group had been synchronous, i.e. that the rank correlations between all pairs were 1

| Group/species | Code | $m$ | $CV$ (%) | Maximum $CV$ (%) |
|---|---|---|---|---|
| Total Collembola | — | 7·49 | 6·1 | 9·2 |
| *Onychiurus absoloni* | Oa | 6·12 | 13·9 | |
| *Isotoma sensibilis* | Is | 5·80 | 7·2 | |
| *Friesea mirabilis* | Fm | 5·67 | 12·6 | |
| *Folsomia quadrioculata* | Fq | 5·46 | 11·0 | |
| *Tullbergia callipygos* | Tc | 4·73 | 19·5 | |
| Total Cryptostigmata | — | 7·76 | 2·5 | 6·4 |
| *Tectocepheus velatus* | Tv | 6·83 | 6·6 | |
| *Platynothrus peltifer* | Pp | 5·42 | 11·2 | |
| *Steganacarus magnus* | Sm | 5·39 | 6·8 | |
| *Ceratoppia bipilis* | Cb | 4·70 | 8·2 | |
| *Adoristes poppei* | Ap | 4·70 | 6·1 | |
| Total Mesostigmata | — | 6·36 | 4·4 | 11·9 |
| *Olodiscus minimus* | Om | 4·97 | 18·3 | |
| *Veigaia nemorensis* | Vn | 4·17 | 17·2 | |
| *Trachytes pyriformis* | Tp | 3·98 | 13·9 | |
| *Parazercon radiatus* | Pr | 3·84 | 28·3 | |
| *Pergamasus lapponicus* | Pl | 3·64 | 13·5 | |

The total community of Cryptostigmata remains remarkably constant throughout the year, since the individual species tend to complement each other. Similar results can be seen for the Mesostigmata, where the coefficient of variation for the whole community is less than those for the five most abundant species. The fact that the individual species of mites are not at their most abundant at the same time of the year invites speculation that this is the result of either extant interspecific competition or temporal separation evolved to avoid it.

The situation in the Collembola is different. Many authors have referred to a pattern of autumnal maximum and vernal minimum in collembolan population sizes. This is seen in Figure 17.2, where there is a considerable degree of synchrony between the five most abundant species. Why the mite species appear to be asynchronous and the Collembola species to be synchronous is a question that cannot yet be answered. An answer will probably lie in the generality or specificity of the feeding habits of the different groups.

*Vertical distribution*

Reviews of the vertical distribution of species of Collembola are given by Christiansen (1964) and Hale (1967), and of mites by Wallwork (1967). These reviews indicate two basic approaches to the study of vertical distribution. In one approach the measurement of depth is absolute: for example, Athias (1975), working on a West African savannah soil, divided the soil into 0–15 cm and 15–40 cm depths, and Usher (1970, 1971b, 1975b) used three slices, each 1 cm thick. In the other approach, depth is relative to the horizons of the soil profile; thus, Poole (1961) recorded Collembola in the litter, humus and sub-humus horizons, and Hurlbutt (1964) observed *Veigaia* mites in the litter and fermentation, humus and $a_1$ horizons.

Again using the data discussed by Usher (1970, 1971b, 1975b), a mean annual depth, weighted by population size, has been calculated for the five most abundant species in each of three arthropod groups (Fig. 17.3). The Collembola are well spaced out on the depth axis, except for *Friesea mirabilis*

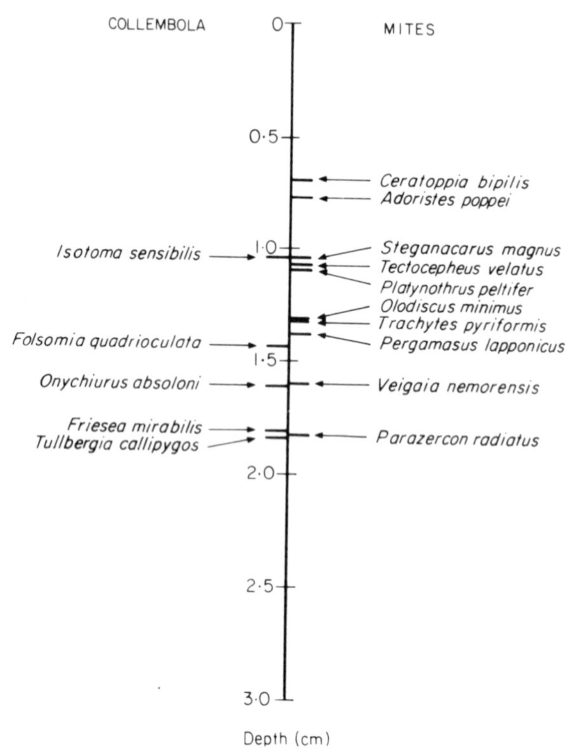


FIG. 17.3. The weighted average annual mean depth of the fifteen species of soil arthropods shown in Figure 17.2. The data are reworked from Usher (1970, 1971b, 1975b).

and *Tullbergia callipygos*, which are unlikely to compete for food since *F. mirabilis* has mouth parts adapted for sucking whilst *T. callipygos* will ingest solid food which it is able to grind. The mites tend to cluster rather more on the vertical axis. All the Cryptostigmata species occur near the surface of the soil, while all the Mesostigmata species are deeper. However, the inadequacy of the sort of data shown in Figure 17.3 is clear for two reasons. First, the method of sampling in 1-cm slices, although thinner than used in most studies of the vertical distribution, is still far too crude, and it can at best only give a rough approximation to the vertical distribution. Secondly, since the arthropods are likely to be associated with particular stages in the decomposition of dead plant material, it is uncertain whether the observed distribution reflects the vertical distribution of the decomposition process and hence of potential food, or the vertical distribution of the species reaction to environmental gradients.

Many of these problems have been overcome by adopting a radically different sampling process (Anderson 1971; Pande & Berthet 1975) in which thin sections of the organic horizon of the soil are cut. The vertical distribution of the species can thus be related both to depth and to the stage of decomposition. Pande and Berthet's study of Cryptostigmata, in 40 horizontal slices of the surface 6 cm of the soil, clearly shows that there is a spatial separation of the species. The depth in the soil is thus likely to be another factor which can be exploited by the species of the soil micro-arthropod community to avoid interspecific competition.

### *Horizontal distribution*

Soil biologists have been fascinated by the aggregated nature of the distribution of soil arthropods, and many studies have been directed solely at demonstrating this form of distribution (reviewed by Christiansen 1964 and Butcher, Snider & Snider 1971). Controversy raged as to whether the distribution was nearer the negative binomial, the Neyman type A, or some other form (for example, Hartenstein 1961); but, with hindsight, it seems that three important features of the aggregated distribution were, at least initially, overlooked.

First, why did the arthropods aggregate where they were observed? Reviewing the literature, Usher (1976) concluded that the location of food and the local conditions of moisture were the two most likely stimuli. Studies by Joosse (1970) have indicated that there may be social reasons for aggregating behaviour: in animals with no courtship, or direct sperm transfer, aggregation brings the two sexes into contact, increasing the probability of fertilization. It has been speculated by Joosse & Verhoef (1974) that the aggregated distribution has an important survival value during the moult. This will be particularly effective if the arthropods are able jointly to provide a defence mechanism against predators. Laboratory studies, as indicated later, suggest that *Hypogastrura* might gain protection by aggregating to moult (Strebel 1932). It seems less likely that species in family Isotomidae, Entomobryidae or Sminthuridae (*sensu* Gisin 1960) could gain protection in this way, although the warning of a predator's proximity afforded by an attack on one invididual may have a survival value to the remainder. The recent detection by Verhoef, Nagelkerke & Joosse (1977) of an aggregation pheromone is of considerable importance, especially if the pheromone causes a species to aggregate and predators to avoid that aggregation.

Secondly, is there any kind of relation between aggregation behaviour and population density? Usher (1969, 1971a, 1975a) attempted to investigate this question by recognizing three distinct kinds of relations, which are outlined in Table 17.2. Using the commoner species shown in Figures 17.2 and 17.3, most of the Collembola and some of the Mesostigmata showed a type I response, whereas the Cryptostigmata showed either a type II or type III response. As a population increases, the Collembola and Mesostigmata are able to found new aggregations, while the Cryptostigmata increased the size of their aggregations. These observations suggest that the Collembola have wider niches, and are therefore able to establish new foci for aggregations, while the Cryptostigmata have narrower niches, and therefore increase their population density within an existing aggregation.

Thirdly, is an aggregation a single-species or a multi-species phenomenon? Attempts to compare indices based on the variances of numbers of the single species with the index of all species pooled together are unconvincing, though they have been used to suggest that aggregations are multi-specific (e.g. Poole 1961). Soil cores are probably far too insensitive to allow for a detailed study of whether several species can join together in an aggregation, although the new techniques of soil sectioning probably offer the best chance of investigating this aspect of soil micro-arthropod communities. If aggregations are found to be multi-specific, one will need to ask if and how the species avoid competition

TABLE 17.2. $r_n$ is defined as the correlation between population density and the number of aggregations per unit of soil volume; $r_s$ as the correlation between density and the mean number of organisms per aggregation. Depending on the significance or non-significance of $r_n$ and $r_s$, the table defines three types of aggregation behaviour, designated I, II and III, with possible variants on each of these basic types (designated I?, etc.). A further three of the cells are unlikely to occur (designated –)

| | $r_n$ | | |
|---|---|---|---|
| $r_s$ | Positive | Non-significant | Negative |
| Positive | III | II | II? |
| Non-significant | I | III? | – |
| Negative | I? | – | – |

within the aggregation, and whether the chemical defences of one species are used by other species to gain protection against predators.

### *Multivariate analyses—an indication of the niche?*

One early empirical attempt to relate micro-arthropod distribution to the environment was that of Haarløv (1960), though Poole (1961) carried out a more extensive analysis using correlation and partial correlation coefficients. Due to the correlation between the environmental variables measured, such attempts have generally proved relatively unsuccessful. Curry's (1973, 1976) studies take the analysis of the relation between soil arthropods and their environment further, by comparing the mean numbers of arthropods occurring beneath different plants, and then using a principal components analysis to identify associations.

One should ask what a principal components or principal co-ordinates analysis is likely to reveal. If a set of samples or observations (species, environmental variables, or both) are ordinated on the principal axes accounting for the greatest variance, then the existence of clusters will indicate communities or associations. If the axes have meaning as resources when the samples drawn from a single cluster or association are analysed, then the location of a species on the diagram gives an indication of its niche in relation to those resources. This approach has been attempted by a principal component analysis of the data for the fifteen species of soil micro-arthropods shown in Figures 17.2 and 17.3. Since the 1 119 samples were taken over a period of one year, they have been divided into three approximately equally sized groups representing the cold period of the year (November to February) and periods of increasing (March to July) and decreasing (August to October) temperature. Having transformed the data by $\ln(x + 1)$, the results of these three analyses, plotted on the same ordination diagram, are shown in Figure 17.4. If the principal axes represent the niches of the species in some abstract manner, a triangle formed by the three analyses could be considered to approximate the area of the niche space occupied by a species. In general, there is remarkably little overlap of the triangles, except for that of *Isotoma sensibilis* which overlaps three of the Cryptostigmata species. The spacing of the species on this diagram suggests that there is some separation of their niches.

However, to understand these niches, the principal axes would need to be identified with a measurable property of the ecosystem. Comparison of the first axis with the depth diagram in Figure 17.3 indicates that species with a small mean depth tend to be at the left of the illustration. No interpretation of the second principal axis can be offered. Depth was a factor included in the sampling since one-third of the samples were taken from each of three layers. To eliminate depth the three counts were added together to give a 'core count', which, after transformation, was analysed (Fig. 17.5). The first two principal

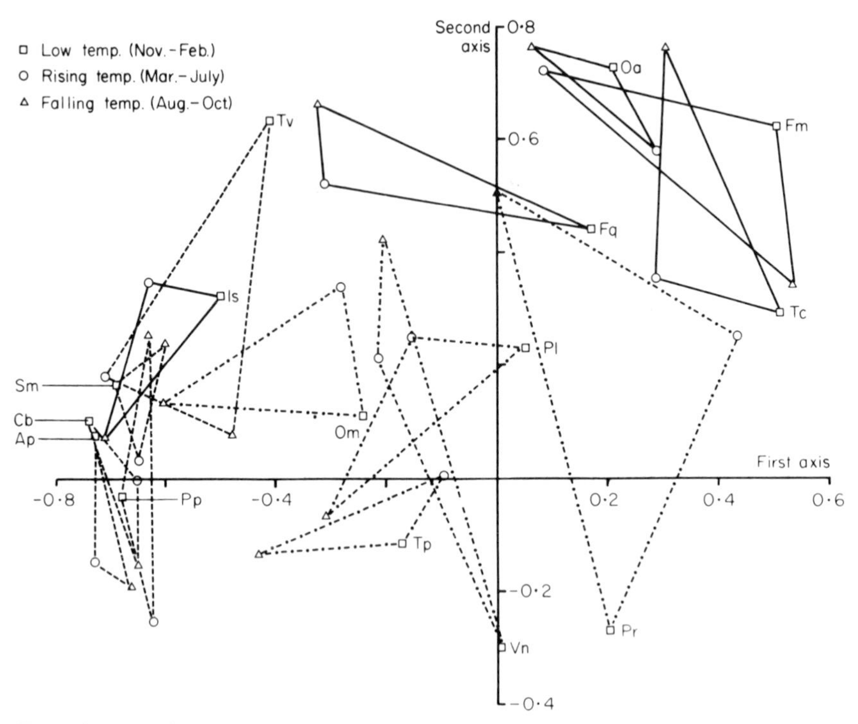


FIG. 17.4. A plot of the first two axes of principal component analyses of the fifteen species (Table 17.1) in 1 119 samples, divided into three groups depending upon the temperature. The positions of a species in the three analyses are joined by a triangle, continuous lines representing Collembola, dashed lines Cryptostigmata and dotted and dashed lines Mesostigmata.

axes are now not interpretable in terms of known environmental variables, though the third axis tends to correlate with depth. The species in Figure 17.5 are all well separated, indicating their niche separation in relation to these abstract axes.

These analyses, in which the principal axes cannot be identified, are certainly less satisfactory than that of Miracle (1974), who was able to identify her axes. She found that her first three axes accounted for 58% of the variance, whereas in Figure 17.4 the first three axes account for only 45%. It is possible that soil biologists, whose history of studying communities is shorter than that of aquatic biologists, have not yet identified some of the environmental influences acting upon soil arthropod populations. It is also possible, and perhaps more likely, that multivariate analyses are not going to shed any light upon the niche structure of communities. Maxwell (1977) indicates that in the behavioural sciences there were great expectations of multivariate techniques which have, in

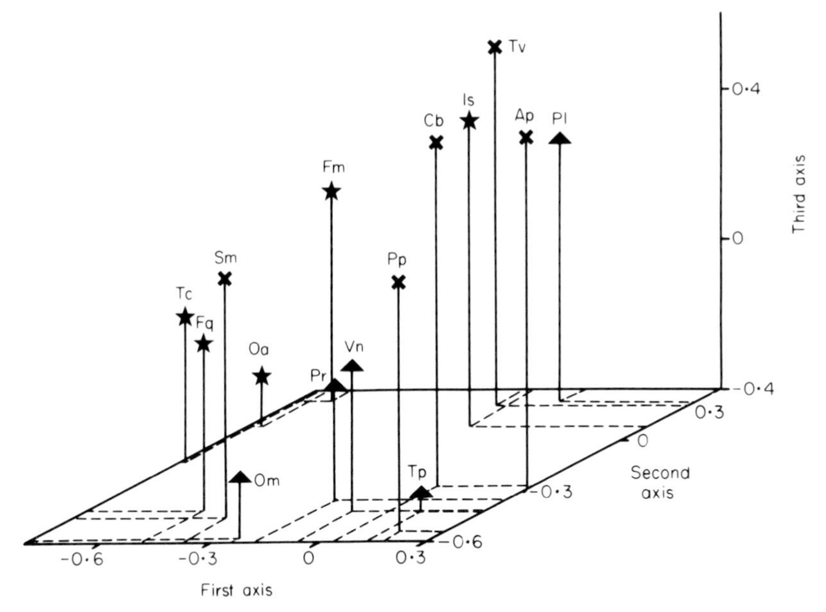


FIG. 17.5. Principal components analysis of the complete data shown in Figure 17.4, but summed so that cores are analysed rather than series of samples taken at three different depths. Stars, crosses and triangles represent Collembola, Cryptostigmata and Mesostigmata respectively.

general, not been realized due to the low level of correlation existing between different categories of 'tests'. The problem of extremely small correlation frequently occurs in ecological data.

## THE AGGLOMERATIVE APPROACH

Immediately one thinks of investigating the dynamics of single-species populations, or two species interacting, the problem of applying the experimental findings needs to be considered. Experiments are usually carried out in the laboratory, where the diurnal temperature cycle has a smaller amplitude than that in the field, and other environmental variables are kept more or less constant. The environment of laboratory cultures is considerably more predictable than that of field populations. Since the soil environment is buffered against extremes of temperature, there is the small consolation that laboratory conditions more closely approach soil conditions than above-ground conditions. Also, soil provides the arthropods with an opaque three-dimensional environment which, if it is recreated in a laboratory, presents the experimenter with problems both of taking a census and of observation. The majority of laboratory studies have reduced the environment to an approximately flat surface, and therefore density cannot be easily compared between populations in cultures and populations in soil.

### *Studies of single species*

The simplest way to investigate the dynamics of single species is to maintain cultures in the laboratory; this has yielded information on the relation between fecundity and density (Green 1964; Usher & Stoneman 1977), mortality (Usher & Stoneman 1977), inhibition of oviposition and oophagy (Waldorf 1971a,b) and life history, speed of development, duration and number of instars, etc. (these have been reviewed by Butcher, Snider & Snider 1971). Another class of techniques relies on field observations or on inference based on field samples. Thus, after nine seasons of field work, Wallace (1967) was able to demonstrate the nature of the control of populations of *Sminthurus viridis* by environmental factors, by a predatory mite and by a density-dependent factor (eating of apparently toxic bodies of dead individuals by early instar insects). Studies by Lebrun (1970) and Joosse (1969) have inferred such population parameters as fecundity, mortality and speed of development, quoting their results as yearly averages or on a seasonal basis. The dilemma facing anyone experimenting with single-species populations is that field studies relate the population parameters to the seasonal cycle in the environment, whereas laboratory studies relate these parameters to the density of the animals; there seems to be no satisfactory method of relating the population parameters both to seasonality and to density.

Another aspect of laboratory studies that is of relevance to community ecology is the productivity of an exploited population and the exploitation that can be tolerated before the species is ruinously heavily exploited. Experiments described by Usher, Longstaff & Southall (1971) on *Folsomia candida* distinguish numerical production of a population from biomass production. Although there were no significant differences between the numerical productions of cultures with excess food in which 30%, 40%, 50% and 60% of the population were removed every 14 days, biomass production was at a maximum when exploitation was 30%. More detailed investigations of *Onychiurus armatus* (*s. lat.*) by Costigan (1975) indicate that a maximum numerical production occurred when between 40% and 60% of the population was removed every 18 days (a model of his experimental system indicated the actual maximum at 48%). These laboratory experiments show that the Collembola can produce large numbers of offspring in the presence of continued predation, though it is unlikely that field population densities approach those of laboratory cultures when production is reduced by intraspecific competition. Whether it is important to consider biomass production (heavily weighted in favour of large individuals) or numerical production (heavily weighted in favour of young, newly emerged individuals) when considering the effects of predators in a natural community is unknown.

### *Competition between species*

Studies of the competitive relations between species of soil arthropods are rare, though more is understood about phytophagous mites (see Huffaker, Vrie & McMurtry 1969). Christiansen (1967) investigated only 13 of the possible 55 pairwise interactions between the 11 species of Collembola that he had available; in only one of his experiments, that of *Sinella coeca* with *S. curviseta*, did he get an indication of long-term co-existence. Inference based on field sampling of *Veigaia* and *Asca* mites has been attempted by Hurlbutt (1968), who suggested that species of moderately high anatomical similarity are more likely to co-exist than species that are either extremely similar or of low similarity. This concept of similarity seems not to have been widely used in ecology; its problem in application lies in the finding of a satisfactory measure of similarity that can be used in many groups. The observation presumably reflects a compromise between character divergence in co-existing generalist species (Lawlor & Maynard Smith 1976) and the constraints imposed by a common environment.

Longstaff (1974) investigated both intra- and inter-specific competition in three species of Collembola, *Hypogastrura denticulata*, *Onychiurus armatus* (*s. lat.*) and *Sinella coeca*, and constructed models of the system (Longstaff 1977). Full details of the culturing and census techniques are given by Longstaff (1974). At laboratory temperature (mean 23°C, range 15–30°C) there was no evidence of co-existence between species. In competition with either *S. coeca* or *O. armatus*, *H. denticulata* usually had become extinct before the end of the 24-week experimental period, irrespective of its initial proportion. *O. armatus* was always eliminated from cultures with *S. coeca*. Of the 36 cultures, only four still had both species surviving together after 24 weeks (these were all cultures of *O. armatus* and *H. denticulata* initiated with equal numbers of the two species or with excess of *H. denticulata*).

In contrast, at a constant temperature of 16°C, *H. denticulata*, in competition with *O. armatus*, always either remained or became the dominant species, though *O. armatus* never become extinct. In cultures with *S. coeca*, *H. denticulata* similarly increased its population size, though numerical dominance appeared to be retained by the species in the greatest abundance in the initiating culture. At this temperature it appears that *H. denticulata* is able to co-exist with either of the other species. *O. armatus*, although persisting with *H. denticulata*, usually became extinct in cultures with *S. coeca*. The fecundity of *O. armatus* was initially increased above the level in single-species cultures by the presence of *H. denticulata*, though in the presence of *S. coeca* the fecundity was either reduced or the animals failed to breed. *S. coeca* populations were apparently unaffected by the presence of *O. armatus*. Of the 36 cultures maintained at 16°C, only in nine of them had one or other species become extinct within 24 weeks. In relation to field conditions, this result could indicate either that it takes longer for competitive exclusion to occur at lower temperatures or that soil arthropods are more able to co-exist at temperatures more similar to, though still higher than, those in their natural environment. That few of the populations at 16°C declined supports the latter hypothesis, though it leaves unanswered the question of what mechanism allows co-existence.

Longstaff's experiments are summarized in Figure 17.6, in which the total productivity (area under the population size curve and hence in units of 'insect weeks') of the different species is compared. If competition is defined as occurring when two or more species all experience depressed fitness due to their mutual presence together (Emlen 1973), Figure 17.6 indicates that there is little genuine competition. At laboratory temperatures only one species of any pair showed a depressed productivity in the presence of the other and hence these interactions are probably best termed amensalism (cf. Longstaff 1976). At 16°C

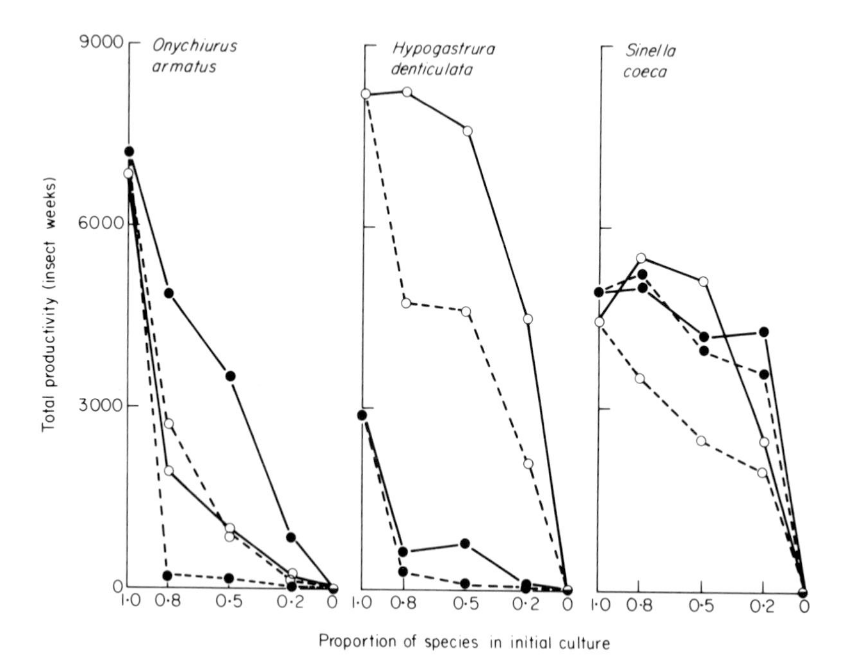


FIG. 17.6. Summaries of the results of competition (Longstaff 1976) by replacement series diagrams. Open and filled circles indicate cultures at 16°C and in the fluctuating laboratory environment respectively. On the *O. armatus* diagram the continuous and dashed lines indicate performance of the species in the presence of *H. denticulata* and *S. coeca*; on the *H. denticulata* diagram the lines represent the presence of *O. armatus* and *S. coeca*; and on the *S. coeca* diagram the lines represent the presence of *O. armatus* and *H. denticulata*, all respectively.

two of the interactions resulted in the depression of the productivity of only one species, while the interaction between *H. denticulata* and *S. coeca* depressed neither species and hence is likely to lead to long-term co-existence of both species.

Without identifying the mechanisms of competition/amensalism, and with totally unpredictable interactions such as the increased fecundity of *O. armatus* in the presence of *H. denticulata*, Longstaff was unable to predict the outcome of the three species interactions from the data he had collected from single-species populations and all their pairwise interactions.

### *Predation by one species on another*

Many groups of animals in the soil ecosystem have been identified as predators: spiders, beetles (especially Carabidae, Staphylinidae), pseudo-scorpions and mites (particularly the Mesostigmata). Whilst predators in all these groups (e.g. Christiansen 1971) will feed on the soil micro-arthropods, there is increasing evidence that the predatory mites exert the largest predatory influence on the species of soil micro-arthropods (e.g. the review by Huffaker, Vrie & McMurtry (1969) for the control of phytophagous mites and Wallace's (1967) study of the control of *Sminthurus viridis*). The importance of predation is indicated by the occurrence of defence and escape mechanisms in the potential prey. Collembola occurring in the litter layer have a well-developed jumping apparatus, the furcula, whilst those deeper in the soil appear to have a chemical defence mechanism which is at least partially effective in deterring predators (Usher & Balogun 1966).

Harris (1974b) studied the feeding of one species of mite, *Pergamasus longicornis*, on three species of Collembola and demonstrated that most of the life-history phenomena were dependent upon the population density of their prey. The time taken to find and capture a prey is dependent upon the density of prey for both juvenile (Fig. 17.7a) and adult mites (Fig. 17.7b). The sexes have different capture rates when the mites are juvenile at which stage the sexes are superficially identical, whereas when the mites are adult (and can be sexed by eye) the capture rates are more closely similar (regression lines are shown in Figs 17.7a,b). The speed with which the mites developed from larva to adult in a constant temperature of 20°C is linearly related to the rate with which prey were caught (Fig. 17.7c). Although the females capture prey quicker than males, there is no difference between the sexes in the relation of development time to capture rate. The implication of the regression line in Figure 17.7c,

$$D = 10{\cdot}3 + 18{\cdot}2t$$

where $D$ is the development time and $t$ is the time per capture in days, is that there is an upper ceiling to a mite's development rate, and that the slower the mite develops the less total prey it needs to catch for development to adulthood.

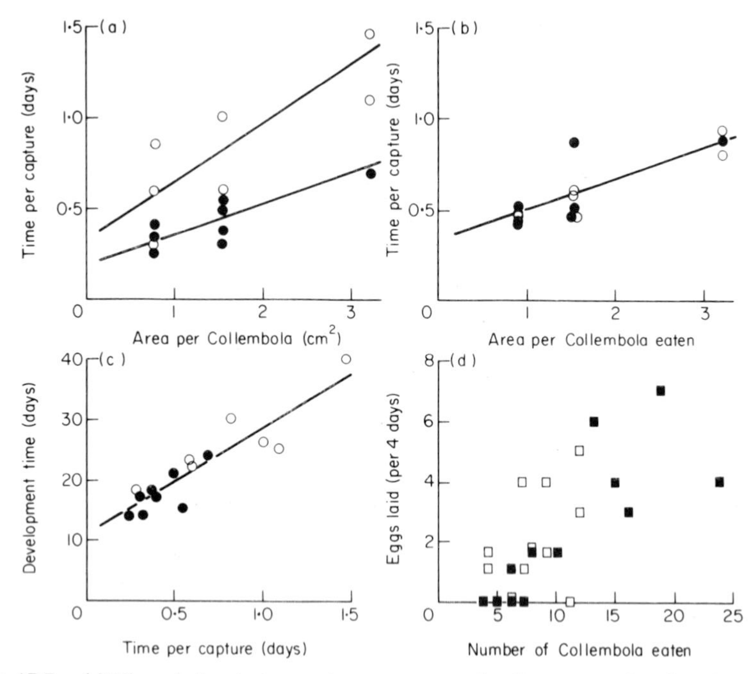


FIG. 17.7. (a) The relation between time per capture by *Pergamasus longicornis* and prey (*Folsomia fimetaria*) density over the period from the predator hatching from an egg till the final moult (Harris 1974b). (b) Similar data for adult *P. longicornis*. (c) The relation between the total development time of *P. longicornis* and the time required to capture and handle one prey individual. In these three diagrams open and filled circles represent males and females respectively. (d) The relation between the number of eggs laid per 4 days and the number of prey eaten by female *P. longicornis*. Open and closed squares indicate maximum prey densities of 0·65 and 1·30 *F. fimetaria* $cm^{-2}$ respectively.

The decrease in the number of prey required for development as *t* increases is contrary to what might be expected if maintenance of the mite took a significant proportion of the food it consumed. Similarly, there appears to be some upper asymptote to the relation between the number of eggs laid and the rate of ingestion of prey, though there is considerable variation in these data (Fig. 17.7d).

An interesting feature of *P. longicornis* is the variety of its effective functional responses. A least squares fit of an integrated version of the Holling disc equation (Harris 1974a) to the observed consumption of *F. fimetaria*, *H. denticulata* and *S. coeca* yielded the curves shown in Figure 17.8. The ranges of prey densities depicted probably span those encountered by the mite in the field (Harris 1974b). The type of the functional response in this predator clearly depends upon the prey on which it is feeding. When feeding on *F. fimetaria*, there is a typical type II response by both male and female mites, whilst both

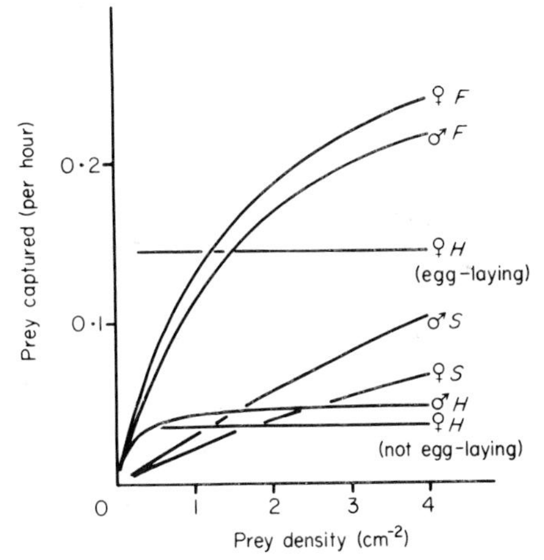


FIG. 17.8. The functional response of *P. longicornis* to prey density, the experimental data being smoothed by Harris' (1974a) model. The symbols indicate first the sex of the mite and second the prey species: *F*, *S* and *H* representing *F. fimetaria*, *S. coeca* and *H. denticulata* respectively. Female mites preying upon *H. denticulata* are divided into two sets according to whether or not they laid eggs.

sexes seem to show a linear response, without a plateau being reached, when feeding on *S. coeca*. Predation by either sex upon *H. denticulata* is effectively independent of prey density, the rise towards the plateau presumably occurring on lower densities than those included in the experiment. These results led Harris & Usher (1978) to speculate that, under predation by *P. longicornis*, populations of *S. coeca* are likely to be more stable than those of *F. fimetaria*, whilst populations of *H. denticulata* are likely to be very unstable. This expectation seems to be in accord with reported observations of swarming in the Collembola.

### *Predation by one species on more than one other*

Davis (1978) investigated predation upon two species of Collembola, *Hypogastrura denticulata* and *Sinella coeca*, at a constant temperature of 16°C. As *Pergamasus longicornis* had displayed excessive cannibalism under Harris' culture conditions, it was replaced by another predatory Mesostigmatid mite, *Hypoaspis aculeifer*. Cultures were set up to investigate each pairwise interaction between the three species, and the results were used to construct a three-species population model. The predictions of this model were compared with data gained from an experiment conducted over a nine-month period in which 56 different initial conditions, representing different proportions of the two-prey species and the life-history stages of the mite, were investigated.

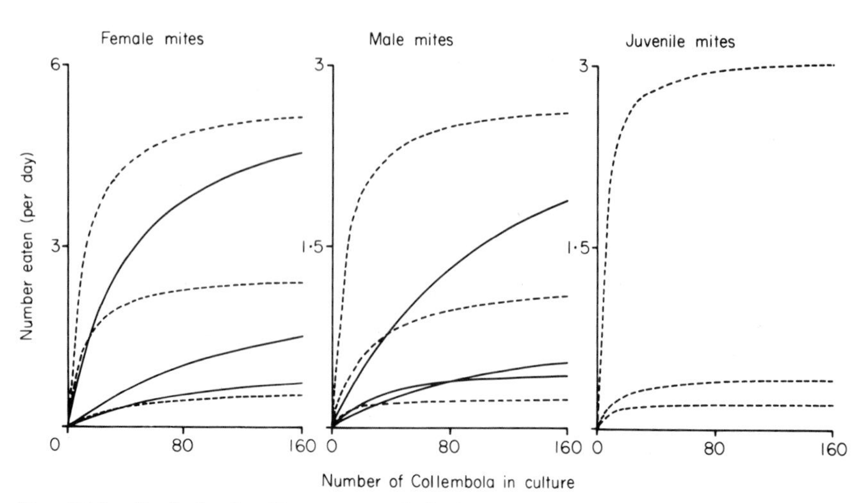


FIG. 17.9. Predation by *Hypoaspis aculeifer* (separated into male, female and juvenile) upon *S. coeca* (continuous lines) and *H. denticulata* (dashed line). In each illustration the upper curve represents predation on small prey, the centre curve on medium-sized prey and the lower curve on large prey. Since juvenile mites were largely unable to catch *S. coeca*, that family of curves has been omitted from the diagram. The curves were derived by Davis (1978).

In the presence of a single-prey species, *H. aculeifer* always displayed a type II response when preying upon one size class of either *S. coeca* or *H. denticulata* (see Fig. 17.9). Whilst the number of eggs laid was linearly related to an approximate measure of the biomass (the cube of the length of the size class concerned) of *S. coeca* eaten, *H. aculeifer* could hardly reproduce on a diet of only *H. denticulata*. When both species were present, maximum daily fecundity was in excess of that which could be expected from the amount of *S. coeca* eaten, and so it would appear that a diet of *H. denticulata* is deficient in some manner and that the presence of *S. coeca* in the diet overcomes this deficiency (Table 17.3). If such interactions between prey species in their effect upon the

TABLE 17.3. The fecundity of *Hypoaspis aculeifer* when feeding upon *Sinella coeca*, *Hypogastrura denticulata*, and both species in mixed culture. In all three experiments there were 40 prey individuals in the culture

| Prey | Number of females | Eggs per female (over a 12-day period) | Maximum number of eggs per female per day |
|---|---|---|---|
| 40 *S. coeca* | 18 | 7·72 | 2 |
| 40 *H. denticulata* | 18 | 0·67 | 1 |
| 20 of each species | 27 | 7·19 | 4 |

predator are common, the aim of the agglomerative approach to build simple studies into complex community studies will be impossible to achieve.

Davis found that *H. aculeifer* females, when offered a choice of prey, displayed switching only when they had been starved. The data were gained from a 12-day experiment, before which the predators were starved for one day, during which counts were made every four days. A significantly higher level of predation was recorded in the first four-day period, during which significant switching occurred towards the preferred prey species, *S. coeca* (using an assumption of a random predator, the observed kill of *S. coeca* was 6·67 individuals above that predicted by the model, whilst the kill of *H. denticulata* was only 0·41 above the prediction; for individuals in the second two four-day periods the figures were 1·01 and 0·94 respectively). The experiments also indicated that there was a significant amount of switching to the smaller size classes of *S. coeca* only during the first four-day period. There was no evidence of switching between size classes when *H. denticulata* was the only prey species. These results contrast to the predictions of the optimum feeding theory (MacArthur & Pianka 1966) and the findings of Ernsting (1977). It is possible that the results for *H. aculeifer* are due to increased activity of starved predators disturbing the more mobile prey species or the more mobile size class of the species, in which case the switching would be a reflection only of the physical interactions which occur in the simple experimental culture and which may not occur in the field. It is possible that *H. aculeifer*, operating in an environment in which food is scarce, and hence also hungry, would not exhibit switching.

Without the inclusion of switching, Davis' predation model always predicted a disturbance of co-existence at densities at which co-existence was to be expected, and an increase in the speed of competitive exclusion when that was the most likely result of the interaction between two prey species in the absence of predation. The results of the experimental cultures were variable (see Davis 1978), but the model's predictions were borne out in all but one of the 56 cultures assessed. It is possible that predators in the soil will experience a much lower density of prey than in culture, and it might therefore be predicted that predators would be hungry more frequently. Also, a diurnal rhythm of temperature and light may well tend to produce hungry predators at various times of the day. If switching by *H. aculeifer* occurs in the field it may result in opposite conclusions with respect to the co-existence of prey species. Although the requirement of simplicity has forced stenophagy upon the majority of predators used in laboratory studies, there is little or no evidence that species of predatory mites are stenophagous in the soil ecosystem and certainly none to refute the belief that the generalist predator is the more common.

A further observation is that *H. aculeifer* populations persisted longer in the total absence of prey than in the presence of only *H. denticulata* and, similarly, juvenile *P. longicornis* suffered a high mortality when feeding on this species (Harris 1974b). *H. denticulata* thus appears to be toxic to these predatory mites.

This may be interpreted as an evolutionary response to its life deeper in the soil than *S. coeca*, where it is less able to avoid predation by jumping.

### *Environmental heterogeneity*

The well-known experiments of Huffaker (1958) indicated that the persistence of simple communities of predatory and non-predatory mites was dependent upon the heterogeneity of the environment in which they live. However, there has been virtually no consideration of this factor in laboratory studies of the soil micro-arthropods. Culver (1974) introduced patchiness into his laboratory cultures, in which, over a 68-day period, *Folsomia candida* and *Lepidocyrtus cinereus* co-existed. However, due to the experimental design one cannot state whether co-existence is a phenomenon of these two species or of the less homogeneous culture conditions. It is true that soil biologists have been slow to experiment with concepts of patchy environments stabilizing either competitive relationships or predator–prey interactions (Murdoch 1977; Hastings 1977); this seems a fruitful field for future research.

## DISCUSSION

Attempts at understanding the structure of soil arthropod communities have been relatively unsuccessful. The review of approaches broadly grouped as 'divisive', which aimed at sampling the community and thereby unravelling the role of each species, has indicated that they have generally proved unsatisfactory. This could be because the approach does not deal in sufficient detail with the role of the individual species; it could be because it is like a 'black box' in which the results of interactions are seen but the nature of the interaction is unknown; it could be because the experimental (as opposed to sampling) and analytical techniques are not yet sufficiently developed; or it could be because the nature of the community does not allow for valid hypotheses to be made and drawn. The problem is essentially statistical in nature; efficient sampling and experimental designs might help to provide more interpretable data. More concentration on the methodology of multivariate analysis, to overcome the problem of relatively weak correlation in field data, might yield results. The factor analysis model, including as it does $k$ common factors as well as unique or residual variances, might give an approximation to niches in $k$-dimensional space. The availability of factor analysis models that are not linear, although not developed from a computational point of view, may prove attractive to ecologists.

Similarly, the 'agglomerative' approaches, aiming to predict ecosystem behaviour from single-species studies and pairwise interactions, have proved unsatisfactory because the interactions of three species together cannot be predicted from a knowledge of all the pairwise interactions; experiments indicate that 'the whole is more than the sum of the parts'. If this proves to be the general rule, it is obvious that the agglomerative approach has no future as a technique in community ecology, though it is essential for population ecology. Associated with the agglomerative approach are both statistical design and model building; the criterion for a successful experiment is the construction of a model that predicts accurately the outcome of the next experiment.

Neither the divisive nor the agglomerative approach is particularly useful in understanding the diversity of the community. Both approaches will inevitably fix on the more abundant species, and produce data on them. It is unlikely that sampling techniques will be suitable for estimating population parameters of both the abundant and the rare species, though one set of samples is used to do both tasks. No-one has yet used a sampling scheme which aims to sample the rare species of the soil fauna (usually more than half the total number of species) efficiently. Similarly, the acquisition of living material of the rare species in sufficient quantity is a difficult barrier to overcome before attempting to study their population parameters in the laboratory.

The realism of any study has also to be considered. Divisive approaches are usually working with field communities, and hence the question does not arise, though the sampling periods may be atypical (drought, colder than average, etc.) unless the period extends over several years. Laboratory studies have all been done in exceptionally homogeneous environments, none of which approaches field conditions. The effects of environmental heterogeneity need to be known if the results of laboratory studies are ever to be applied to field communities.

If neither the divisive nor the agglomerative approaches are satisfactory, what are to be used in their place? Five possibilities can be suggested.

First, no-one has sampled a community in the field whilst, at the same time, designing experiments with small numbers of species from the same community in the laboratory. Only when both approaches are tried on the same community will one know by how much they fail to satisfy the criteria of understanding the structure and functioning of the complete community.

Secondly, it would seem appropriate to attempt a combined divisive/agglomerative approach on a simple soil arthropod community, such as those of the Antarctic and sub-Antarctic (see, for example, Tilbrook 1973), where there is a chance that all species of arthropods can be encompassed both by the field sampling and by the laboratory population studies. A community of such low diversity thus has the advantage that both approaches might meet, but it has the disadvantage that results are probably not generally applicable to more diverse communities.

Thirdly, a greater knowledge of the trophic structure of the arthropod communities is required. Anderson (1975) lists some of the studies on feeding habits of soil animals, and he comments upon the apparent predominance of generalized feeding. In some instances food webs have been constructed, but the detail is usually insufficient for the studies of community dynamics. The recent use of radioactive traces, $^{14}C$ by Gifford (1967), $^{32}P$ by Ernsting & Joosse (1974) and $^{45}Ca$ by Kowal & Crossley (1971), provides means whereby the connections in the food web can be identified, and possibly whereby the flow of energy and nutrients quantified. Whether such studies will agree with Anderson's statement about generality of feeding (i.e. a high degree of connectivity) or whether they demonstrate that connectivity is less than we currently think (as indicated by the studies of Mills & Sinha 1971 and Visser & Whittaker 1977) remains to be seen.

Fourthly, the study of soil communities that are not in a steady state might indicate community phenomena that were previously unknown or unsuspected. Thus, the study of a community undergoing successional change may demonstrate the characteristics of species involved in invasion and extinction processes (e.g. Parr 1978).

And finally, perturbation experiments will demonstrate how the remaining species in a community react to a major disturbance. Such experiments have frequently been tried, especially by the application of insecticides. However, the majority of these studies have recorded only the number of 'mites', 'Collembola', etc., on treated and control plots, and do not identify the groups to specific level. Other perturbations, such as mowing, raking, grazing of different intensities, silvicultural practices, the addition or subtraction of litter or the application of fungicides, have been tried, but the majority of papers in this field do not record organisms at the species level.

Amongst these observational and experimental approaches to the community, it seems essential to include the parallel development of ecological theory. The problems of whether complexity begets or reflects stability, and the effect of a perturbing factor (Goh 1975), as happens with the periodic litter fall on top of the soil arthropod community, have relevance to understanding the structure of micro-arthropod communities. It is possible that the more stable soil environment and the less complex soil arthropod community have contributed to the biological stability shown in Figure 17.2.

## ACKNOWLEDGMENTS

We should like to acknowledge the financial support of the Natural Environment Research Council (to P.R.D.) and the Science Research Council (to J.R.W.H. and B.C.L.), and permission to work on sites at Rannoch (Forestry Commission), Givendale (Mrs B. Jackson) and Wharram Quarry (Yorkshire Naturalists' Trust Ltd.).

## REFERENCES

**Anderson J.M. (1971)** Observations on the vertical distribution of Oribatei (Acarina) in two woodland soils. *IV Colloquium Pedobiologiae* (Ed. by INRA) pp. 257–272. Proceedings of the 4th International Colloquium of Soil Zoology, Paris, 1970.

**Anderson J.M. (1975)** The enigma of soil animal species diversity. *Progress in Soil Zoology* pp. 51–58. Proceedings of the 5th International Colloquium of Soil Zoology, Prague, 1973. Academic, Czechoslovak Academy of Sciences.

**Athias F. (1975)** Donnée complémentaires sur l'abondance et la distribution verticale des microarthropodes de la savane de Lamto (Côte d'Ivoire). *Bulletin du Muséum National d'Histoire Naturelle, Écologie Générale*, **24**, 1–28.

**Butcher J.W., Snider R. & Snider R.J. (1971)** Bioecology of edaphic Collembola and Acarina. *Annual Review of Entomology*, **16**, 249–288.

**Christiansen K. (1964)** Bionomics of Collembola. *Annual Review of Entomology*, **9**, 147–178.

**Christiansen K. (1967)** Competition between collembolan species in culture jars. *Revue d'Écologie et de Biologie du Sol*, **4**, 439–462.

**Christiansen K. (1971)** Factors affecting predation on Collembola by various arthropods. *Annales de Spéléologie*, **26**, 97–106.

**Costigan P.A. (1975)** Experimental and computer studies on the response to exploitation of laboratory populations of *Onychiurus quadriocellatus* (Insecta: Collembola). *Unpublished B.A. thesis, University of York.*

**Culver D. (1974)** Competition between Collembola in a patchy environment. *Revue d'Écologie et de Biologie du Sol*, **11**, 533–540.

**Curry J.P. (1973)** The arthropods associated with the decomposition of some common grass and weed species in the soil. *Soil Biology and Biochemistry*, **5**, 645–657.

**Curry J.P. (1976)** The arthropod communities of some common grasses and weeds of pasture. *Proceedings of the Royal Irish Academy, Series B*, **76**, 641–665.

**Davis P.R. (1978)** Approaches towards modelling competition and predation in a simple community of soil arthropods. *Unpublished D.Phil. thesis, University of York.*

**Drift J. van der (1959)** Field studies on the surface fauna of forests. *Bijdragen tot de Dierkunde*, **29**, 79–103.

**Emlen J.M. (1973)** *Ecology: An Evolutionary Approach.* Addison-Wesley, Reading, Massachusetts.

**Ernsting G. (1977)** Effects of food deprivation and type of prey on predation by *Notiophilus biguttatus* F. (Carabidae) on springtails (Collembola). *Oecologia (Berlin)*, **31**, 13–20.

**Ernsting G. & Joosse E.N.G. (1974)** Predation on two species of surface dwelling Collembola: a study with radio-isotope labelled prey. *Pedobiologia*, **14**, 222–231.

**Ghilarov M.S. (1977)** Why so many species and so many individuals can coexist in the soil. *Ecological Bulletins (Stockholm)*, **25**, 593–597.

**Gifford D.R. (1967)** An attempt to use $^{14}C$ as a tracer in a Scots pine (*Pinus sylvestris* L.) litter decomposition study. *Secondary Productivity of Terrestrial Ecosystems* (Ed. by K. Petrusewicz) pp. 687–693. Panstwowe Wydawnictwo Naukowe, Warsaw & Krakow.

**Gisin H. (1960)** *Collembolenfauna Europas.* Museum d'Histoire Naturelle, Geneva.

**Goh B.S. (1975)** Stability, vulnerability and persistence of complex ecosystems. *Ecological Modelling*, **1**, 105–116.

**Green C.D. (1964)** The effect of crowding upon the fecundity of *Folsomia candida* (William) var. *distincta* (Bagnall) (Collembola). *Entomologia Experimentalis et Applicata*, **7**, 62–70.

**Haarløv N. (1960)** Microarthropods from Danish soils: ecology, phenology. *Oikos, Supplement No.* **3**, 1–176.

**Hale W.G. (1966)** A population study of moorland Collembola. *Pedobiologia*, **6**, 65–99.

**Hale W.G. (1967)** Collembola. *Soil Biology* (Ed. by A. Burges & F. Raw) pp. 397–411. Academic Press, London.

**Harris J.R.W. (1974a)** The kinetics of polyphagy. *Ecological Stability* (Ed. by M.B. Usher & M.H. Williamson) pp. 123–139. Chapman & Hall, London.

**Harris J.R.W. (1974b)** Aspects of the kinetics of predation with reference to a soil mite, *Pergamasus longicornis* Berlese. *Unpublished D.Phil. thesis, University of York.*

**Harris J.R.W. & Usher M.B. (1978)** Laboratory studies of predation by the grassland mite *Pergamasus longicornis* Berlese and their possible implications for the dynamics of populations of Collembola. *Scientific Proceedings of the Royal Dublin Society, Series A*, **6**, 143–153.

**Hartenstein R. (1961)** On the distribution of forest soil microarthropods and their fit to 'contagious' distribution functions. *Ecology*, **42**, 190–194.

**Hastings A. (1977)** Spatial heterogeneity and the stability of predator–prey systems. *Theoretical Population Biology*, **12**, 37–48.

**Huffaker C.B. (1958)** Experimental studies on predation: dispersion factors and predator–prey oscillations. *Hilgardia*, **27**, 343–383.

**Huffaker C.B., Vrie M. van de & McMurtry J.A. (1969)** The ecology of tetranychid mites and their natural control. *Annual Review of Entomology*, **14**, 125–174.

**Hurlbutt H.W. (1964)** Structure and distribution of *Veigaia* (Mesostigmata) in forest soils. *Acarologia* (*1st International Congress of Acarology*), 150–152.

**Hurlbutt H.W. (1968)** Coexistence and anatomical similarity in two genera of mites, *Veigaia* and *Asca*. *Systematic Zoology*, **17**, 261–271.

**Hutchinson G.E. (1957)** Concluding remarks. *Cold Spring Harbor Symposium in Quantitative Biology*, **22**, 425–427.

**Joosse E.N.G. (1969)** Population structure of some surface dwelling Collembola in a coniferous forest soil. *Netherlands Journal of Zoology*, **19**, 621–634.

**Joosse E.N.G. (1970)** The formation and biological significance of aggregations in the distribution of Collembola. *Netherlands Journal of Zoology*, **20**, 299–314.

**Joosse E.N.G. & Verhoef H.A. (1974)** On the aggregational habits of surface dwelling Collembola. *Pedobiologia*, **14**, 245–249.

**Kowal N.E. & Crossley D.A. (1971)** The ingestion rates of micro-arthropods in pine mor, estimated with radioactive calcium. *Ecology*, **52**, 444–452.

**Lawlor L.R. & Maynard Smith J. (1976)** The coevolution and stability of competing species. *American Naturalist*, **110**, 79–99.

**Lebrun Ph. (1970)** Écologie et biologie de *Nothrus palustris* (C.L. Koch 1839)—Acarien, Oribate. IV. Survivance, fécondité, action d'un prédateur. *Acarologia*, **12**, 827–848.

**Lions J.C. (1973)** Application du concept de la diversité spécifique à la dynamique de trois populations d'oribates (acariens) de la forêt de la Sainte-Baume (Var). *Écologie Méditerranea*, **1**, 165–192.

**Longstaff B.C. (1974)** Experimental and computer studies of competition in soil Collembola. *Unpublished D.Phil. thesis, University of York.*

**Longstaff B.C. (1976)** The dynamics of collembolan populations: competitive relationships in an experimental system. *Canadian Journal of Zoology*, **54**, 948–962.

**Longstaff B.C. (1977)** The dynamics of collembolan populations: a matrix model of single species population growth. *Canadian Journal of Zoology*, **55**, 314–324.

**MacArthur R.H. & Pianka E.R. (1966)** On optimal use of a patchy environment. *American Naturalist*, **100**, 603–609.

**MacArthur R.H. & Levins R. (1967)** The limiting similarity, convergence, and divergence of coexisting species. *American Naturalist*, **101**, 377–385.

**Macfadyen A. (1952)** The small arthropods of a *Molinia* fen at Cothill. *Journal of Animal Ecology*, **21**, 87–117.

**Maxwell A.E. (1977)** *Multivariate Analysis in Behavioural Research.* Chapman & Hall, London.

**Mills J.T. & Sinha R.N. (1971)** Interactions between a springtail, *Hypogastrura tullbergi*, and soil-borne fungi. *Journal of Economic Entomology*, **64**, 398–401.

**Miracle M.R. (1974)** Niche structure in freshwater zooplankton: a principal components approach. *Ecology*, **55**, 1306–1316.

**Moritz M. (1963)** Über Oribatidengemeinschaften (Acari: Oribatei) norddeutscher Laubwaldböden, unter besonderer Berücksichtigung der die Verteilung regelnden Milieubedingungen. *Pedobiologia*, **3**, 142–243.

**Morris M.G. (1969)** Differences between invertebrate faunas of grazed and ungrazed chalk grassland. III. The heteropterous fauna. *Journal of Applied Ecology*, **6**, 475–487.

**Morris M.G. (1971)** Differences between invertebrate faunas of grazed and ungrazed chalk grassland. IV. Abundance and diversity of Homoptera-Auchenorhyncha. *Journal of Applied Ecology*, **8**, 37–52.

**Murdoch W.W. (1977)** Stabilizing effects of spatial heterogeneity in predator–prey systems. *Theoretical Population Biology*, **11**, 252–273.

**Niijima K. (1971)** Seasonal changes in collembolan populations in a warm temperate forest of Japan. *Pedobiologia*, **11**, 11–26.

**Pande Y.D. & Berthet P. (1975)** Observations on the vertical distribution of soil Oribatei in a woodland soil. *Transactions of the Royal Entomological Society of London*, **127**, 259–275.

**Parr T.W. (1978)** An analysis of soil micro-arthropod succession. *Scientific Proceedings of the Royal Dublin Society, Series A*, **6**, 185–196.

**Poole T.B. (1961)** An ecological study of the Collembola in a coniferous forest soil. *Pedobiologia*, **1**, 113–137.

**Strebel O. (1932)** Beiträge zur Biologie, Ökologie und Physiologie einheimischer Collembolen. *Zeitschrift für Morphologie und Ökologie der Tiere*, **25**, 31–153.

**Tamura H. (1976)** Population studies on *Folsomia octoculata* (Collembola: Isotomidae) in a subalpine coniferous forest. *Revue d'Écologie et de Biologie du Sol*, **13**, 69–91.

**Tilbrook P.J. (1973)** The signy Island terrestrial reference sites. I. An introduction. *British Antarctic Survey Bulletin*, **33 & 34**, 65–76.

**Usher M.B. (1969)** Some properties of the aggregations of soil arthropods: Collembola. *Journal of Animal Ecology*, **38**, 607–622.

**Usher M.B. (1970)** Seasonal and vertical distribution of a population of soil arthropods: Collembola. *Pedobiologia*, **10**, 224–236.

**Usher M.B. (1971a)** Properties of the aggregations of soil arthropods, particularly Mesostigmata (Acarina). *Oikos*, **22**, 43–49.

**Usher M.B. (1971b)** Seasonal and vertical distribution of a population of soil arthropods: Mesostigmata. *Pedobiologia*, **11**, 27–39.

**Usher M.B. (1975a)** Some properties of the aggregations of soil arthropods: Cryptostigmata. *Pedobiologia*, **15**, 355–363.

**Usher M.B. (1975b)** Seasonal and vertical distribution of a population of soil arthropods: Cryptostigmata. *Pedobiologia*, **15**, 364–374.

**Usher M.B. (1976)** Aggregation responses of soil arthropods in relation to the soil environment. *The Role of Terrestrial and Aquatic Organisms in Decomposition Processes* (Ed. by J.M. Anderson & A. Macfadyen) pp. 61–94. Blackwell Scientific Publications, Oxford.

**Usher M.B. & Balogun R.A. (1966)** A defence mechanism in *Onychiurus* (Collembola, Onychiuridae). *Entomologists' Monthly Magazine*, **102**, 237–238.

**Usher M.B., Longstaff B.C. & Southall D.R. (1971)** Studies on populations of *Folsomia candida* (Insecta: Collembola). The productivity of populations in relation to food and exploitation. *Oecologia (Berlin)*, **7**, 68–79.

**Usher M.B. & Stoneman C.F. (1977)** *Folsomia candida*—an ideal organism for population studies in the laboratory. *Journal of Biological Education*, **11**, 83–90.

**Verhoef H.A., Nagelkerke C.J. & Joosse E.N.G. (1977)** Aggregation pheromones in Collembola (Apterygota): a biotic cause of aggregation. *Revue d'Écologie et de Biologie du Sol*, **14**, 21–25.

**Visser S. & Whittaker J.B. (1977)** Feeding preference of certain litter fungi by *Onychiurus subtenuis* (Collembola). *Oikos*, **29**, 320–325.

**Waldorf E. (1971a)** Selective egg cannibalism in *Sinella curviseta* (Collembola: Entomobryidae). *Ecology*, **52**, 673–675.

**Waldorf E.S. (1971b)** Oviposition inhibition in *Sinella curviseta* (Collembola: Entomobryidae). *Transactions of the American Microscopical Society*, **90**, 314–325.

**Wallace M.M.H. (1967)** The ecology of *Sminthurus viridis* (L.) (Collembola). I. Processes influencing numbers in pastures in Western Australia. *Australian Journal of Zoology*, **15**, 1173–1206.

**Wallwork J.A. (1967)** Acari. *Soil Biology* (Ed. by A. Burges & F. Raw) pp. 363–395. Academic Press, London.

**Wood T.G. (1967)** Acari and Collembola of moorland soils from Yorkshire, England. I. Description of the sites and their populations. *Oikos*, **18**, 102–117.

# 18. THE STRUCTURE AND DYNAMICS OF ECOLOGICAL COMMUNITIES

ROBERT M. MAY

*Biology Department, Princeton University,*
*Princeton, N.J. 08540, U.S.A.*

This concluding chapter is divided into two distinct parts. The first part is a brief review of what I conceive to be some of the main themes that have emerged recently from theoretical and empirical studies of the dynamics of single populations. These themes have been sounded in many of the earlier chapters in this book, and they serve to draw together a large and diverse amount of information about the way real populations behave in the laboratory and in the field. Much of this work, at the level of single populations, has the satisfying property of putting theoretical curves through, or at least near, collections of data points.

In contrast, the second part is almost wholly speculative. Here mathematical models are used to explore questions about the way multispecies systems are organized. The models are necessarily greatly oversimplified, and at best bear a metaphorical relation to real ecosystems. This second part begins with a review of the various studies suggesting that, as a mathematical generality, increasing complexity (more species, a richer web of interactions) leads to diminishing dynamical stability (less ability to withstand, or recover from, disturbances). It goes on to review several mechanisms whereby the structure of real ecosystems may reconcile special kinds of complexity with dynamical stability.

## THE DYNAMICS OF SINGLE POPULATIONS

Three ideas that run through this volume are recapitulated: the richness of dynamical behaviour latent in the simplest of non-linear equations; the occurrence of more than one stable population value; the effects of spatial 'roughness' or 'patchiness'. Looking ahead, some serious problems arising from the unpredictable nature of most environments are foreshadowed.

### *Stable points, stable cycles, chaos*

Most real populations consist of many distinct but overlapping age classes. Two extremes that are relatively easy to handle are where there is only one discrete age class (no overlap between generations) and where generations overlap continuously (so that population growth is a continuous process). In the former case, we have a sequence of discrete population values, $N_t$ (the population $N$ in generation $t$), described by a first-order difference equation; in the latter case, we have a continuous population variable, $N(t)$, described by a first-order differential equation.

For non-overlapping generations the population growth equation takes the form

$$N_{t+1} = F(N_t) \tag{18.1}$$

where $F(N)$ is in general some non-linear or density-dependent function of $N$. Some typical forms for $F(N)$ are shown in Figure 18.1. The essential features are a tendency at low densities for populations to grow from one generation to the next, at high densities for them to decrease and for the severity of this non-linear response (the steepness of the hump in Fig. 18.1) to be 'tuned' by one or more biological parameters.

The dynamical behaviour of a population described by eqn (18.1) depends on how steeply non-linear are the density-dependent mechanisms embodied in $F(N)$; that is, on how steep is the hump in Figure 18.1. For modestly non-linear $F(N)$ the population will tend to settle stably to a constant value (a stable point). As the hump steepens, this stable point gives way to stable cycles in

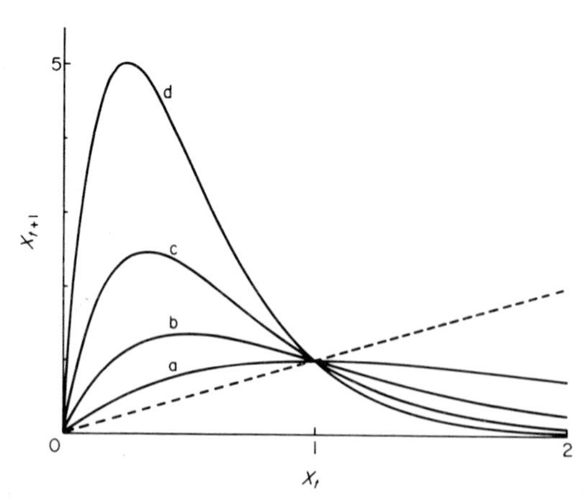


FIG. 18.1. The solid curves (a)–(d) represent some typical functions $F(X)$ relating the population in generation $t$, $X_t$, to that in generation $t + 1$, $X_{t+1}$, cf. eqn (18.1). The dashed line at 45° represents zero population growth, $X_{t+1} = X_t$; possible equilibrium population values lie at its intersection with the function $F(X)$. Specifically, curves (a) and (b) here describe populations with a stable equilibrium point; curve (c) gives stable cycles; and curve (d) gives chaotic dynamics. For a more full discussion of such figures see May & Oster (1976).

which the population alternates up and down, with successive generations assuming values above and below the stable point value. As the hump steepens yet further, manifesting an increasingly severe non-linearity, these stable cycles give way in turn to dynamic behaviour which, despite the simple and rigidly deterministic nature of eqn (18.1), is effectively indistinguishable from the sample function of a random process.

This progression from a stable point, through stable cycles to chaotic behaviour, is reviewed by May & Oster (1976), where both the mathematical details and the biological implications are developed more fully.

For a population with continuously overlapping generations population growth is continuous and obeys an equation of the form

$$dN/dt = f(N) \tag{18.2}$$

Here, as before, $f(N)$ is some non-linear or density-dependent function; the most commonly assumed form is the simple logistic, $f(N) = rN(1 - N/K)$. More realistically, the density-dependent mechanisms that bear upon population growth rates will incorporate time delays arising, for example, from the time vegetation takes to recover from grazing, or from the time before the effects of crowding show up in decreased numbers of adults. As a result, the function $f$ on the right-hand side of eqn (18.2) depends not just on $N(t)$ but on past values of $N$ as well.

One simple and thoroughly explored form that exhibits this is the so-called 'time-delayed logistic' (Hutchinson 1948; May 1973),

$$dN/dt = rN[1 - N(t - T)/K] \tag{18.3}$$

If the time delay ($T$) in the regulatory mechanism is short compared to the intrinsic biological response time of this system (the inverse of the growth rate, $1/r$), that is if $rT$ is small (specifically, if $rT < \pi/2$), there is a stable equilibrium value at $N = K$. However, if $rT$ is large (specifically, if $rT > \pi/2$), the system oscillates stably in 'limit cycles' whose period and amplitude are set by the biological parameters $r$ and $T$; the amplitude or ratio between maximum and minimum populations in such cycles can be very severe as $rT$ increases, but the period of the cycles remains around $4T$. Although eqn (18.3) exhibits only a stable point, or stable cycles, other more complicated forms for the density-dependent function $f(N)$ can, once time lags are present, unfold the full spectrum of stable points, stable cycles and chaos (Mackey & Glass 1977; MacDonald 1978).

This dramatic array of dynamical behaviour that lurks in deceptively simple, deterministic models for single populations (be they discrete or continuous) has several implications.

The detailed mathematical structure of these phenomena (as reviewed by

May 1976a) is of intrinsic fascination. As stressed by several people, the transition to the chaotic regime has analogies with the problem of turbulence in fluid dynamics. There is a legitimate worry that attempts to use simple deterministic models to understand the behaviour of populations may often be frustrated by the intractable technical difficulties that turbulent behaviour poses in other disciplines (May & Oster 1976; Guckenheimer, Oster & Ipaktchi 1976).

Taking a more cheerful view, it does seem that much of the observed behaviour of field and laboratory populations can be understood on the basis of the spectrum of dynamical possibilities sketched above. The papers by Lawton and McNeill and by Hassell in this volume provide two examples. A more deliberate attempt to use the theory in a quantitative way to interpret many laboratory studies is in May *et al.* (1974) and May (1976b, ch. 2). A fresh example is given in Figure 18.2. The task of reviewing field studies in this light is complicated by the fact that most natural populations are embedded in a multi-species complex, to which violence may be done by the assumption that we can study single-species dynamics in isolation (Hassell, Lawton & May 1976). One broad insight that can be usefully extracted is a detail-independent explanation of 'wildlife's four-year cycle' (Elton 1942): density-dependent regulatory mechanisms in boreal environments are likely to operate with characteristic time lags ($T$) of one year; for mice, voles, lemmings and other relatively small and fecund animals we expect relatively large $r$, and thence relatively large values of $rT$; the upshot will be stable cycles whose amplitude may be very variable but whose period is of the order of $4T$ or four years (May 1976b, ch. 2).

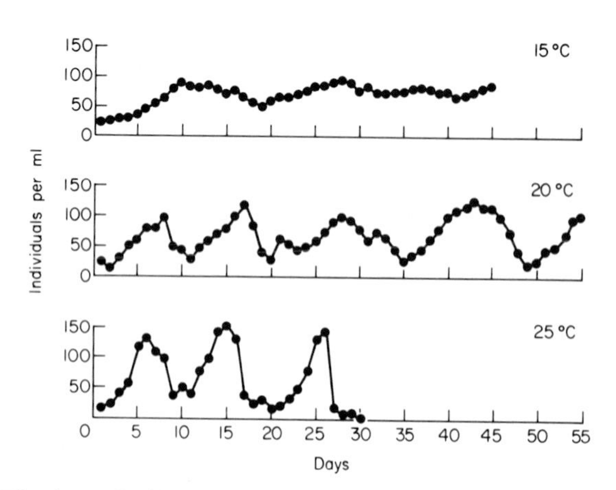


FIG. 18.2. The dynamical behaviour of laboratory populations of rotifers are shown as functions of time for three different temperatures (from Halbach 1979). A plausible interpretation is that increasing temperature leads to faster biological growth rates $r$, and thus to increasing values of $rT$ in some population growth equation of the general form of eqn (18.2); as a result, the dynamics progresses from a stable point to stable limit cycles of increasing amplitude.

*Alternative stable states: thresholds and breakpoints*

Figure 18.1 illustrates the simplest situation for a single population in which there is only one possible equilibrium point (which may or may not be stable). In many circumstances, however, we have the situation depicted in Figure 18.3. Here there is a regime (corresponding, for example, to intermediate levels of predation) in which the function $F(N)$ in eqn (18.1) has not one but two humps, giving rise to two alternative possible equilibrium points.

The existence of a regime with two such equilibrium points has been documented recently in several contexts: for vegetation biomass (as a function of stock density in a managed pasture or rangeland; Noy-Meir 1975); for fish and whale populations (as a function of harvesting intensity; Clark 1974; Gulland 1975; Peterman 1979 this volume); for near-shore marine communities (as a function of the densities of sea otter populations; Simenstad, Estes & Kenyon 1978); for insect pest populations (as a function of predation levels; Southwood 1977a; Southwood & Comins 1976; Ludwig, Jones & Holling 1978; Peterman 1979 this volume); and for various host–parasite systems (MacDonald 1965; Bradley & May 1978; Anderson 1979 this volume).

Mechanisms common to all these examples are illustrated in Figures 18.4 and 18.5. In Figure 18.4 the solid line depicts the intrinsic biological growth

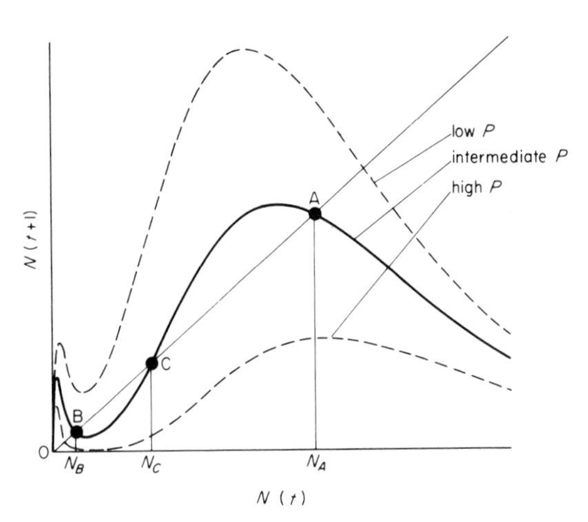


FIG. 18.3. This is a more complicated version of Figure 18.1, where now the curve relating the population in generation $t$, $N(t)$, to that in generation $t + 1$, $N(t + 1)$, can have two humps and three intersections with the 45° line corresponding to zero population growth. As discussed in the text, at low levels of predation there is only one possible equilibrium point; at intermediate levels of predation there can be two (possibly stable) equilibrium points at A and B, separated by an unstable point at C; at high levels of predation there is again a unique possible equilibrium point. (From May 1977, where details are given.)

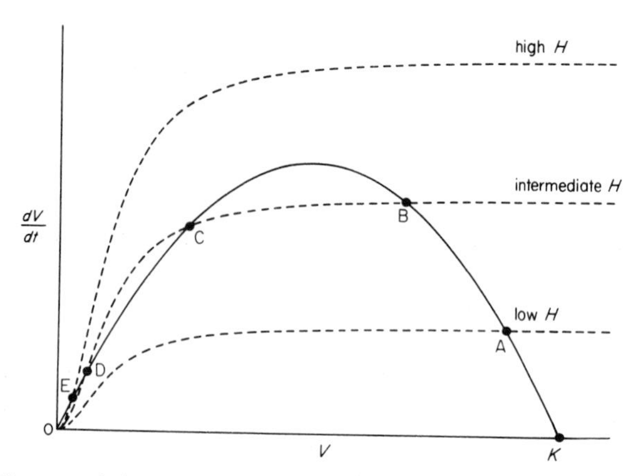


FIG. 18.4. The rate of change of vegetation biomass (or of fish population or insect population), $dV/dt$, is shown as a function of $V$. The solid curve is the natural, ungrazed vegetation growth rate. The dashed curves are loss rates due to grazing (of type III pattern) at high, intermediate and low herbivore densities, $H$. Where the solid curve lies above the dashed one, the net growth rate is positive; where the solid curve lies below the dashed one, the net growth rate is negative; the points of intersection of the curves correspond to possible equilibrium points. For further discussion see the text. (From May 1977.)

rate of the population in question (vegetation, fish, insect pests) as a function of its density. To this logistic style of intrinsic growth is added mortality from predation (which may be herbivores on the vegetation, or fishers on the fish, or predators on the insect pests). The predation is 'type III', rising faster than linearly at low prey density but saturating to a constant per predator at high

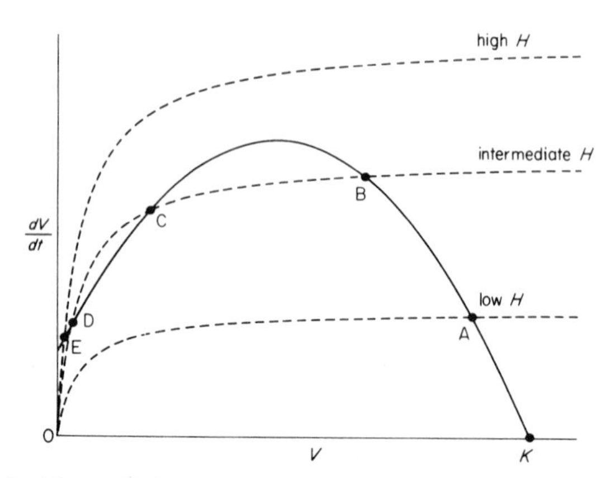


FIG. 18.5. As for Figure 18.4, except here the natural growth rate is finite even for $V = 0$ (corresponding to immigration of seeds, or fish, or insects), and the loss rate due to grazing (or harvesting, or predation) is of type II.

densities. The outcome is clearly a unique equilibrium point at high and at low predation levels, with an intermediate level of predation at which there are two possible stable equilibrium points (at B and D) separated by an unstable point (at C). Figure 18.5 illustrates an equivalent situation where the intrinsic biological growth curve for the population has a finite value at zero population density (corresponding to gains coming from immigration of seeds, or fish, or insect pests) and where the superimposed mortality from predation is of 'type II' form (rising linearly at low prey density). Again the upshot is a unique state at high and at low predation levels, with the possibility of two stable equilibrium states at intermediate levels of predation.

The net result is shown schematically in Figure 18.6, which displays the possible equilibrium values of the vegetation biomass (or fish population, or insect pest population) as a function of the level of predation by grazing stock (or harvesting fishermen, or natural predators). At low predation levels there is a unique equilibrium state, corresponding roughly to that set by environmental resources. At some *threshold* level of predation, $T_1$, there discontinuously appears a second alternative stable state. Which of the two states the system settles towards depends on the initial conditions; the system will recover from small disturbances, but sufficiently large disturbances (crossing the dashed '*breakpoint*' line in Fig. 18.6) will carry the system from one stable

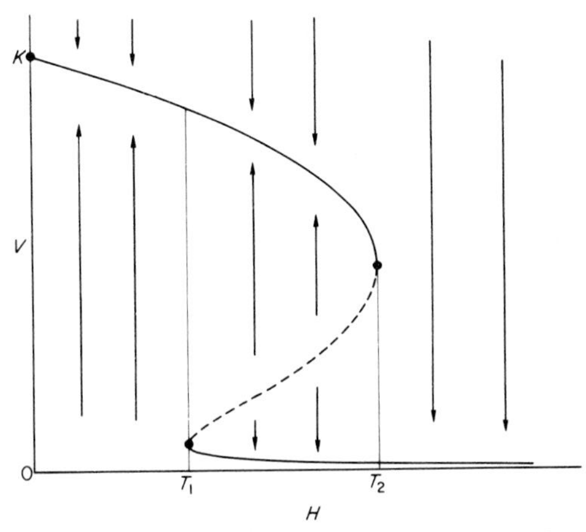


FIG. 18.6. The equilibrium values of the vegetation biomass (or fish population, or insect prey population), $V$, are shown as a function of the stocking density (or harvesting rate, or predator density), $H$. For a fixed value of $H$ below the lower threshold at $T_1$, or above the upper threshold at $T_2$, there is a unique equilibrium value of $V$; any initial $V$ value will move to this equilibrium, as indicated by the arrows. For $H$ between $T_1$ and $T_2$ there are two alternative equilibria for $V$; as shown by the arrows, the system will move to the upper or lower equilibrium, depending on whether the initial value of $V$ lies above or below the dashed 'breakpoint' curve.

state to the other. Finally, as the predation level increases beyond a second threshold level, at $T_2$, the original equilibrium state disappears discontinuously and the system again possesses a unique equilibrium state, now set predominantly by predation.

This qualitative scheme for the appearance of alternative stable states has been applied by the authors listed above to give a quantitative account of observed data on grazing ecosystems (mainly sheep), on various fisheries, on alternate near-shore communities in the Aleutian Islands, on the outbreaks of the eucalyptus psyllid *Cardiaspina albitextura* in Australia, on the introduction of the European sawfly *Diprion hercyniae* in Canada and on the behaviour of the spruce budworm in Canada. An attempt to draw these threads together in a unifying review is by May (1977).

Two additional comments are worth making.

First, note that the dynamics in the neighbourhood of one or both of the alternative possible equilibrium points may be a stable point, or a stable cycle, or chaos, just as for the simpler models. This leads to complications that have not yet been fully discussed.

Second, some of the above remarks can be recast, *post hoc*, in the language of catastrophe theory. I think nothing is gained thereby. (In mathematical jargon catastrophe theory is abbreviated as CT, which initials, by a felicitous coincidence, have a lewder meaning in American demotic; in biology catastrophe theory has so far proved indeed a tease, all promise and no fulfillment.)

### *Spatial 'roughness' or 'patchiness'*

Another theme that has pervaded this book (in the chapters by Taylor & Taylor; Steele; Lawton & McNeill; and Hassell) is the influence of spatial inhomogeneity on population dynamics.

The roles that spatial 'roughness' or 'patchiness' can play, particularly in enabling the overall persistence of vagile, 'boom-and-bust' populations, has recently been the subject of important reviews by Levin (1976), Southwood (1977b), Taylor & Taylor (1977) and Levandowsky & White (1977). I shall not attempt to recapitulate or summarize these reviews here, but just state what I think is the main moral to emerge: the type of analysis outlined in the previous sections assumes spatial homogeneity, which may apply to laboratory populations; but for field populations complications stemming from inhomogeneities will be the usual case. It is, therefore, heartening to see the studies of insect host–parasitoid (Hassell 1979 this volume), host–parasite (Anderson 1979 this volume) and general plant–herbivore or insect prey–predator (Lawton & McNeill 1979 this volume; Beddington, Free & Lawton 1978), which include some of the best quantitative data available to ecologists, attempting to deal with these complications.

### *Stochastic versus deterministic worlds*

The above themes have been largely developed within settings where the biological and environmental parameters are deterministic. This is, furthermore, true of other central ideas in contemporary evolutionary ecology: the search for optimal life-history strategies (represented in this volume by the chapter by Law) takes place in a basically deterministic framework; computations of 'evolutionarily stable strategies' (ess) (Dawkins 1976; Maynard Smith 1976; Hamilton & May 1977; Cowie & Krebs 1979 this volume) are done for models with well-defined environmental parameters; the goal of optimizing sustained yield (with or without economic constraints; Beddington, Steele & Peterman 1979 this volume) is basically a deterministic one.

Many environments, however, are subject to unpredictable fluctuations which introduce ineluctably stochastic elements into any calculation. At one level such continual environmental perturbations enable history and accident to promote variety and diversity (as emphasized by Connell 1979 this volume) or to set a chancy stage upon which a limited repertoire of patterned plays are acted out (as in Diamond 1979 this volume).

In a deeper sense such stochasticity can undercut any notion of *the* optimum sustained yield, or *the* optimum life-history strategy, or *the* ess. As is clear in the management of investment portfolios, trade-offs between average yield and acceptable levels of risk admit no unique solution, but rather depend on the diverse goals of different investors. In more technical language environmental unpredictability and concomitant historical accidents are not consistent with the existence of a 'gradient field', and without an underlying gradient field the concepts of unique optimum sustained yield, or ess, or optimal life-history strategies lose their meaning. Attention has been called to this worrisome set of general issues by Oster (1979).

## THE ORGANIZATION OF MULTISPECIES SYSTEMS

Throughout the bulk of this volume we have seen mathematical models that are fairly tightly tied to data. This is no longer possible when we move on to multispecies models, at least not if the models are to be sufficiently simple for us to learn anything general from them.

Oversimplified caricatures of reality nevertheless have a use, if only in posing questions and sharpening discussion. In this second part of my chapter I hope to show how such multispecies models are useful as one among many tools for working on the question of the relationship between stability and complexity in real ecosystems. Beginning with a survey of past work based on very general models for complex ecosystems, I review several suggestions that have been made as to tactical tricks whereby real ecosystems may reconcile special non-random complexity with dynamical stability.

### *Stability versus complexity in general models*

As discussed at length elsewhere (May 1974), by stability I mean either the ability to recover from some particular perturbation or the (related) ability of the system to persist in the face of the continual stream of environmental disturbances to which real communities are subjected. Complexity is loosely taken to increase as the number of species increases and/or as the number of interconnections among species increases and/or as the average strength of the interactions increases.

In these senses Elton (1958) has argued that increased complexity begets increased stability, a notion re-echoed by many people in the 1960's. One ingredient of Elton's web of arguments draws upon the dynamical instability that characterizes many simple prey–predator models.

However, if simple one prey–one predator models (of the kind introduced by Lotka and Volterra) are compared with the analogous *n* prey–*n* predator models, it becomes clear that the multispecies models are increasingly likely to be grossly unstable as *n* increases (May 1974). This relatively early piece of work was for the specially antisymmetric Lotka–Volterra prey–predator equation. Several later studies have used more general versions of the Lotka–Volterra equations to explore the relation between stability and complexity. If $N_i(t)$ is the population of the $i$th species ($i = 1, 2, 3, \ldots, S$ if $S$ species are present) at time $t$, these equations are

$$dN_i/dt = r_i N_i \left[1 - \sum_{j=1}^{S} a_{ij} N_j\right] \qquad (18.4)$$

Here $a_{ij}$ gives a measure of the effect of the presence of the $j$th species upon the growth rate of the $i$th species; the specially (and unrealistically) antisymmetric prey–predator models have $a_{ij} = -a_{ji}$. Assigning the coefficients $a_{ij}$ random values that are equally likely to be positive or negative (except for the diagonal elements, $a_{ii}$, which are kept negative), Roberts (1974) has shown numerically that the probability to get a 'feasible' equilibrium, at which all $S$ species are present with positive population densities, $N_i(\text{equil}) > 0$ for all $i$, decreases as $S$ increases; in the unlikely event that a feasible equilibrium does exist in one of Roberts' models, it will usually be stable against disturbances. Goh & Jennings (1977) have provided analytic insights into Roberts' studies, showing that the probability to have a feasible equilibrium in his 'randomly assembled' $S$-species ecosystems is roughly $2^{-S}$. Gilpin (1975) has further shown that if the assumption that all species are autotrophs (expressed as $a_{ii} < 0$ for all $i$) is relaxed, then not only does the probability to find a feasible equilibrium decrease as $S$ increases, but also the probability for such a feasible equilibrium to be stable decreases with increasing $S$. A particularly nice numerical study of these Lotka–Volterra ecosystems is due to Gilpin & Case (1976), who show that the typical 'randomly assembled' $S$-species system will collapse to one or other of many alternative simpler systems, usually comprising only two to four species. Their numerical results suggest that the number of alternative stable states scales exponentially with $S$; the particular simpler state to which the system settles after the extinction of most species depends on the initial conditions (that is, on the initial species mix). This is a very brief summary of a large and growing literature, all bringing the message that the more species that are stirred into the pot in these abstractly general models the harder it is to end up with a stable ecosystem.

A different and interesting approach to the stability–complexity question is by Caswell (1976). He frames a 'neutral hypothesis' by studying the likely distribution of relative abundance of individuals among $S$ species, assuming *no* biological interactions among them. Comparing the 'neutral' distributions of species' relative abundance with real ones for birds, fish, insects and plant species in tropical and temperate zones, Caswell finds the real communities to be less diverse (either in the sense of fewer species or of greater dominance by a few common species) than would be the case in the absence of interspecific interactions. The discrepancy is greatest in the tropics, where biotic effects are most pronounced. Caswell concludes 'the diversity of natural communities may be maintained in spite of, rather than because of, such [biological] interactions [between species]'.

A very abstract way of approaching the question is to eschew particular dynamical models, such as the Lotka–Volterra system of eqn (18.4), and ask only about the dynamical behaviour of the populations in the neighbourhood of their equilibrium point. That is, we write $X_i(t)$ to denote the fluctuation about equilibrium in the $i$th population, $X_i(t) = N_i(t) - N_i(\text{equil})$, and study the linear dynamical system,

$$dX_i/dt = \sum_{j=1}^{S} b_{ij} X_i \tag{18.5}$$

Here the coefficients $b_{ij}$ measure the influence of the $j$th species upon the rate of change of the $i$th in the neighbourhood of equilibrium. Numerical (Gardner & Ashby 1970; McMurtrie 1975) and analytic (May 1972) studies of randomly assembled food webs of this kind involve three parameters: the number of species, $S$; the average connectance of the web (the number of food links in the web as a fraction of the total number of topologically possible links), $C$; the average magnitude of the interaction between linked species, $b$. If all self-regulatory terms are taken to be $b_{ii} = -1$, for large $S$ these systems will tend to be stable if

$$b\sqrt{(SC)} < 1 \tag{18.6}$$

and unstable otherwise. Thus, increasing complexity, in the sense of an increasing number of species $S$, or of increasing connectance $C$, or of increasing average interaction strength $b$, works against dynamical stability. Broadly similar conclusions follow from qualitative considerations of food web topology (Levins 1975).

### *Conclusions based on general models*

The above work all points to increased complexity being associated with decreased dynamical stability for general 'randomly constructed' model ecosystems.

Two intertwined conclusions emerge (May 1974, pp. 3–4 and 75–77).

Other things being equal, we could expect increasing numbers of species in a community to be associated with increased dynamical fragility and diminished ability to withstand a given level of environmental disturbance. Thus, a relatively stable or predictable environment may permit that fragile thing—complexity and species richness—while a relatively unstable or unpredictable environment requires a dynamically robust and therefore relatively simple ecosystem.

On the other hand, real communities are not assembled randomly. They are the winnowed products of the long workings of evolutionary processes. What special structural features of real ecosystems may help reconcile species richness and apparent complexity with dynamical stability? This question has recently been pursued in various directions.

### *Loosely coupled subsystems*

Suppose that in the food web described by eqn (18.5) the connections in the $b_{ij}$ matrix are not made at random but are constrained to be mainly in relatively small blocks. As observed by May (1972) and McMurtrie (1975), such non-randomly connected webs can then be significantly more stable than would be the case for the same values of $S$, $C$ and $b$ in a random web.

Siljak (1974, 1975a,b) has developed these ideas more fully and rigorously, introducing the notion of 'connective stability'. If a system can be disassembled (in a specified formal sense) into loosely coupled subsystems, it can be stable even with strong interactions or with time-dependent changes in the interactions among the components of the system. Goh (1978) has refined Siljak's analysis to account for the fact that state variables in ecological systems must be non-negative. Drawing this and other work together in a masterly review, Goh (1978) considers the stability of ecosystem models 'subjected to (a) large perturbations of the initial state, (b) certain finite changes in system parameters, (c) continual disturbances of the system dynamics and (d) invasions and extinction of non-endemic species. The main conclusion is that a complex ecosystem model is robust relative to all types of perturbations if (a) it is a collection of subsystems each of which is self-regulating and (b) from the total system point of view the interactions between subsystems are weaker than the self-regulating interactions of the subsystems.'

Given that assembly as a set of loosely coupled subsystems enables a community to reconcile species richness with dynamical robustness, the question remains whether real ecosystems avail themselves of this trick. Gilbert (1975) has suggested that his *Helioconus–Passiflora* systems play this game. Lawton & Pimm (1978) and Beddington & Lawton (1978) have argued the case more broadly, observing, for example, that most insect herbivores are monophagous or oligophagous, giving rise to relatively discrete food chains even in species-rich plant communities.

McNaughton (1978) has recently endeavoured to test some of these ideas in the field. He collected data on plant species in 17 grassland stands in Serengeti National Park (during May and June 1977) and estimated the parameters pertinent to eqn (18.5) as follows: $S$ is the number of species in the stand; $b_{ij}$ is the botanist's point correlation coefficient for nearest neighbour data (being +1 if two species always occur together and −1 if they never do); $b$ is the average magnitude of $|b_{ij}|$; and the connectance $C$ is the proportion of all $b_{ij}$ that are significantly different from zero. By studying a system where species compete within a single trophic level, McNaughton avoids some of the objections to using the model (18.5) which ignores much of the structure of food webs. Both the average interaction strength $b$ and the connectance $C$ manifest significant declines as the species richness $S$ of the grasslands increases. Particularly relevant to the notion that some communities may be organized into 'blocks' of strongly interacting species is McNaughton's observation that the product $SC$ was remarkably constant (at $4{\cdot}7 \pm 0{\cdot}7$, 95% confidence interval) for his 17 stands. Finally, McNaughton tested the idea that the combination $b\sqrt{(SC)}$ might be a very crude measure of the system's ability to withstand disturbance, as in eqn (18.6), with relatively small values implying relatively high stability. In four places he was able to compare the species' relative abundance patterns in unprotected grassland with plots that had been protected from grazing for more than 10 years; there is a significant tendency for grazing to cause smaller changes in the plots with small values of $b\sqrt{(SC)}$. Indeed, in view of the crudities of the model and the difficulties of the field study, their accord is probably largely coincidence.

The skeptic's view has been well put by Murdoch (1979), who notes that many of the most intensively studied natural communities (such as the rocky intertidal and freshwater communities) do not appear to be comprised of loosely coupled subunits.

*Number of trophic levels*

An important property of natural ecosystems is that food chains are typically short, rarely consisting of more than four or five trophic levels (Hutchinson 1959).

This point has recently been emphasized by Pimm & Lawton (1977). Cohen (1978) and Pimm (private communication) have given it quantitative buttressing by analysing data compiled by Cohen for 19 food webs containing 102 top predators (species themselves free from predation); the 19 webs embrace terrestrial, freshwater and marine examples. Cohen and Pimm have independently traced out all the food chains connecting top predators to basal species (plants, detritus, or arthropods falling into freshwater systems). The number of trophic levels is fairly consistently around three, and for only one of the 102 top predators does Pimm find a food chain involving more than six species (five links).

Such steady patterns in the number of trophic levels are in pronounced contrast to the great variability in the amount of energy flowing through different ecological systems. Primary productivity varies over three to four orders of magnitude in both terrestrial and aquatic ecosystems, and the productivity of fish and of terrestrial animal populations varies over five or more orders of magnitude. There are further variabilities in the efficiency of energy transfer from one level to the next, with such efficiencies typically being much lower for warm-blooded than for other animals.

The conventional explanation for the number of trophic levels is that they are determined by energy flow. If only, say, 10% of the energy entering one level is effectively transferable to the level above it, the number of levels is clearly limited. However, as Pimm and Lawton emphasize, this explanation is not easily reconciled with the number of trophic levels being essentially independent of enormous variations in the amount of energy flow and in the transfer efficiencies: 'Food chains are not noticeably shorter in barren Arctic and Antarctic terrestrial ecosystems compared with a productive tropical savannah or the fish guilds of a tropical coral reef.'

Pimm and Lawton alternatively suggest the explanation may lie in the dynamics of the various populations in the community. They use numerical studies of the stability properties of variously structured Lotka–Volterra models to argue that long food chains may typically result in population fluctuations so severe as to make it hard for top predators to exist. Saunders (1978) has supplied analytic insight into Pimm and Lawton's numerical studies, observing that because stabilizing density-dependent effects enter via the primary producers (at the foot of the trophic ladder), the stabilization is attenuated as food chains lengthen. De Angelis *et al.* (1978) make the additional point that, within Pimm and Lawton's Lotka–Volterra models, long food chains may be kept relatively stable provided the transit time of a molecule or a unit of energy through the web is fast. This suggestion is an interesting one; a corollary (Janetos private communication) is that food chains involving endothermic species may be longer than those for ectotherms, because transit times for passage of energy units through the web are undoubtedly shorter in the former systems than in the latter. I am not aware of a review of the available data from these viewpoints.

Although Pimm and Lawton's idea that the number of trophic levels may be limited by considerations of dynamical stability is broadly supported by these studies on Lotka–Volterra models, the notion must remain subject to mistrust, as deriving from special mathematical models.

In short, this discussion of the length of food chains evokes familiar echoes: the empirical patterns are widespread and abundantly documented, but there is no compelling explanation.

## *Other constraints and patterns in food webs*

In addition to having a limited number of trophic levels, and possibly being organized into loosely coupled blocks, the interrelations among species in real food webs can be constrained in many ways that make the system very different from a randomly assembled one. This point, first made by May (1974, pp. 76–77), has been trenchantly re-emphasized by Lawlor (1978).

Lawlor suggests that if an interaction matrix $b_{ij}$ is to represent a biologically reasonable system, there are several constraints on its sign structure, including: (i) not more than five to seven trophic levels (which, *pace* previous section, he attributes to thermodynamics); (ii) there should be no food loops in which species are consumed by other species lower in the trophic hierarchy (a constraint which ignores the role of decomposers); (iii) there must be primary producers, with all matrix elements either zero or negative (or, if positive, mutualistic). As $S$ becomes large the fraction of randomly connected matrices that are in this sense reasonable becomes negligible. For $S \gg 1$ the fraction of randomly connected matrices with no three species food loops goes roughly as $\exp(-m^3C^6/192)$. This analysis is too simple, as it overlooks the often crucial role of decomposers, but it is indicative of the need to study the effects of biologically realistic constraints. A miscellany of such studies will now be catalogued.

### (*a*) *Donor control*

De Angelis (1975) has observed that in many systems there is an intimate relation between the diagonal and off-diagonal elements in the matrix of interaction coefficients; if the net biomass flow is conserved (so that none is lost from the system), then the rate of change of perturbed biomass of one species is balanced by compensating rates of change in other species. De Angelis has argued that such 'donor-dependent' effects, coupled with assimilation efficiencies that are explicitly less than 100%, can contribute negative (and therefore stabilizing) components to the diagonal matrix elements that *increase* as the number of species $S$ and connectance $C$ increase. I think that a rough qualitative understanding of De Angelis' numerical results follows from the observation that his donor control effects add to the typical diagonal matrix element in eqn (18.5) an extra term of the order of

$$\Delta b_{ii} \sim \tfrac{1}{2}SC[\gamma R_1 - R_2] \tag{18.7}$$

Here $\gamma$ is the overall assimilation efficiency ($\gamma < 1$) for converting the consumed species into biomass of the consuming species. $R_1$ and $R_2$ are quantities defined by De Angelis, and are very roughly related to the average magnitude of non-zero column and row elements in eqn (18.5). $C$ enters because only the fraction of interactions that are non-zero contributes to the factor in the square brackets. Note that if $R_1 \sim R_2 \sim b$, with both $R_1$ and $R_2$ independent of $S$ and $C$, these effects make a stabilizing negative contribution to the diagonal elements, and the stability criterion (18.6) is very crudely modified to read

$$b\sqrt{(SC)} < 1 + \tfrac{1}{2}SCb(1 - \gamma) \tag{18.8}$$

This explains the essential features of De Angelis' numerical results, whereby relatively large values of $S$ and $C$ can enhance stability.

One reservation about this work is that losses from the system can undercut the donor control arguments that lead essentially to eqn (18.8). Another reservation is that, as the number of species and the connectance increase, the average non-zero interaction strengths ($R_1$ and $R_2$) may well decrease; if you interact with many species, you may do so relatively weakly with each one. Some further discussion of this and other related models is in May (1974, pp. 221–223). A comprehensive review of the literature on stability in donor-controlled model ecosystems, and the beginnings of a discussion of the consequences of 'acceptor control', is due to Mazanov (1978). Much of Mazanov's discussion is, however, particular to systems that are exactly linear, which he justifies by appeal to ideas introduced by Patten (1975). The formal elegance of Patten's work notwithstanding, I find the idea that real ecosystems have exactly linear dynamics to be too idiosyncratic to warrant serious attention.

### (*b*) *Predator functional responses*

Nunney (1978) has recently drawn attention to the possibility that predators with stabilizing functional responses can contribute to ecosystem stability, with the contribution increasing with $S$ and $C$ in a manner akin to that discussed by De Angelis. Specifically, if the functional response for the average predator upon its average prey species, $N$, be denoted by $G(N)$, then there is added to the diagonal matrix elements in the ecosystem model of eqn (18.5) a contribution of the order of

$$\Delta b \sim \tfrac{1}{2}SC[G - N\,dG/dN] \qquad (18.9)$$

As discussed in general terms by Murdoch & Oaten (1975), for predators with 'type III' functional responses we can have $dG/dN > G/N$, and hence stabilizing (negative) effects on the dynamics of the system. Moreover, such type III predation patterns increasingly seem to be the rule rather than the exception (Oaten & Murdoch 1977; Hassell, Lawton & Beddington 1977). Thus, if the factor in square brackets in eqn (18.9) has some constant negative value independent of $S$ and $C$, the stability criterion of eqn (18.6) is again modified to the general form of eqn (18.8) and the dynamical stability of the system can *increase* as $S$ and $C$ increase. However, as Nunney points out, in homogeneous ('coarse-grained') environments we may expect the interaction of the predator with any one prey species to decrease as more and more species are added to its diet; that is, for the magnitude of $G$ and $dG/dN$ to decrease as $S$ and $C$ increase. In this event the earlier results of pp. 394–396 remain intact. On the other hand, for an inhomogeneous ('fine-grained') environment a predator could conceivably add new items to its diet (increasing $S$ and $C$) without diminishing the average $G$, whereupon stability increases as $S$ and $C$ increase.

Rounding out his general insights with numerical simulations, Nunney concludes 'there exists an underlying trend causing stability to decrease with complexity, but [the simulations] also show that other factors can reverse this trend'.

### (c) *Omnivory*

The effects of various patterns of omnivory upon the dynamics of ecosystems modelled by Lotka–Volterra equations have been explored by Pimm & Lawton (1978). Broadly speaking, their numerical results lead to the conclusion that omnivory and overall dynamical stability are easier to reconcile if the omnivores and their variegated prey are of similar sizes and population densities, a situation that most commonly pertains to insect parasitoids. This point has also been made on more abstract grounds (May 1978), in an attempt to account for the diversity of insects in general and the diversity of parasitoids in particular.

Pimm and Lawton show that the patterns of omnivory that they would expect to be most likely to persist on an evolutionary time scale are, indeed, consistent with real food webs (although they note that the data are less full and reliable than they would wish).

### (*d*) *The physical size of species*

Elton (1927) has emphasized that 'size has a remarkably great influence on the organization of animal communities. We have already seen how animals form food chains in which the species become progressively larger in size or, in the case of parasites, smaller in size [or, I interpose, in the case of parasitoids, roughly the same size].' Cousins (1978) has recently revived this theme, seeking to show how many community patterns can be understood as arising from the interplay of thermodynamics and physical aspects of plant structure and animal size. A more empirical approach has been adopted by May (1978), who has compiled a lot of information about the numbers of species of various kinds of terrestrial animals as functions of their physical size.

It remains my impression that the whole question of the relation between community structure and the sizes of the constituent species remains wide open.

### (*e*) *Some other patterns*

Two further empirical patterns, both somewhat enigmatic, have recently been documented by Cohen (1977, 1978).

Drawing together data from food webs in 14 different communities, Cohen (1977) shows that the ratio of number of kinds of prey to number of kinds of predators is remarkably constant at around 3/4. Here the 'kind of organism' may be 'a stage in the life cycle, or a size class within a single species, or a collection of functionally or taxonomically related species according to the practice of the original report'. Within this admittedly fuzzy frame, the number of prey kinds, $n$, is related to the number of predator kinds, $m$, by the linear regression

$$m = 1{\cdot}8 + 0{\cdot}71n \qquad (18.10)$$

The regression coefficient has standard deviation of 0·07.

Cohen (1978) has also presented an analysis of some 30 food webs aimed at determining the *minimum* number of dimensions of (abstract) niche space that are necessary to represent, in a strict if formal sense, the overlaps among observed trophic niches. The food webs are all for communities in habitats with limited physical and temporal heterogeneity ('single habitats'), and Cohen finds that the overlap of species along feeding dimensions can almost always be represented as one-dimensional within his scheme. As Cohen stresses, the concepts have ambiguities, the data are poor and the feeding relationships are usually reported in binary form (present or absent) rather than quantitatively. Even so, the result is intriguing.

If these two phenomenological relationships of Cohen's are accepted at face value, they cry out for explanation in terms of some fundamental theory of the dynamics of real ecosystems.

## CONCLUSIONS

The two parts of this chapter are quite dissonant. The first part argued that a good understanding of the observed dynamics of single populations is currently emerging, despite complications in the range of dynamical behaviour and in the number of alternative stable states that can be exhibited. In contrast, the survey of multispecies systems in the second part could only list some empirical patterns that are as yet largely unexplained, and some abstractly theoretical speculations about mechanisms that may help stabilize real ecosystems.

In the natural world (as opposed to the laboratory) there are, however, few species that can be legitimately treated as isolated populations. Most single-species dynamics is embedded in a multispecies context. This fact has practical implications in such diverse areas as the management of crop pests, the design of conservation regions and the control of vectors of tropical diseases. There are growing doubts about the practicality of managing fish stocks as single species rather than as part of a multispecies complex (Steele 1979 this volume; Beddington 1979 this volume; Horwood 1978; Getz 1978). Gulland (1977) has indeed suggested that just as the fisheries literature has pioneered, and given the best quantitative data, on single-species populations, so too it may lead the way in multispecies studies.

This volume thus ends on a suitably hortatory note: some answers for the dynamics of single populations; a list of questions about the structure and dynamics of multispecies systems; and a clear need for these questions to be addressed.

## ACKNOWLEDGMENTS

I have been helped by too many people to list. This work was supported in part by the NSF under grant DEB77-01565.

## REFERENCES

**Anderson R.M. (1979)** The influence of parasitic infection on the dynamics of host population growth. *Population Dynamics* (Ed. by R.M. Anderson, B.D. Turner & L.R. Taylor) pp. 245–281. Blackwell Scientific Publications, Oxford.

**Beddington J.R. (1979)** Harvesting and population dynamics. *Population Dynamics* (Ed. by R.M. Anderson, B.D. Turner & L.R. Taylor) pp. 307–320. Blackwell Scientific Publications, Oxford.

**Beddington J.R. & Lawton J.H. (1978)** On the structure and behaviour of ecosystems. *Journal de Physique*, **39c** 5–39.

**Beddington J.R., Free C.A. & Lawton J.H. (1978)** Modelling biological control: on the characteristics of successful natural enemies. *Nature, London*, **273**, 513–519.

**Bradley D.J. & May R.M. (1978)** Consequences of helminth aggregation for the dynamics of schistosomiasis. *Proceedings of the Royal Society of Tropical Medicine and Hygiene* (in press).

**Caswell H. (1976)** Community structure: a neutral model analysis. *Ecological Monographs*, **46**, 327–354.

**Clark C.W. (1974)** Possible effects of schooling on the dynamics of exploited fish populations. *Journal du Conseil. Conseil permanent international pour l'exploration de la mer*, **36**, 7–14.

**Cohen J.E. (1977)** Ratio of prey to predators in community food webs. *Nature, London*, **270**, 165–167.

**Cohen J.E. (1978)** *Food Webs and Niche Space*. Princeton University Press, Princeton.

**Connell J.H. (1979)** Tropical rain forests and coral reefs as open non-equilibrium systems. *Population Dynamics* (Ed. by R.M. Anderson, B.D. Turner & L.R. Taylor) pp. 141–163. Blackwell Scientific Publications, Oxford.

**Cousins S.H. (1978)** A trophic continuum derived from plant structure, animal size and a detritus cascade (in press).

**Cowie R.J. & Krebs J.R. (1979)** Optimal foraging in patchy environments. *Population Dynamics* (Ed. by R.M. Anderson, B.D. Turner & L.R. Taylor) pp. 183–205. Blackwell Scientific Publications, Oxford.

**Dawkins R. (1976)** *The Selfish Gene*. Oxford University Press, Oxford.

**De Angelis D.L. (1975)** Stability and connectance in food web models. *Ecology*, **56**, 238–243.

**De Angelis D.L., Gardner R.H., Mankin J.B., Post W.M. & Carney J.H. (1978)** Energy flow and the number of trophic levels in ecological communities. *Nature, London*, **273**, 406–407.

**Diamond J.M. (1979)** Community structure; is it random, or is it shaped by species differences and competition? *Population Dynamics* (Ed. by R.M. Anderson, B.D. Turner & L.R. Taylor) pp. 165–181. Blackwell Scientific Publications, Oxford.

**Elton C. (1927)** *Animal Ecology*. Sidgwick & Jackson, London.

**Elton C. (1942)** *Voles, Mice and Lemmings: Problems in Population Dynamics*. Oxford University Press, Oxford.

**Elton C.S. (1958)** *The Ecology of Invasions by Animals and Plants*. Methuen & Co., London.

**Gardner M.R. & Ashby W.R. (1970)** Connectance of large dynamical (cybernetic) systems: critical values for stability. *Nature, London*, **228**, 784.

**Getz W.M. (1978)** On harvesting two competing populations (in press).

**Gilbert L.E. (1975)** Ecological consequences of a coevolved mutualism between butterflies and plants. *Coevolution of Animals and Plants* (Ed. by L.E. Gilbert & P.H. Raven) pp. 210–240. University of Texas Press, Austin, Texas.

**Gilpin M.E. (1975)** Stability of feasible predator–prey systems. *Nature, London*, **254**, 137–138.

**Gilpin M.E. & Case T. (1976)** Multiple domains of attraction in competition communities. *Nature, London*, **261**, 40–42.

**Goh B.S. (1978)** Robust stability concepts for ecosystem models. *Theoretical Systems Ecology* (Ed. by E. Halfon). Academic Press, New York.

**Goh B.S. & Jennings L.S. (1977)** Feasibility and stability in randomly assembled Lotka–Volterra models. *Ecological Modelling*, **3**, 63–71.

**Guckenheimer J., Oster G.R. & Ipaktchi A. (1976)** The dynamics of density-dependent population models. *Journal of Mathematical Biosciences*, **4**, 101–147.

**Gulland J.A. (1975)** The stability of fish stocks. *Journal du Conseil. Conseil permanent international pour l'exploration de la mer*, **37**, 199–204.

**Gulland J.A. (1977)** The analysis of data and development of models. *Fish Population Dynamics* (Ed. by J.A. Gulland) pp. 67–95. Wiley, London.

**Halbach U. (1979)** Introductory remarks. *Fortschritte der Zoologie* (in press).

**Hamilton W.D. & May R.M. (1977)** Dispersal in stable habitats. *Nature, London*, **269**, 578–581.

**Hassell M.P. (1979)** The dynamics of predator–prey interactions: polyphagous predators, competing predators and hyperparasitoids. *Population Dynamics* (Ed. by R.M. Anderson, B.D. Turner & L.R. Taylor) pp. 283–306. Blackwell Scientific Publications, Oxford.

**Hassell M.P., Lawton J.H. & Beddington J.R. (1977)** Sigmoid functional responses by invertebrate predators and parasitoids. *Journal of Animal Ecology*, **46**, 249–262.

**Hassell M.P., Lawton J.H. & May R.M. (1976)** Patterns of dynamical behaviour in single-species populations. *Journal of Animal Ecology*, **45**, 471–486.

**Horwood J.W. (1978)** Management and models of marine multispecies complexes. *Dynamics in Large Mammals* (Ed. by C. Fowler & T. Smith). Wiley, New York (in press).

**Hutchinson G.E. (1948)** Circular causal systems in ecology. *Annals of the New York Academy of Science*, **50**, 221–246.

**Hutchinson G.E. (1959)** Homage to Santa Rosalia, or why are there so many kinds of animals? *American Naturalist*, **93**, 145–159.

**Law R. (1979)** Ecological determinants in the evolution of life histories. *Population Dynamics* (Ed. by R.M. Anderson, B.D. Turner & L.R. Taylor) pp. 81–103. Blackwell Scientific Publications, Oxford.

**Lawlor L.R. (1978)** A comment on randomly constructed model ecosystems. *American Naturalist*, **112**, 445–447.

**Lawton J.H. & Pimm S.L. (1978)** Population dynamics and the length of food chains. *Nature, London*, **272**, 190.

**Lawton J.H. & McNeill S. (1979)** Between the devil and the deep blue sea: on the problem of being a herbivore. *Population Dynamics* (Ed. by R.M. Anderson, B.D. Turner & L.R. Taylor) pp. 223–244. Blackwell Scientific Publications, Oxford.

**Levandowsky M. & White B.S. (1977)** Randomness, time scales, and the evolution of biological communities. *Evolutionary Biology*, **10**, 69–161.

**Levin S.A. (1976)** Population dynamics in heterogeneous environments. *Annual Review of Ecology and Systematics*, **7**, 287–310.

**Levins R. (1975)** Evolution in communities near equilibrium. *Ecology of Species and Communities* (Ed. by M.L. Cody & J.M. Diamond) pp. 16–50. Harvard University Press, Cambridge, Massachusetts.

**Ludwig D., Jones D.D. & Holling C.S. (1978)** Qualitative analysis of insect outbreak systems: the spruce budworm and forest. *Journal of Animal Ecology*, **47**, 315–332.

**MacDonald G. (1965)** The dynamics of helminth infections, with special references of schistosomes. *Transactions of the Royal Society of Tropical Medicine and Hygiene*, **59**, 489–506.

**MacDonald N. (1978)** The prevalence of chaos. *Nature, London*, **271**, 305–306.

**Mackey M.C. & Glass L. (1977)** Oscillation and chaos in physiological control systems. *Science*, **197**, 287–289.

**McMurtrie R.E. (1975)** Determinants of stability of large, randomly connected systems. *Journal of Theoretical Biology*, **50**, 1–11.

**McNaughton S.J. (1978)** Stability and diversity of ecological communities. *Nature, London*, **274**, 251–253.

**May R.M. (1972)** Will a large complex system be stable? *Nature, London*, **238**, 413–414.

**May R.M. (1973)** Time-delay versus stability in population models with two and three trophic levels. *Ecology*, **54**, 315–325.

**May R.M. (1974)** *Stability and Complexity in Model Ecosystems*, 2nd edition. Princeton University Press, Princeton.

**May R.M. (1976a)** Simple mathematical models with very complicated dynamics. *Nature, London*, **261**, 459–467.

**May R.M. (Ed.) (1976b)** *Theoretical Ecology: Principles and Applications*. Saunders, Philadelphia, and Blackwell Scientific Publications, Oxford.

**May R.M. (1977)** Thresholds and breakpoints in ecosystems with a multiplicity of stable states. *Nature, London*, **269**, 471–477.

**May R.M. (1978)** The dynamics and diversity of insect faunas. *Diversity of Insect Faunas* (Ed. by L.A. Mound & N. Waloff) pp. 188–204. Royal Entomological Society Symposium, Sept. 1977. Blackwell Scientific Publications, Oxford.

**May R.M., Conway G.R., Hassell M.P. & Southwood T.R.E. (1974)** Time delays, density dependence, and single species oscillations. *Journal of Animal Ecology*, **43**, 747–770.

**May R.M. & Oster G.F. (1976)** Bifurcations and dynamic complexity in single ecological models. *American Naturalist*, **110**, 573–599.

**Maynard Smith J. (1976)** Evolution and the theory of games. *American Scientist*, **64**, 41–45.

**Mazanov A. (1978)** Acceptor control in model ecosystems. *Journal of Theoretical Biology*, **71**, 21–38.

**Murdoch W.W. (1979)** Predation and the dynamics of prey populations. *Fortschritte der Zoologie* (in press).

**Murdoch W.W. & Oaten A. (1975)** Predation and population stability. *Advances in Ecological Research*, **9**, 2–131.

**Noy-Meir I. (1975)** Stability of grazing systems: an application of predator–prey graphs. *Journal of Ecology*, **63**, 459–481.

**Nunney L. (1978)** The stability of complex model ecosystems. *American Naturalist* (in press).

**Oaten A. & Murdoch W.W. (1977)** More on functional response and stability. *American Naturalist*, **111**, 383–386.

**Oster G.F. (1979)** Optimization models in ecology. *Fortschritte der Zoologie* (in press).

**Patten B.C. (1975)** Ecosystem linearization: an evolutionary design problem. *American Naturalist*, **109**, 529–539.

**Peterman R.M. (1979)** The dynamics of resilience: shifting stability domains in fish and insect systems. *Population Dynamics* (Ed. by R.M. Anderson, B.D. Turner & L.R. Taylor) pp. 321–341. Blackwell Scientific Publications, Oxford.

**Pimm S.L. & Lawton J.H. (1977)** Number of trophic levels in ecological communities. *Nature, London*, **268**, 329–331.

**Pimm S.L. & Lawton J.H. (1978)** On feeding on more than one trophic level. *Nature, London*, **275**, 542–544.

**Roberts A.P. (1974)** The stability of a feasible random ecosystem. *Nature, London*, **251**, 607–608.

**Saunders P.T. (1978)** Population dynamics and the length of food chains. *Nature, London*, **272**, 189–190.

**Siljak D.D. (1974)** Connective stability of complex systems. *Nature, London*, **249**, 280.

**Siljak D.D. (1975a)** When is a complex ecosystem stable? *Mathematical Biosciences*, **25**, 25–50.

**Siljak D.D. (1975b)** Connective stability of competitive equilibrium. *Automatica*, **11**, 389–400.

**Simenstad C.A., Estes J.A. & Kenyon K.W. (1978)** Aleuts, sea otters, and alternate stable-state communities. *Science*, **200**, 403–411.

**Steele J.H. (1979)** Interactions in marine ecosystems. *Population Dynamics* (Ed. by R.M. Anderson, B.D. Turner & L.R. Taylor) pp. 343–357. Blackwell Scientific Publications, Oxford.

**Southwood T.R.E. (1977a)** The relevance of population dynamic theory to pest status. *The Origins of Pest, Parasite, Disease and Weed Problems* (Ed. by J.M. Cherrett & G.R. Sagar) pp. 35–54. Blackwell Scientific Publications, Oxford.

**Southwood T.R.E. (1977b)** Habitat, the templet for ecological strategies? *Journal of Animal Ecology*, **46**, 337–366.

**Southwood T.R.E. & Comins H.N. (1976)** A synoptic population model. *Journal of Animal Ecology*, **45**, 949–965.

**Taylor L.R. & Taylor R.A.J. (1977)** Aggregation, migration and population mechanics. *Nature, London*, **265**, 415–421.

**Taylor R.A.J. & Taylor L.R. (1979)** A behavioural model for the evolution of spatial dynamics. *Population Dynamics* (Ed. by R.M. Anderson, B.D. Turner & L.R. Taylor) pp. 1–27. Blackwell Scientific Publications, Oxford.

# AUTHOR INDEX

Figures in italics refer to pages where full references appear

# SUBJECT INDEX

Related names appear in parentheses